B/NA

Symmetric and *G*-algebras

Mathematics and Its Applications

Volume 60

Symmetric and *G*-algebras

With Applications to Group Representations

by

Gregory Karpilovsky

Department of Mathematics,
California State University, Chico,
U.S.A.

KLUWER ACADEMIC PUBLISHERS
DORDRECHT / BOSTON / LONDON

MATH
cep/2e

Library of Congress Cataloging in Publication Data

Karpilovsky, Gregory, 1940-
   Symmetric and G-algebras : with applications to group
representations / Gregory Karpilovsky.
      p.   cm. -- (Mathematics and its applications ; v.60)
   ISBN 0-7923-0761-5 (alk. paper)
   1. Frobenius algebras. 2. Representations of groups. 3. Group
algebras.   I. Title. II. Title: G-algebras. III. Series:
Mathematics and its applications (Kluwer Academic Publishers) ; v
60.
QA251.5.K36   1990
512'.24--dc20
                                                    90-34936
                                                       CIP

ISBN 0-7923-0761-5

Published by Kluwer Academic Publishers,
P.O. Box 17, 3300 AA Dordrecht, The Netherlands.

Kluwer Academic Publishers incorporates
the publishing programmes of
D. Reidel, Martinus Nijhoff, Dr W. Junk and MTP Press.

Sold and distributed in the U.S.A. and Canada
by Kluwer Academic Publishers,
101 Philip Drive, Norwell, MA 02061, U.S.A.

In all other countries, sold and distributed
by Kluwer Academic Publishers Group,
P.O. Box 322, 3300 AH Dordrecht, The Netherlands.

Printed on acid-free paper

Printed in the Netherlands

To my children
SUZANNE and ELLOITT

'Et moi, ..., si j'avait su comment en revenir,
je n'y serais point allé.'

　　　　　　　　　Jules Verne

The series is divergent; therefore we may be
able to do something with it.

　　　　　　　　　O. Heaviside

One service mathematics has rendered the
human race. It has put common sense back
where it belongs, on the topmost shelf next
to the dusty canister labelled 'discarded non-
sense'.

　　　　　　　　　Eric T. Bell

Mathematics is a tool for thought. A highly necessary tool in a world where both feedback and non-linearities abound. Similarly, all kinds of parts of mathematics serve as tools for other parts and for other sciences.

Applying a simple rewriting rule to the quote on the right above one finds such statements as: 'One service topology has rendered mathematical physics ...'; 'One service logic has rendered computer science ...'; 'One service category theory has rendered mathematics ...'. All arguably true. And all statements obtainable this way form part of the raison d'être of this series.

This series, *Mathematics and Its Applications*, started in 1977. Now that over one hundred volumes have appeared it seems opportune to reexamine its scope. At the time I wrote

"Growing specialization and diversification have brought a host of monographs and textbooks on increasingly specialized topics. However, the 'tree' of knowledge of mathematics and related fields does not grow only by putting forth new branches. It also happens, quite often in fact, that branches which were thought to be completely disparate are suddenly seen to be related. Further, the kind and level of sophistication of mathematics applied in various sciences has changed drastically in recent years: measure theory is used (non-trivially) in regional and theoretical economics; algebraic geometry interacts with physics; the Minkowsky lemma, coding theory and the structure of water meet one another in packing and covering theory; quantum fields, crystal defects and mathematical programming profit from homotopy theory; Lie algebras are relevant to filtering; and prediction and electrical engineering can use Stein spaces. And in addition to this there are such new emerging subdisciplines as 'experimental mathematics', 'CFD', 'completely integrable systems', 'chaos, synergetics and large-scale order', which are almost impossible to fit into the existing classification schemes. They draw upon widely different sections of mathematics."

By and large, all this still applies today. It is still true that at first sight mathematics seems rather fragmented and that to find, see, and exploit the deeper underlying interrelations more effort is needed and so are books that can help mathematicians and scientists do so. Accordingly MIA will continue to try to make such books available.

If anything, the description I gave in 1977 is now an understatement. To the examples of interaction areas one should add string theory where Riemann surfaces, algebraic geometry, modular functions, knots, quantum field theory, Kac-Moody algebras, monstrous moonshine (and more) all come together. And to the examples of things which can be usefully applied let me add the topic 'finite geometry'; a combination of words which sounds like it might not even exist, let alone be applicable. And yet it is being applied: to statistics via designs, to radar/sonar detection arrays (via finite projective planes), and to bus connections of VLSI chips (via difference sets). There seems to be no part of (so-called pure) mathematics that is not in immediate danger of being applied. And, accordingly, the applied mathematician needs to be aware of much more. Besides analysis and numerics, the traditional workhorses, he may need all kinds of combinatorics, algebra, probability, and so on.

In addition, the applied scientist needs to cope increasingly with the nonlinear world and the extra mathematical sophistication that this requires. For that is where the rewards are. Linear models are honest and a bit sad and depressing: proportional efforts and results. It is in the non-linear world that infinitesimal inputs may result in macroscopic outputs (or vice versa). To appreciate what I am hinting at: if electronics were linear we would have no fun with transistors and computers; we would have no TV; in fact you would not be reading these lines.

There is also no safety in ignoring such outlandish things as nonstandard analysis, superspace and anticommuting integration, p-adic and ultrametric space. All three have applications in both electrical engineering and physics. Once, complex numbers were equally outlandish, but they frequently proved the shortest path between 'real' results. Similarly, the first two topics named have already provided a number of 'wormhole' paths. There is no telling where all this is leading - fortunately.

Thus the original scope of the series, which for various (sound) reasons now comprises five subseries: white (Japan), yellow (China), red (USSR), blue (Eastern Europe), and green (everything else), still applies. It has been enlarged a bit to include books treating of the tools from one subdiscipline which are used in others. Thus the series still aims at books dealing with:

- a central concept which plays an important role in several different mathematical and/or scientific specialization areas;
- new applications of the results and ideas from one area of scientific endeavour into another;
- influences which the results, problems and concepts of one field of enquiry have, and have had, on the development of another.

Symmetry is important. It does not necessarily involve groups, and can take many forms. And, if the past two centuries are anything to go by, the concept and its generalizations will become even more important.

In algebra 'symmetry' manifests itself for instance in the theory of Frobenius algebras, symmetric algebras and G-algebras. The theory of these has seen rapid growth in the last 15 years or so and has reached a level where a first systematic and comprehensive account becomes both possible and desirable. Moreover, there are important applications, particularly to group representation theory as the present book amply shows.

The shortest path between two truths in the real domain passes through the complex domain.

> J. Hadamard

La physique ne nous donne pas seulement l'occasion de résoudre des problèmes ... elle nous fait pressentir la solution.

> H. Poincaré

Never lend books, for no one ever returns them; the only books I have in my library are books that other folk have lent me.

> Anatole France

The function of an expert is not to be more right than other people, but to be wrong for more sophisticated reasons.

> David Butler

Amsterdam, 25 April 1990   Michiel Hazewinkel

Preface

In this volume we present the theory of symmetric and G-algebras together with a number of applications to modular group representation theory and block theory of group algebras. The theory of symmetric and G-algebras has experienced a rapid growth in the last ten to fifteen years, acquiring mathematical depth and significance and leading to new insights in group representation theory. By now a stage has clearly been reached when a systematic account has become both possible and desirable. We hope to help the reader early on to an understanding of the basic structure of the theory, and to motivate at each successive stage the introduction of new concepts and new tools.

This book is written on the assumption that the reader has had the equivalent of a standard first-year graduate algebra course. Apart from a few specific results and the general knowledge we have presupposed, the book is entirely self-contained. A systematic description of the material is supplied by the introduction to individual chapters. There is a fairly large bibliography of works which are either directly relevant to the text or offer supplementary material of interest. An attempt has been made to give credit for some of the major methods and theorems, but we have stopped far short of trying to trace each theorem to its source.

A word about notation. As is customary, Theorem 2.4.2 denotes the second result of Section 4 of Chapter 2; however, for simplicity, all references to this result within Chapter 2 itself, are designated as Theorem 4.2.

The following is a brief description of the content of the book. In Chapter 1 we review some basic ring-theoretic results. Many readers may wish to glance briefly at the contents of this chapter, referring back to the relevant sections when they are needed later.

In Chapter 2 we provide an extensive account of the theory of Frobenius and symmetric algebras. After recording some basic facts, we show that if A is a Frobenius algebra over a field F and G a finite group, then any crossed product F-algebra $A*G$ is Frobenius. We then establish an important result due to Harris (1988) which asserts that if A is semisimple, then any crossed product F-algebra $A*G$ is symmetric. Turning to modules over group algebras, we apply the above results to demonstrate that if V is a finitely generated semisimple FN-module (N is a normal subgroup of G), then the F-algebra $End_{FG}(V^G)$ is symmetric. As a byproduct of our investigations, we establish a number of important results pertaining to modular representation theory of groups and block theory of group algebras.

The aim of Chapter 3 is twofold: first to investigate symmetric local algebras A over an algebraically closed field F with $dim_F Z(A) \leq 5$ and second to apply the results obtained to modular representation theory of finite groups. More specifically, assume that F is of prime characteristic p and B is a block of a group algebra FG with exactly one simple FG-module. Denote by $k(B)$ the number of irreducible complex characters in B and by D a defect group of B. As an application of our results on symmetric local algebras, we demonstrate that if $k(B) \leq 4$, then $k(B) = |D|$ and if $k(B) = 5$, then either $|D| = 5$ or D is nonabelian of order 8. The results above are relevant to the Brauer's conjecture which states that for any B, $k(B) \leq |D|$.

Chapter 4, the final chapter, provides an extensive account of the theory of G-algebras. Much of the material presented is based on works of Green (1968), Puig (1981), Broué and Robinson (1986) and Ikeda (1987). As one of the many applications, we present a criterion for an indecomposable two-sided ideal of RG to be a block of RG. Among other applications, we mention a remarkable result due to Robinson (1983, Corollary 1) regarding defect groups of blocks.

I would like to express my gratitude to my wonderful wife for the tremendous help and constant encouragement which she has given me in the preparation of this book. My grateful thanks to my daughter, Suzanne, and my son Elliott, for bearing my great despair, forgiving my inattentions, and denying me pity.

Contents

Notation

Number systems

$\mathbb{N}$	the natural numbers
$\mathbb{Z}$	the rational integers
$\mathbb{Q}$	the rational numbers
$\mathbb{R}$	the real numbers
$\mathbb{C}$	the complex numbers

Set theory

$\subset$	proper inclusion
$\subseteq$	inclusion
$|X|$	the cardinality of the set X
$X - Y$	the complement of Y in X
$\emptyset$	the empty set

Number theory

$a|b$	a divides b
$a \nmid b$	a does not divide b
(a, b)	greatest common divisor of a and b
n_p	the p-part of n
$n_{p'}$	the p'-part of n

Group theory

F^*	the multiplicative group of a field F
$< X >$	the subgroup generated by X
$\mathbb{Z}_n$	the cyclic group of order n
$G_1 \times G_2$	direct product of G_1 and G_2
$H \lhd G$	H is a normal subgroup of G
$C_G(X)$	the centralizer of X in G
$N_G(X)$	the normalizer of X in G
$[x, y]$	$= x^{-1}y^{-1}xy$
$[X, Y]$	$=< [x, y] \| x \in X, y \in Y >$
$G' = [G, G]$	the commutator subgroup of G
$Z(G)$	the center of G
$\Phi(G)$	the Frattini subgroup of G
S_n	symmetric group of degree n
A_n	alternating group of degree n
$GL(V)$	the group of all nonsingular transformations of the vector space V
$Cl(G)$	the set of conjugacy classes of G
$Syl_p(G)$	the set of Sylow p-subgroups of G
$H \subseteq_G K$	H is G-conjugate to a subgroup of K
$H =_G K$	H and K are G-conjugate
$\delta(C)$	defect group of $C \in Cl(G)$
$0_{p'}(G)$	the largest normal p'-subgroup of G
$0_p(G)$	the largest normal p-subgroup of G
$0_{p',p}(G)$	defined by $0_{p',p}(G)/0_{p'}(G) = 0_p(G/0_{p'}(G))$

Rings, modules and characters

$Z(R)$	the center of R
$A \otimes B$	tensor product
$char R$	the characteristic of R
$\oplus_{i \in I} R_i$	direct sum of rings
$\prod_{i \in I} R_i$	direct product of rings
RG	the group algebra of G over R
$F^\alpha G$	the twisted group algebra of G over F

$R * G$	crossed product of G over R
$J(R)$	the Jacobson radical of a ring R
$J(V)$	the radical of an R-module V
$Supp\, x$	support of x
$End_R(V)$	the algebra of R-endomorphisms of V
$M_n(R)$	$n \times n$ matrices with entries in R
V^G	induced module
V_E	$= E \otimes_F V$
A_E	$= E \otimes_F A$
$ann(V)$	the annihilator of V
$P(V)$	projective cover of V
${}^gV$	G-conjugate of V
V^*	contragredient module
$i(V, W)$	the intertwining number
$dim_F V$	dimension of V over F
X^G	the induced character
$J(B)$	the Jacobson radical of a block B
$I(G)$	the augmentation ideal of RG
$aug(x)$	the augmentation of x
$Hom_R(V, W)$	R-homomorphisms from V to W
$C_A(X)$	the centralizer of X in A
R°	opposite ring
$[A, A]$	the commutator subspace of A
C^+	$= \sum_{x \in C} x$
$Soc(V)$	the socle of V
$l(X)$	left annihilator of X
$r(X)$	right annihilator of X
${}^\perp X$	$= \{a \in A \mid \psi(aX) = 0\}$
$X^\perp$	$= \{a \in A \mid \psi(Xa) = 0\}$
$T(A)$	$= \{a \in A \mid a^{p^n} \in [A, A] \quad \text{for some} \quad n \geq 1\}$
A^{p^n}	$= \{a^{p^n} \mid a \in A\}$
$P_n(A)$	$= FA^{p^n} + [A, A]$
$T_n(A)$	$= \{a \in A \mid a^{p^n} \in [A, A]\}$
$X^{p^{-n}}$	$= \{g \in G \mid g^{p^n} \in X\}$
$Rey(A)$	the Reynolds ideal of A
$Tr_H^G : A^H \to A^G$	the trace map

A_H^G $= Im(Tr_H^G)$

$A(K)$ $= A^K/(J(R)A^K + \sum_{H \subset K} A_H^K)$

$Br_K : A^K \to A(K)$ the Brauer morphism

Field theory

E/F	field extension
$(E : F)$	degree of E over F
$Gal(E/F)$	Galois group of E over F
$K(S)$	the smallest subfield containing S and K
$\hat{F}$	the algebraic closure of F

Block theory

$B = B(e)$	the block containing e
$V \in B$	V lies in the block B
$\delta(e)$	defect group of e
$\delta(B)$	defect group of B

Cohomology theory

$Z^2(G, A)$	the group of all A-valued 2-cocycles of G
δg	a coboundary
$B^2(G, A)$	the set of all coboundaries
$H^2(G, A)$	$= Z^2(G, A)/B^2(G, A)$

Chapter 1

Preliminaries

The main purpose of this chapter is to review some basic ring-theoretic results, as well as to fix conventions and notations for the rest of the book. Because we presuppose a familiarity with various elementary group-theoretic and ring-theoretic terms, only a brief description of them is presented. Many readers may wish to glance briefly at the contents of this chapter, referring back to the relevant sections when they are needed later.

1.   Notation and terminology

It is important to establish at the outset various conventions that we shall use throughout the book.

We shall write $A - B$ for the complement of the subset B in the set A, while $A \subset B$ will mean that A is a proper subset of B. The symbol for a map will be written before the element affected, and consequently, when the composition $f \circ g$ of maps is indicated, g is the first to be carried out.

All rings in this book are *associative* with $1 \neq 0$ and subrings of a ring R are assumed to have the same identity element as R. Each ring homomorphism will be assumed to preserve identity elements.

Let R be a ring. An element x in R is called *nilpotent* if $x^n = 0$ for some integer $n \geq 1$. An ideal J in R is *nil* if every element of J is nilpotent, while J is *nilpotent* if there is an integer $n \geq 1$ such that $J^n = 0$, where J^n is the product of J with itself n times. The ring of all $n \times n$-matrices over R will be denoted by $M_n(R)$.

Let R be a ring. An element e in R is an *idempotent* if $e^2 = e$. Two idempotents u, v of R are *orthogonal*, if $uv = vu = 0$. A nonzero idempotent is *primitive* if it cannot be written as a sum of two nonzero orthogonal idempotents. We denote by $Z(R)$ the centre of R. An idempotent e is called *centrally* primitive if e is a primitive idempotent of $Z(R)$.

Let R be a ring. From this we can construct a new ring R°, called the *opposite ring* of R. Both the underlying set and the additive structure of R° are just those of R. But the multiplication, denoted by $\circ$, is given by

$$x \circ y = yx \qquad \text{for all} \quad x, y \in R$$

It is straightforward to verify that R° is a ring with these operations. Clearly, $Z(R) = Z(R^\circ)$ and R is commutative if and only if $R = R^\circ$.

Suppose that A is a ring, R a commutative ring, and $f : R \to Z(A)$ is a ring homomorphism. The resulting system (A, R, f) is called an *R-algebra*. It will be convenient to suppress the f and speak of A as an R-algebra or as an algebra over R. Thus A is an R-algebra (with respect to some f) if and only if there is an ideal I of R with R/I isomorphic to a subring of $Z(R)$. Setting $ra = f(r)a, r \in R, a \in A$, it follows that A is an R-module such that

$$r(xy) = x(ry) = (rx)y \qquad \text{for all} \quad r \in R, x, y \in A$$

Conversely, if A is an R-module for which the above equalities hold, then the map $R \to Z(A), r \mapsto r \cdot 1$ is a ring homomorphism and hence A is an R-algebra. By a *homomorphism of R-algebras*, we understand a ring homomorphism which is also a homomorphism of R-modules.

Let $(R_i), i \in I$, be a family of rings and let R be the direct product set $\prod_{i \in I} R_i$. One can define addition and multiplication on R by the rules

$$(x_i) + (y_i) = (x_i + y_i), \qquad (x_i)(y_i) = (x_i y_i) \qquad (x_i, y_i \in R_i)$$

An immediate verification shows that R is a ring; we shall refer to R as the *direct product* of the family $(R_i), i \in I$.

We next record some conventions pertaining to modules. Let R be a ring. By an R-module, we always understand a unital R-module; the symbols $_RM$ or M_R will be used to underline the fact that M is a left or right R-module, respectively. We usually consider *left* R-modules, and, in this case, speak simply about an R-module.

A nonzero R-module V is *indecomposable* if 0 and V are the only direct summands of V. A nonzero R-module V is *simple* if 0 and V are the only submodules of V.

Let V and W be R-modules. We denote by $Hom_R(V, W)$ the additive abelian group of all R-homomorphisms of V into W. The same notation will be employed in case V and W are right R-modules. In case $V = W$, we write $End_R(V)$ instead of $Hom_R(V, V)$. Observe that $End_R(V)$ is a ring under the multiplication

$$(fg)(v) = f(g(v)) \qquad \text{for all} \quad f, g \in End_R(V), v \in V$$

Let V be an R-module. A submodule W of V is called *maximal* if W is a *proper* submodule (i.e. $W \neq V$) amd W is not contained in any proper submodule of V.

Given a pair of rings R, S we say that V is an (R, S)-*bimodule* if V is a left R-module and a right S-module, with the actions of R and S on V commuting

$$(rv)s = r(vs) \qquad (r \in R, s \in S, v \in V)$$

An R-module V is said to have an R-*basis* $(v_i), i \in I$, if there exist elements $v_i \in V$ such that each $v \in V$ can be written as a finite sum $v = \sum r_i v_i$ with uniquely determined coefficients $r_i \in R$. Such modules V are called *free* R-modules. By a *projective* module, we understand a direct summand of a free module.

Unless explicitly stated otherwise, all groups are assumed to be *multiplicative*. The multiplicative group of a field is denoted by F^*.

Let G be a group and let S be an arbitrary set. An *action* of G on S is a homomorphism θ of G into the permutation group of S. When the action is understood, we shall suppress the symbol θ and instead write $s \mapsto {}^g s$ for $\theta(g), g \in G, s \in S$. The *stabilizer* $G(s)$ of $s \in S$ is defined by

$$G(s) = \{g \in G | {}^g s = s\}$$

and the orbit $^G s$ of $s \in S$ is defined by

$$^G s = \{\, ^g s | g \in G \}$$

We say that G acts *transitively* if $^G s = S$ for some (and hence all) $s \in S$. One readily verifies that $G(s)$ is a subgroup of G and that

$$|\, ^G s| = (G : G(s))$$

Throughout the book, S_n and A_n denote symmetric and alternating groups of degree n, respectively, while $\mathbb{Z}_n$ is a cyclic group of order n.

If S is a subset of a group G, $< S >$ will denote the subgroup of G generated by S, while

$$C_G(S) \quad \text{and} \quad N_G(S)$$

is the *centralizer* and *normalizer* of S in G, respectively. As usual, $Z(G)$ denotes the centre of G.

Let H be a subgroup of G. A subset of G containing just one element from each left coset xH is called a *left transversal* for H in G, and right transversals are defined correspondingly.

Let g be an element of a finite group G and let p be a prime. We say that g is a p-element if g has order equal to a power of p; we say that g is a p'-element (or is p-regular) if its order is prime to p. Each $g \in G$ can be written in a unique way $g = g_p g_{p'}$, where g_p is a p-element, $g_{p'}$ is a p'-element, and g_p and $g_{p'}$ commute. The elements g_p and $g_{p'}$ are called the p and p'-parts of g, respectively. A conjugacy class C of G is called p-regular (respectively, p-singular) if the order of the elements of C is not divisible by p (respectively, divisible by p).

2. Artinian, noetherian and semisimple modules

Throughout this section, R denotes an arbitrary ring.

Let M be an R-module. We say that M is *artinian* (respectively, *noetherian*) if every descending (respectively, ascending) chain of submodules of M terminates. The ring R itself is called *artinian* (respectively, *noetherian*) if the left regular R-module $_R R$ is artinian (respectively, noetherian).

Let M be an R-module. We say that M is *finitely generated* if $M = \sum_{i=1}^{n} Rv_i$ for some finitely many $v_1, \ldots, v_n$ in M. The module M is said to be *finitely cogenerated* if for every family $\{M_i | i \in I\}$ of submodules of M with $\cap_{i \in I} M_i = 0$, there is a finite subset J of I such that

$$\cap_{j \in J} M_j = 0$$

2.1. Proposition. *For any R-module M, the following conditions are equivalent:*

(i) M is artinian.

(ii) Every nonempty subset of submodules of M has a minimal element.

(iii) Every factor-module of M is finitely cogenerated.

Proof. (i) $\Rightarrow$ (ii): Suppose that there is a nonempty set S of submodules of M such that S does not have a minimal element. Then, for any given $N \in S$, the set $\{N' \in S | N' \subset N\}$ is nonempty. Applying the axiom of choice, there is a function $N \mapsto N'$ with $N \supset N'$ for any N in S. By choosing any N in S, we therefore obtain an infinite descending chain

$$N \supset N' \supset N'' \supset \cdots$$

of submodules of M.

(ii) $\Rightarrow$ (iii): Let N be any submodule of M and let $\{M_i | i \in I\}$ be a family of submodules of M with $\cap_{i \in I} M_i = N$. It suffices to show that $\cap_{j \in J} M_j = N$ for some finite subset J of I. To this end, put

$$S = \{\cap_{k \in K} M_k | K \subseteq I \quad \text{is finite}\}$$

By hypothesis, S has a minimal element, say $\cap_{j \in J} M_j$ with $J \subseteq I$ and J finite. It is clear that $N = \cap_{j \in J} M_j$, as required.

(iii) $\Rightarrow$ (i): Consider a descending chain $M_1 \supseteq M_2 \supseteq \cdots \supseteq M_n \supseteq \cdots$ of submodules of V. If $N = \cap_{k \geq 1} M_k$, then by hypothesis M/N is finitely cogenerated. Thus $N = M_n$ for some $n \geq 1$. But then $M_{n+i} = M_n$ for all $i \geq 1$, as desired. ∎

2.2. Corollary. *Let $M \neq 0$ be an artinian module. Then M has a simple submodule.*

Proof. Apply Proposition 2.1 for the nonempty set of all nonzero submodules of M. ∎

2.3. Proposition. *For any R-module M, the following conditions are equivalent:*
(i) M is noetherian.
(ii) Every nonempty set of submodules of M has a maximal element.
(iii) Every submodule of M is finitely generated.

Proof. The proof of this result is dual to that of Proposition 2.1 and therefore will be omitted. ∎

We next record the following useful property.

2.4. Proposition. *Let M be a finitely generated R-module. Then every proper submodule of M is contained in a maximal submodule (in particular, if $M \neq 0$, then M has a maximal submodule).*

Proof. By assumption, $M = Rv_1 + \cdots + Rv_m$ for some $v_1, \ldots, v_m$ in M. Denote by N a proper submodule of M and by S the set of all proper submodules of M which contain N. Then $N \in S$ and if (N_i) is a chain in S, then $\cup_i N_i$ is a submodule. If $M = \cup_i N_i$, then $v_j \in N_{k_j}$ for some $k_j, 1 \leq j \leq m$. Let N_k be the largest of the modules $N_{k_1}, \ldots, N_{k_m}$. Then N_k contains $v_1, \ldots, v_m$ and hence $N_k = M$, a contradiction. Thus S is inductive and so, by Zorn's lemma, it has a maximal element. This gives us a maximal submodule of M containing N, as required. ∎

2.5. Proposition. *Let $0 \rightarrow U \rightarrow V \rightarrow W \rightarrow 0$ be an exact sequence of R-modules. Then V is artinian (noetherian) if and only if both U and W are artinian (noetherian).*

Proof. Assume that V is artinian. Then, since U and W are isomorphic to a submodule and factor module of V, respectively, it is clear from the definition that both U and W are artinian.

Conversely, suppose that both U and W are artinian. To show that V is artinian, we may assume that $U \subseteq V$ and $W = V/U$. Consider a descending chain

$$V_1 \supseteq V_2 \supseteq \cdots \supseteq V_n \supseteq \cdots$$

of submodules of V. Because V/U is artinian, there is an integer m such that

$$V_m + U = V_{m+i} + U \qquad (i = 1, 2, \ldots)$$

Since U is artinian, there is an integer $n \geq m$ such that

$$V_n \cap U = V_{n+i} \cap U \qquad (i = 1, 2, \ldots)$$

Taking into account modularity and the fact that $V_n \supseteq V_{n+i}$, we have for each $i = 1, 2, \ldots$

$$
\begin{aligned}
V_n &= V_n \cap (V_n + U) = V_n \cap (V_n + U) = V_{n+i} + (V_n \cap U) \\
&= V_{n+i} + (V_{n+i} \cap U) = V_{n+i}
\end{aligned}
$$

Thus V is artinian. The proof of the noetherian case is dual. ■

2.6. Corollary. *Let* $V = \oplus_{i=1}^n V_i$. *Then* V *is artinian (noetherian) if and only if each module* V_i *is artinian (noetherian).*

Proof. The case $n = 2$ follows from Proposition 2.5. The general case is a straightforward induction on n. ■

2.7. Corollary. *Let* R *be a ring. Then* R *is artinian (noetherian) if and only if every finitely generated* R-*module is artinian (noetherian).*

Proof. Suppose that R is artinian. If V is a finitely generated R-module, then V is a factor module of a free R-module W of finite rank. By Corollary 2.6, W is artinian, hence so is V, by Proposition 2.5. Conversely, if every finitely generated R-module is artinian, then so is R, since R is generated by 1. The proof of the noetherian case is similar. ■

2.8. Corollary. *Let* A *be an algebra over a commutative ring* R *such that* A *is a finitely generated* R-*module. If* R *is artinian (noetherian), then* A *is artinian (noetherian).*

Proof. This is a direct consequence of Corollary 2.7. ∎

We next examine direct decompositions of modules.

2.9. Proposition. *Let R be any ring and let $V \neq 0$ be an R-module which is either artinian or noetherian. Then V is a finite direct sum of indecomposable submodules.*

Proof. For any nonzero R-module M that is not a finite direct sum of indecomposable submodules, choose a proper decomposition $M = M' \oplus N'$ where M' is not a finite direct sum of indecomposable submodules. Assume by way of contradiction that V cannot be written as a finite direct sum of indecomposable submodules. Then

$$V = V' \oplus W', \quad V' = V'' \oplus W'', \ldots$$

is a sequence of proper decompositions. Thus there exist infinite chains

$$W' \subset W' \oplus W' \subset \cdots \text{ and } \quad V \supset V' \supset V'' \supset \cdots$$

proving that V is neither artinian nor noetherian. ∎

2.10. Proposition. *Let $_R R = V_1 \oplus \cdots \oplus V_n$ with $V_i \neq 0$ for $i = 1, \ldots, n$ and write $1 = e_1 + \cdots + e_n$ with $e_i \in V_i$. Then $\{e_1, \ldots, e_n\}$ is a set of pairwise orthogonal idempotents in R and $V_i = Re_i$, $1 \leq i \leq n$. Conversely, if $\{e_1, \ldots, e_n\}$ is a set of pairwise orthogonal idempotents in R, then*

$$R(\sum_{i=1}^{n} e_i) = \oplus_{i=1}^{n} Re_i$$

Proof. Given $r \in R$, we have $r = r \cdot 1 = 1 \cdot r = re_1 + \cdots + re_n$. Hence, in particular,

$$e_i = e_i e_1 + \cdots + e_i e_n$$

which shows that $e_i e_j = \delta_{ij} e_i$ for all i, j. Furthermore, $Re_i \subseteq V_i$ and $_R R = Re_1 \oplus \cdots \oplus Re_n$, which implies that $V_i = Re_i$ for all i.

Conversely, assume that $\{e_1, \ldots, e_n\}$ is a set of pairwise orthogonal idempotents and put $e = \sum_{i=1}^{n} e_i$. Then $e^2 = e$ and $ee_i = e_i e = e_i$

for all i. Hence $Re = \sum_{i=1}^{n} Re_i$. If $\sum r_i e_i = 0$, then multiplication on the right by e_j gives $r_j e_j = 0$ for all j. Therefore, $Re = \oplus_{i=1}^{n} Re_i$, as required. ■

2.11. Corollary. *Assume that a ring R is either artinian or noetherian. Then there exists a set $\{e_1, \ldots, e_n\}$ of pairwise orthogonal primitive idempotents of R with $e_1 + \cdots + e_n = 1$. Moreover,*

$$_R R = Re_1 \oplus \cdots \oplus Re_n$$

where each Re_i is an indecomposable R-module.

Proof. Apply Propositions 2.9 and 2.10. ■

We shall refer to the R-modules $Re_1, \ldots, Re_n$ in Corollary 2.11 as the *principal indecomposable R-modules*.

3. Semisimple modules

In this section, R denotes an arbitrary ring. An R-module V is said to be *semisimple* if every submodule of V is a direct summand of V. Thus, by definition, 0 is a semisimple R-module. The name "semisimple" will be justified by proving that a nonzero R-module V is semisimple if and only if V is a direct sum of simple modules. As a preliminary, we first establish the following result.

3.1. Proposition. *Let V be a semisimple R-module.*
(i) Every submodule of V is isomorphic to a homomorphic image of V and every homomorphic image of V is isomorphic to a submodule of V.
(ii) Every submodule and every homomorphic image of V is semisimple.
(iii) If $V \neq 0$, then V contains a simple submodule.

Proof. (i) If W is a submodule of V, then $V = W \oplus W'$ for some submodule W' of V. Then $W \cong V/W'$ and $V/W \cong W'$, as asserted.
(ii) Let W be a submodule of V and let $U = V/W$. By (i), it suffices

to show that U is semisimple. Suppose that $U_1 = V_1/W$ is a submodule of U. We may choose V_2 with $V = V_1 \oplus V_2$. Then

$$U = U_1 \oplus (V_2 + W)/W$$

and hence U is semisimple.

(iii) Fix a nonzero v in V. By Zorn's lemma, there is a submodule W of V maximal with respect to the property that $v \notin W$. Write $V = W \oplus W'$ for some submodule W' of V. If W' is not simple, then $W' = W_1 \oplus W_2$ for some nonzero submodules W_1, W_2 of V. Since

$$(W \oplus W_1) \cap (W \oplus W_2) = W$$

it follows that $v \notin W \oplus W_i$ for $i = 1$ or $i = 2$. But this contradicts the maximality of W and thus W' is simple, as asserted. ∎

We are now ready to provide the following characterizations of semisimple modules.

3.2. Proposition. *For any nonzero R-module V, the following conditions are equivalent:*
(i) V is semisimple.
(ii) V is a direct sum of simple submodules.
(iii) V is a sum of simple submodules.

Proof. (i) $\Rightarrow$ (ii): Consider the collection of sets of simple submodules of V whose sum is direct. Owing to Proposition 3.1(iii), it is nonempty and, by Zorn's lemma, there is a maximal element, say $\{V_i\}$, in this collection. Put $W = \oplus V_i$ and write $V = W \oplus W'$. If $W' \neq 0$, then by Proposition 3.1(ii), (iii), W' contains a simple submodule V'. Hence $W + V' = V' \oplus (\oplus V_i)$, contrary to the maximality of $\{V_i\}$. Accordingly, $W' = 0$ and $V = W$, as asserted.

(ii) $\Rightarrow$ (iii): This is obvious.

(iii) $\Rightarrow$ (i): Let W be a submodule of V. Owing to Zorn's lemma, there is a submodule W' of V maximal with the property that $W \cap W' = 0$. Thus we must have $W + W' = W \oplus W'$. We are therefore left to verify that $V = W \oplus W'$.

Assume by way of contradiction that $W \oplus W' \neq V$ and chose $v \in V$

such that $v \notin W \oplus W'$. By assumption, $v = v_1 + \cdots + v_n$ where $v_i \in V_i$ and V_i is a simple submodule of V, $1 \leq i \leq n$. Hence $v_j \notin W \oplus W'$ for some j which forces

$$V_j \cap (W \oplus W') \neq V_j$$

But V_j is simple, so $V_j \cap (W \oplus W') = 0$ and therefore

$$W \oplus W' + V_j = W \oplus W' \oplus V_j$$

which shows that $W \cap (W' \oplus V_j) = 0$. Because this contradicts the maximality of W', the result is established. ∎

3.3. Corollary. *Let $V \neq 0$ be a semisimple R-module. Then the following conditions are equivalent:*
(i) V is artinian.
(ii) V is noetherian.
(iii) V is a finite direct sum of simple submodules.

Proof. Owing to Corollary 2.6, if $V = \oplus_{i=1}^{n} V_i$ then V is artinian (noetherian) if and only if each V_i is artinian (noetherian). Now apply Proposition 3.2. ∎

A ring R is said to be *semisimple* if the regular module $_R R$ is semisimple.

3.4. Corollary. *A ring R is semisimple if and only if every R-module is semisimple.*

Proof. It suffices to show that if R is semisimple, then so is every R-module. To this end, let V be an R-module and let $v \in V$. Then the map $_R R \to Rv$, $r \mapsto rv$ is a surjective homomorphism of R-modules. Hence, by Proposition 3.1(ii), Rv is semisimple. Since $V = \sum_{v \in V} Rv$, the result follows by virtue of Proposition 3.2. ∎

3.5. Lemma. *Let $V = \oplus V_i$ where each V_i is simple and let W be a simple submodule of V. Then $W \subseteq \oplus V_j$ where j ranges over those i for which $W \cong V_i$.*

Proof. Given a nonzero $w \in W$, we may write $w = \sum v_i$ with $v_i \in V_i$ and with finitely many $v_i \neq 0$. If $v_i \neq 0$, then the map $rw \rightarrow rv_i, r \in R$, is a nonzero homomorphism from $W = Rw$ to $V_i = Rv_i$. Because W, V_i are simple, $W \cong V_i$ and the result follows. ∎

Let V be a semisimple R-module. We say that V is *homogeneous* if V can be written as a sum of isomorphic simple submodules. The sum of all simple submodules of V which are isomorphic to a given simple submodule of V is called a *homogeneous component* of V.

3.6. Proposition. *Let $V = \oplus_{i \in I} V_i$ where each V_i is simple and let $J \subseteq I$ be such that the $V_j, j \in J$, are all representatives of the isomorphic classes of $V_i, i \in I$. For any given $j \in J$, let W_j denote the sum of all V_i with $V_i \cong V_j$. Then*
(i) The $W_j, j \in J$, are all homogeneous components of V.
(ii) $V = \oplus_{j \in J} W_j$.

Proof. (i) Apply Lemma 3.5.
(ii) This is a direct consequence of the definition of W_j. ∎

Let V be any R-module. A submodule W of V is said to be *fully invariant* in V if it is mapped into itself by all R-endomorphisms of V.

3.7. Proposition. *Let $V \neq 0$ be a semisimple R-module. A submodule W of V is fully invariant if and only if W is a sum of certain homogeneous components of V.*

Proof. Keeping the notation of Proposition 3.6, fix $f \in End_R(V)$. If $V_i \cong V_j$ then either $f(V_i) = 0$ or $f(V_i) \cong V_i \cong V_j$, since V_i is simple. Thus $f(W_j) \subseteq W_j$, so each W_j (and hence the sum of some of them) is fully invariant.

Conversely, suppose that W is fully invariant. We need only verify that if P is a simple submodule of W and Q is a simple submodule of V such that $Q \cong P$, then $Q \subseteq W$. If $Q = P$, this is clear; otherwise $P \cap Q = 0$, so $P + Q = P \oplus Q$ and thus $V = P \oplus Q \oplus U$ for a submodule U of V. Let $f : P \rightarrow Q$ be the given isomorphism and define an endomorphism g of V by

$$g(x + y + z) = f(x) + f^{-1}(y) + z \qquad (x \in P, y \in Q, z \in U)$$

Then $W \supseteq g(W) \supseteq g(P) = F(P) = Q$, as required. ∎

We close by recording the following useful observation.

3.8. Corollary. *Let $V \neq 0$ be a semisimple R-module and write V as the direct sum of its homogeneous components : $V = \oplus_{j \in J} W_j$. Then*

$$End_R(V) \cong \prod_{j \in J} End_R(W_j)$$

Proof. It is obvious that any family (f_j) with $f_j \in End_R(W_j)$ determines a unique endomorphism f of V. The map

$$\begin{cases} \prod_{j \in J} End_R(W_j) & \to & End_R(V) \\ (f_j) & \mapsto & f \end{cases}$$

is clearly an injective ring homomorphism. Since each W_j is fully invariant (Proposition 3.7), any endomorphism f of V maps each W_j into W_j. Setting $f_j = f|W_j$, $j \in J$, it follows that $(f_j) \mapsto f$, as required. ∎

4. The radical and socle of modules and rings

Throughout, R denotes an arbitrary ring. Given an R-module V, the *radical $J(V)$* of V is defined to be the intersection of all maximal submodules of V. If V contains no maximal submodules, then by definition $J(V) = V$. Note however that, if $V \neq 0$ is finitely generated, then $J(V) \neq V$ by virtue of Proposition 2.4. The *Jacobson radical $J(R)$* of R is defined by

$$J(R) = J(_R R)$$

Thus, by definition, $J(R)$ is the intersection of all maximal ideals of R. We say that R is *semiprimitive* if $J(R) = 0$.

The sum of simple submodules of an R-module V, written $Soc(V)$, is called the *socle* of V. If V contains no simple submodules, then we put $Soc(V) = 0$. The *socle* of R, written $Soc(R)$, is defined by

$$Soc(R) = Soc(_R R)$$

4.1. Proposition. *For any nonzero R-module V, the following conditions are equivalent:*

(i) V is a finite direct sum of simple submodules.

(ii) V is semisimple artinian.

(iii) V is artinian and $J(V) = 0$.

Proof. (i) $\Rightarrow$ (ii): This is a direct consequence of Corollary 2.6 and Proposition 3.2.

(ii) $\Rightarrow$ (iii): We may harmlessly assume that $V \neq 0$ in which case V is a direct sum of simple submodules, by Proposition 3.2. Hence, by the definition of $J(V)$, $J(V) = 0$.

(iii)$\Rightarrow$(i): Owing to Proposition 2.1, V is finitely cogenerated. Because $J(V) = 0$, it follows that $\cap_{i=1}^{n} V_i = 0$ for some maximal submodules $V_1, \ldots, V_n$ of V. Hence V is isomorphic to a submodule of $\prod_{i=1}^{n}(V/V_i)$. Thus, by Propositions 3.2 and 3.1(ii), V is semisimple. Now apply Corollary 3.3. ∎

4.2. Corollary. *For any ring R, the following conditions are equivalent:*

(i) $_R R$ is a finite direct sum of simple submodules.

(ii) R is semisimple.

(iii) R is semiprimitive artinian.

(iv) Every R-module is semisimple.

Proof. The equivalence of (ii) and (iv) is the content of Corollary 3.4. That (i) implies (ii) is a consequence of Proposition 3.2, while (iii) implies (i) by Proposition 4.1. Finally, assume that R is semisimple. Since 1 lies in the sum of finitely many simple submodules, it follows that $_R R$ is a finite direct sum of simple submodules. Now apply Proposition 4.1(iii) for $V =_R R$. ∎

4.3. Proposition. *(i) If $V \neq 0$ is an artinian module, then $Soc(V) \neq 0$.*

(ii) V is semisimple if and only if $Soc(V) = V$.

(iii) $Soc(V)$ is the unique largest semisimple submodule of V.

Proof. (i) By Corollary 2.2, V has a simple submodule and thus

$Soc(V) \neq 0$.

(ii) We may clearly assume that $V \neq 0$, in which case the required assertion follows by virtue of Proposition 3.2.

(iii) Apply Proposition 3.2. ∎

4.4. Proposition. *Let* $f : W \to V$ *be a homomorphism of* R-*modules.*

(i) $f(J(W)) \subseteq J(V)$ *with equality if* f *is surjective and* $Ker f \subseteq J(W)$.

(ii) $f(Soc(W)) \subseteq Soc(V)$.

Proof. (i) If V has no maximal submodules, then $J(V) = V$ and so $f(J(W)) \subseteq J(V)$. Suppose that M is a maximal submodule of V. Then the map

$$f^* : W \to V/M$$

given by

$$f^*(w) = f(w) + M$$

is an R-homomorphism. Because $Ker f^*$ is a maximal submodule of W, $J(W) \subseteq Ker f^*$ and so $f(J(W)) \subseteq M$. Hence $f(J(W)) \subseteq J(V)$.

Suppose that f is surjective and $Ker f \subseteq J(W)$. If W has no maximal submodules, then so does V, in which case

$$f(J(W)) = f(W) = V = J(V)$$

Assume that the set $\{M_i | i \in I\}$ of all maximal submodules of W is nonempty. By assumption, $Ker f \subseteq M_i$ for all $i \in I$. If X and Y are the lattices of submodules of W containing $Ker f$ and submodules of V, respectively, then the map $U \mapsto f(U)$, $U \in X$, is an isomorphism of X onto Y. Hence $\{f(M_i) | i \in I\}$ is the set of all maximal submodules of V and

$$J(V) = \cap_{i \in I} f(M_i) = f(\cap_{i \in I} M_i) = f(J(W))$$

as asserted.

(ii) Apply Propositions 4.3(iii) and 3.1(ii). ∎

4.5. Corollary. *Let* W *be a submodule of an* R-*module* V.

(i) $J(W) \subseteq J(V)$.

(ii) $J(V/W) \supseteq (J(V) + W)/W$.
(iii) If $W \subseteq J(V)$, then $J(V/W) = J(V)/W$.
(iv) $Soc(W) = W \cap Soc(V)$.

Proof. (i) Apply Proposition 4.4(i) for the case where f is the inclusion map.

(ii) and (iii). Apply Proposition 4.4(i) for the case where $f : V \to V/W$ is the natural homomorphism.

(iv) By Proposition 4.4(ii), applied for the case where f is the inclusion map, we have $Soc(W) \subseteq Soc(V)$ and so $Soc(W) \subseteq W \cap Soc(V)$. The opposite inclusion being a consequence of Propositions 4.3(iii) and 3.2, the result follows. ∎

4.6. Corollary. *(i)* $J(V/J(V)) = 0$ *and* $J(V) \subseteq W$ *for any submodule W of V for which* $J(V/W) = 0$.

(ii) For any given submodule W of V, $W = J(V)$ if and only if $W \subseteq J(V)$ *and* $J(V/W) = 0$.

(iii) For any given submodule W of V, $W = Soc(V)$ if and only if $W \supseteq Soc(V)$ *and* $Soc(W) = W$.

Proof. (i) The first assertion is a consequence of Corollary 4.5(iii), while the second follows from Corollary 4.5(ii).

(ii) Apply (i).

(iii) Apply Proposition 4.3(iii). ∎

4.7. Corollary. *Let I be an ideal of R.*
(i) $I = J(R)$ if and only if $I \subseteq J(R)$ and $J(R/I) = 0$.
(ii) If $I \subseteq J(R)$, then $J(R/I) = J(R)/I$.

Proof. Apply Corollaries 4.6(ii) and 4.5(iii) together with the obvious fact that the radical of the R-module R/I is the same as the Jacobson radical of the ring R/I. ∎

4.8. Proposition. *For any ring R, $Soc(R)$ is an ideal of R.*

Proof. For any given $r \in R$, the right multiplication by r is an endomorphism of $_R R$. Hence, by Proposition 3.1(ii), $Soc(R)r$ is semisimple. Thus $Soc(R)r \subseteq Soc(R)$, by Proposition 4.3(iii). ∎

A submodule W of an R-module V is said to be *superfluous* if for every submodule X of V, $W + X = V$ implies $X = V$.

4.9. **Proposition.** *Let V be an R-module. Then $J(V)$ is the sum of all superfluous submodules in V. Furthermore, if V is finitely generated, then $J(V)$ is the unique largest superfluous submodule of V.*

Proof. Assume that W is a superfluous submodule of V. To prove that $W \subseteq J(V)$, we may assume that V has a maximal submodule M. If $W \not\subseteq M$, then $M + W = V$ and so $M = V$, a contradiction. Thus every superfluous submodule of V is contained in $J(V)$.

Now assume that $x \in V$. If W is a submodule of V with $Rx + W = V$, then either $W = V$ or there is a maximal submodule M of V such that $W \subseteq M$ and $x \notin M$. If $x \in J(V)$, then the latter cannot occur. Hence $x \in J(V)$ implies that Rx is a superfluous submodule of V, proving the first assertion.

Assume that V is finitely generated. By the foregoing, it suffices to verify that $J(V)$ is a superfluous submodule of V. To this end, suppose that $V = W + J(V)$ for a submodule W of V. By Corollary 4.5(ii), we have $J(V/W) = V/W$. Since V/W is finitely generated, it follows from Proposition 2.4 that $V/W = 0$. Hence $V = W$ as desired. ∎

4.10. **Proposition.** *Let an R-module V be generated by $v_1, \ldots, v_n$ and let $v \in V$. Then $v \in J(V)$ if and only if for all $r_i \in R$, $1 \le i \le n$, the elements $v_i + r_i v$, $1 \le i \le n$, generate V.*

Proof. Let $v \in J(V)$ and let W be the submodule of V generated by the elements $v_i + r_i v$, $1 \le i \le n$. Then $W + J(V) = V$ and so, by Proposition 4.9, $W = V$.

Conversely, assume that $v \notin J(V)$. Then there is a maximal submodule W of V such that $v \notin W$. Hence $V = Rv + W$, so for each i there exists $r_i \in R$ such that $v_i \in -r_i v + W$, where $v_i + r_i v \in W$. Therefore the elements $v_i + r_i v, 1 \le i \le n$, do not generate V, thus completing the proof. ∎

Let V be an R-module. The *annihilator* of V, written $ann(V)$, is defined by

$$ann(V) = \{r \in R | rV = 0\}$$

It is clear that $ann(V)$ is an ideal of R and that V may be viewed as an $R/ann(V)$-module. We say that V is *faithful* if $ann(V) = 0$.

An ideal I of R is said to be *primitive* if the ring R/I has a faithful simple module. Obviously, I is primitive if and only if I is the annihilator of a simple R-module.

4.11. Lemma. *(i) An R-module V is simple if and only if $V \cong R/X$ for some maximal left ideal X of R.*

(ii) Every maximal left ideal of R contains a primitive ideal and every primitive ideal is the intersection of the maximal left ideals containing it.

(iii) $J(R)$ is the intersection of the annihilators of simple R-modules. In particular, $J(R)$ is an ideal of R.

Proof.   (i) It is clear that, for any maximal left ideal X of R, R/X is simple. Conversely, suppose that V is simple and choose $0 \neq v \in V$. Then Rv is a nonzero submodule of V and therefore $V = Rv$. The map $R \to V$, $r \mapsto rv$ is thus a surjective R-homomorphism. Hence $R/X \cong V$ for some left ideal X of R. Since V is simple, X is maximal.

(ii) Let X be a maximal left ideal of R. By (i), R/X is a simple R-module. Hence the annihilator of R/X is a primitive ideal contained in X.

Let I be a primitive ideal of R and let V be a simple R-module whose annihilator is I. Given a nonzero x in V, put $V_x = \{r \in R | rx = 0\}$. Then V_x is a left ideal of R. Since V is simple, $V = Rx \cong R/V_x$ and so V_x is a maximal left ideal of R. Since I is the intersection of all V_x with $0 \neq x \in V$, (ii) is established.

(iii) Apply (i) and (ii). ■

4.12. Proposition.   *Let V be an R-module.*

(i) $J(R)V \subseteq J(V)$ with equality if $R/J(R)$ is artinian.

(ii) Assume that $R/J(R)$ is artinian. Then V is semisimple if and only if $J(R)V = 0$.

Proof. (i) If V has no maximal submodules, then $J(V) = V$ and so $J(R)V \subseteq J(V)$. Let M be a maximal submodule of V. Then V/M is a simple R-module. Since $J(R)$ annihilates V/M (Lemma 4.11(iii)), it follows that $J(R)V \subseteq M$. Thus $J(R)V \subseteq J(V)$.

Assume that $R/J(R)$ is artinian. Owing to Corollary 4.6(ii), it suffices to verify that $J(V/J(R)V) = 0$. Because $J(R/J(R)) = 0$ (Corollary 4.7(i) and Lemma 4.7(iii)) and $R/J(R)$ is artinian, every $R/J(R)$-module is semisimple (Corollary 4.2). In particular, the $R/J(R)$-module $V/J(R)V$ is semisimple. Hence $V/J(R)V$ is a semisimple R-module and therefore $J(V/J(R)V) = 0$.

(ii) If V is semisimple, then $J(R)V = 0$ by Lemma 4.11(iii). Conversely, assume that $J(R)V = 0$. Then V can be regarded as a module over the semiprimitive artinian ring $R/J(R)$. Hence, by Corollary 4.2, V is a semisimple $R/J(R)$-module. Thus V is a semisimple R-module, as required. ∎

4.13. Proposition. *(Nakayama's lemma). Let W be a submodule of a finitely generated R-module V. If $W + J(R)V = V$, then $W = V$.*

Proof. Owing to Proposition 4.12, $J(R)V \subseteq J(V)$ and hence $W + J(V) = V$. But, by Proposition 4.9, $J(V)$ is a superfluous submodule of V, hence the result. ∎

We next examine some ring-theoretic properties of $J(R)$. An element x of R is called a *left* (respectively, *right*) unit if there exists $y \in R$ such that $yx = 1$ (respectively, $xy = 1$). By a *unit*, we understand an element x in R which is both a left and a right unit. Hence $x \in R$ is a unit if and only if there exists $y \in R$ (denoted by x^{-1}) such that $xy = yx = 1$.

4.14. Proposition. *Let x be an element of a ring R. Then $x \in J(R)$ if and only if for all $r \in R$, $1 - rx$ is a left unit. In particular, $J(R)$ contains no nonzero idempotents.*

Proof. We first observe that $1 - rx$ is a left unit if and only if $R(1-rx) = R$. Hence the first assertion is a consequence of Proposition 4.10.

Suppose that e is an idempotent of $J(R)$. Then, by the above $x(1 - e) = 1$ for some $x \in R$. Hence $e = 1 \cdot e = x(1 - e)e = 0$, as required. ∎

4.15. Corollary.   *Let $(R_i), i \in I$, be a family of rings. Then*

$$J\left(\prod_{i \in I} R_i\right) = \prod_{i \in I} J(R_i)$$

Proof.   An element $(r_i) \in \prod_{i \in I} R_i$ is a left unit if and only if r_i is a left unit of R_i, for all $i \in I$. Now apply Proposition 4.14. ∎

4.16. Proposition.   *For any ring R, $J(R)$ is the unique largest ideal I of R such that $1 - rx$ is a unit of R for all $r \in R, x \in I$.*

Proof.   Owing to Proposition 4.14, it suffices to show that $1 - x$ is a unit for all $x \in J(R)$. Since, by Proposition 4.14, $1 - x$ is a left unit, we have $y(1 - x) = 1$ for some $y \in R$. Therefore $z = 1 - y = -yx$ is in $J(R)$ and so $1 = y'(1 - z) = y'y$ for some $y' \in R$. Thus y is a unit and $1 - x = y^{-1}$ is also a unit. ∎

4.17. Corollary.   *For any ring R, $J(R^\circ) = J(R)$ and, in particular, $J(R)$ is the intersection of all maximal right ideals of R.*

Proof.   Apply Proposition 4.16. ∎

4.18. Corollary.   *Let I be a left or right nil ideal of R. Then $I \subseteq J(R)$.*

Proof.   Assume that I is a left nil ideal of R and let $x \in I$. Then, for any given $r \in R, rx \in I$ and so $(rx)^n = 0$ for some $n \geq 1$. Applying the identity

$$(1 - y)(1 + y + \cdots + y^{n-1}) = (1 + y + \cdots + y^{n-1})(1 - y) = 1 - y^n$$

we see that $1 - rx$ is a unit. Thus, by Proposition 4.14, $I \subseteq J(R)$. If I is a right nil ideal, then the same argument applied to R° yields the result. ∎

4.19. Proposition. *For any ring R, the following conditions are equivalent:*

(i) R is artinian.

(ii) R is noetherian, $R/J(R)$ is artinian and $J(R)$ is nilpotent.

Proof. (i)$\Rightarrow$ (ii): Put $J = J(R)$ and consider the chain $J \supseteq J^2 \supseteq \cdots$. By assumption, $J^k = J^{k+1} = \cdots$ for some $k \geq 1$. Setting $I = J^k$ we have $I^2 = I$. Assume by way of contradiction that $I \neq 0$. Then there exists a left ideal X in R with $IX \neq 0$, for example $X = R$. Let M be a minimal element in the set of all such X. Then $I(JM) = (IJ)M = IM \neq 0$ and, since $JM \subseteq M$, we have $JM = M$. By Proposition 4.13, it therefore suffices to show that M is finitely generated. By hypothesis, there exists $x \in M$ with $Ix \neq 0$ and hence $I(Rx) \neq 0$. Because $Rx \subseteq M$, we have $M = Rx$, proving that $J(R)$ is nilpotent.

Since R is artinian, so is $R/J(R)$ by Proposition 2.5. To prove that R is noetherian, we argue by induction on the nilpotency index n of $J(R)$. If $n = 1$, then there is nothing to prove. Consider the exact sequence

$$0 \to J^{n-1} \to R \to R/J^{n-1} \to 0$$

Since $J(R/J^{n-1}) = J/J^{n-1}$ (Corollary 4.7(ii)), it follows from the induction hypothesis that R/J^{n-1} is noetherian. Hence, by Proposition 2.5, it suffices to verify that J^{n-1} is noetherian. But J^{n-1} is artinian (as a submodule of an artinian module $_RR$) and J^{n-1} is semisimple, by Proposition 4.12(ii). Hence, by Corollary 3.3, J^{n-1} is noetherian, as required.

(ii) $\Rightarrow$ (i): We argue by induction on the nilpotency index n of $J(R)$. If $n = 1$, there is nothing to prove. By hypothesis, $R/J(R)^{n-1}$ is artinian. On the other hand, $J(R)^{n-1}$ is noetherian (as a submodule of a noetherian module $_RR$). Because $R/J(R)$ is artinian and $J(R)^n = 0$, it follows from Corollary 3.3 and Proposition 4.12(ii) that $J(R)^{n-1}$ is artinian. Thus, by Proposition 2.5, R is artinian. ∎

4.20. Corollary. *Let R be an artinian ring. Then every finitely generated R-module is both artinian and noetherian.*

Proof. Apply Proposition 4.19 and Corollary 2.7. ∎

4.21. Proposition. *Let $R/J(R)$ be an artinian ring. Then*

$$Soc(R) = \{r \in R | J(R)r = 0\}$$

Proof. Sinc $Soc(R)$ is a semisimple R-module, $J(R)x = 0$ for all $x \in Soc(R)$. Setting $V = \{r \in R | J(R)r = 0\}$, it follows that $Soc(R) \subseteq V$.

Since V can be regarded as an $R/J(R)$-module and $R/J(R)$ is artinian, V is a semisimple $R/J(R)$-module (Corollary 4.2). Hence V is a semisimple R-module and, by Proposition 4.3(iii), $V \subseteq Soc(R)$. ∎

4.22. Proposition. *Let R be a ring such that every simple R-module is isomorphic to a submodule of $_RR$. Then*

$$J(R) = \{r \in R | rSoc(R) = 0\}$$

Proof. The hypothesis on R ensures that $rSoc(R) = 0$ if and only if $rV = 0$ for every simple R-module V. Now apply Lemma 4.11(iii). ∎

5. The Krull-Schmidt theorem

In this section, we shall record a number of results pertaining to the uniqueness of decompositions of modules. As usual, R denotes an arbitrary ring.

A ring R is said to be *local* if $R/J(R)$ is a division ring. In what follows, we write $U(R)$ for the unit group of R.

5.1. Lemma. *For any ring R, the following conditions are equivalent:*
  (i) R is local.
  (ii) R has a unique maximal left ideal.
  (iii) $J(R) = R - U(R)$.
  (iv) The set of nonunits of R is a left ideal.

Proof. (i) $\Rightarrow$ (ii): Let I be a maximal left ideal of R. Then $I/J(R)$ is a proper ideal of $R/J(R)$ and thus $I = J(R)$.

(ii) $\Rightarrow$ (iii): Let J be a unique maximal left ideal of R. Since $J(R)$ is the intersection of all maximal left ideals of R, we have $J(R) = J$, and so J is an ideal. The inclusion $J(R) \subseteq R - U(R)$ being obvious, suppose that $x \in R - U(R)$. If $Rx \neq R$, then Rx lies in a maximal left ideal of R, whence $Rx \subseteq J(R)$, so $x \in J(R)$. On the other hand, if $Rx = R$ then $yx = 1$ for some $y \in R$. Clearly $y \notin J(R)$, otherwise $1 = yx \in J(R)$. Hence $Ry = R$, so $zy = 1$ for some $z \in R$ and hence $z = x$. Thus $x \in U(R)$, a contradiction.

(iii) $\Rightarrow$ (iv): Obvious.

(iv) $\Rightarrow$ (i): Let $x \notin J(R)$ and let $I = R - U(R)$. If M is a maximal left ideal of R, then $M \subseteq I \neq R$, whence $M = I = J(R)$. Hence x is a unit and so $x + J(R)$ is a unit of $R/J(R)$. Consequently, $R/J(R)$ is a division ring, as required. ∎

5.2. Corollary. *Let R be a local ring. Then*
(i) The only idempotents of R are 0 and 1.
(ii) For any ideal I of R, R/I is a local ring.

Proof. (i) Let $e \neq 1$ be an idempotent of R. Then e is a nonunit and hence, by Lemma 5.1, $e \in J(R)$. Thus, by Proposition 4.14, $e = 0$.

(ii) Let J/I be a maximal left ideal of R/I. Then J is a maximal left ideal of R, hence $J = J(R)$ by Lemma 5.1. Thus R/I has a unique maximal left ideal and, by Lemma 5.1, R/I is local. ∎

Let R be an arbitrary ring and let $V \neq 0$ be an R-module. We say that V is *strongly indecomposable* if $End_R(V)$ is a local ring.

5.3. Lemma. *(i) If V is strongly indecomposable, then V is indecomposable.*

(ii) If V is indecomposable and both artinian and noetherian, then every $f \in End_R(V)$ is either a unit or nilpotent (in particular, V is strongly indecomposable).

Proof. (i) Assume that V is strongly indecomposable and that $V = V' \oplus V''$ is a direct decomposition of V. If $\pi : V \to V'$ is the projection map, then π is an idempotent of the local ring $End_R(V)$, so $\pi = 0$ or $\pi = 1$ by Corollary 5.2(i). Hence either $V' = 0$ or $V' = V$, proving that V is indecomposable.

(ii) Assume that V is indecomposable and both artinian and noetherian. Let L_n and X_n be, respectively, the image and the kernel of f^n,

$n \geq 1$. We assert that

$$V = L_n \oplus X_n$$

for a sufficiently large n; if substantiated, it will follow that f is nilpotent, provided f is a nonunit.

Since V is both artinian and noetherian, there exists $n \geq 1$ such that $L_n = L_{2n}$ and $X_n = X_{2n}$. Then, for any $x \in V$, we may find $y \in V$ for which $u(x) = u^2(y)$, where $u = f^n$. Thus

$$x - u(y) \in X_n \quad \text{and} \quad x = (x - f(y)) + f(y) \in L_n + X_n$$

Moreover, given $x \in L_n \cap X_n$, we have $u(x) = 0$ and $x = u(y)$ for some $y \in V$. Then $y \in X_{2n} = X_n$, whence $x = u(y) = 0$, proving that $V = L_n \oplus X_n$.

To prove that V is strongly indecomposable, it suffices, by Lemma 5.1, to verify that every nonunit $f \in E = End_R(V)$ is in $J(E)$. Let g be an arbitrary element of E. Then gf is a nonunit and hence is nilpotent. Thus $1 - gf$ is a unit of E, so by Proposition 4.14, $f \in J(E)$ as required. ∎

The following result is Azumaya's generalization of a classical theorem of Krull and Schmidt.

5.4 Theorem. *Let R be a ring, let V be an R-module and let*

$$V = \oplus_{i=1}^{m} V_i = \oplus_{j=1}^{n} W_j$$

where each of the R-modules V_i, W_j is strongly indecomposable. Then $m = n$ and, after possibly reordering the $W_j, V_i \cong W_i$ for all $i \in \{1, \ldots, m\}$.

Proof. We argue by induction on $min\{m, n\}$. If $m = 1$ or $n = 1$, then the result is obvious since V is indecomposable. Let $e_i : V \to V_i$ and $f_j : V \to W_j$ be the projection maps, $1 \leq i \leq m$, $1 \leq j \leq n$. Observe that $e_1 = \sum_{j=1}^{n} e_1 f_j e_1$ and that, by hypothesis, $End_R(V_1)$ is local. Thus $e_1 f_j e_1$ is a unit of $End_R(V_1)$ for some j. By renumbering the W_j, we may therefore assume that $e_1 f_1 e_1$ is an automorphism of $V_1 = e_1 V$. Hence $f_1 V_1 \subseteq W_1$ and the kernel of f_1 on V_1 is 0.

We now claim that

$$V = f_1 V_1 \oplus (\oplus_{i=2}^m V_i) \tag{1}$$

Indeed, assume that $x \in f_1 V_1 \cap \oplus_{i=2}^m V_i$. Then $e_1 x = 0$ and $x = f_1 e_1 y$ for some $y \in V$. Thus $e_1 f_1 e_1 y = e_1 x = 0$ and so $e_1 y = 0$, since $e_1 f_1 e_1$ is an automorphism on V_1. Hence $x = f_1 e_1 y = 0$, proving that $f_1 V_1 \cap \oplus_{i=2}^m V_i = 0$.

Now fix $x \in V$. Then $e_1 x \in e_1 V = e_1 f_1 e_1 V$ and hence $e_1 x = e_1 f_1 e_1 w$ for some $w \in V$. It follows that $e_1(x - f_1 e_1 w) = 0$ and $x - f_1 e_1 w \in \oplus_{i=2}^m V_i$. Hence

$$x = f_1 e_1 w + (x - f_1 e_1 w) \in f_1 V_1 + \sum_{i=2}^m V_i,$$

proving (1).

Since $f_1 V_1 \subseteq W_1$, it follows from (1) that

$$W_1 = f_1 V_1 \oplus \{W_1 \cap \oplus_{i=2}^m V_i\}$$

But W_1 is indecomposable and $f_1 V_1 \neq 0$, so $W_1 = f_1 V_1 \cong V_1$ and

$$V = W_1 \oplus (\oplus_{i=2}^m V_i)$$

Consequently

$$\oplus_{i=2}^m V_i \cong V/W_1 \cong \oplus_{j=2}^n W_j$$

and therefore, by induction, $m = n$ and after possibly reordering the W_j, we have $V_i \cong W_i$ for all $i \in \{1, \dots, m\}$. ∎

Let V be an R-module. We say that V has the *unique decomposition property* if the following two properties hold:

(i) V is a finite direct sum of indecomposable modules;

(ii) If $V = \oplus_{i=1}^m V_i = \oplus_{j=1}^n W_j$, where each V_i, W_j is indecomposable, then $m = n$ and, after possibly reordering the $W_j, V_i \cong W_i$ for all $i \in \{1, \dots, m\}$.

5.5. Corollary. *Let R be a noetherian ring such that each finitely generated indecomposable R-module is strongly indecomposable. Then* *any finitely generated nonzero R-module has the unique decomposition property.*

Proof.   Because R is noetherian, it follows from Corollary 2.7 that any finitely generated R-module $V \neq 0$ is noetherian. Hence, by Proposition 2.9, V is a finite direct sum of indecomposable modules. The desired conclusion is therefore a consequence of Theorem 5.4. ■

5.6 Corollary.   *(Krull-Schmidt theorem). Let $V \neq 0$ be an R-module which is both artinian and noetherian. Then V has the unique decomposition property.*

Proof.   By Proposition 2.9, V is a finite direct sum of indecomposable modules. Since these indecomposable modules are again artinian and noetherian, it follows from Lemma 5.3(ii) that they are strongly indecomposable. Now apply Theorem 5.4. ■

5.7.   Corollary.   *Let R be an artinian ring. Then any finitely generated R-module $V \neq 0$ has the unique decomposition property.*

Proof.   Since R is artinian, the module V is both artinian and noetherian by virtue of Corollary 4.20. The required assertion is therefore a consequence of Corollary 5.6. ■

5.8.   Lemma.   *Let R be a ring, let V be an R-module and let*

$$V = \oplus_{i=1}^{n} V_i = \oplus_{i=1}^{n} W_i$$

where $V_i \cong W_i$ for all $i \in \{1, \ldots, n\}$. If $\alpha_i : V \to V_i$ and $\beta_i : V \to W_i$ are projection maps, then

$$\theta^{-1}\alpha_i\theta = \beta_i \qquad (1 \leq i \leq n)$$

for some automorphism θ of V.

Proof.   Since $V_i \cong W_i$, $1 \leq i \leq n$, we may choose an automorphism θ of V such that $\theta(W_i) = V_i$, $1 \leq i \leq n$. Then, for any $w_i \in W_i, (\theta^{-1}\alpha_i\theta)(w_i) = w_i$ and for any $j \neq i$, $(\theta^{-1}\alpha_i\theta)(W_j) = 0$. Thus $\theta^{-1}\alpha_i\theta = \beta_i$, $1 \leq i \leq n$, as required. ■

5.9. Proposition. *Let R be a ring. Assume that R possesses two sets $\{e_i | i = 1, \ldots, m\}$ and $\{f_j | j = 1, \ldots, n\}$ of orthogonal primitive idempotents such that each Re_i, Rf_j is strongly indecomposable and $\sum_{i=1}^{m} e_i = \sum_{j=1}^{n} f_j = 1$. Then $m = n$ and if the f_j are suitably ordered, there exists a unit u of R such that $u^{-1}e_iu = f_i, i = 1, \ldots, m$.*

Proof. We have direct decompositions

$$R = \oplus_{i=1}^{m} Re_i = \oplus_{j=1}^{n} Rf_j$$

where each Re_i, Rf_j is strongly indecomposable, by hypothesis. Hence, by Theorem 5.4, $m = n$ and after possibly reordering the Rf_j, we have

$$R = \oplus_{i=1}^{m} Re_i = \oplus_{i=1}^{m} Rf_i$$

with $Re_i \cong Rf_i$, $1 \leq i \leq m$. Let $\alpha_i : R \to Re_i$ and $\beta_i : R \to Rf_i$ be the projection maps. Then $\alpha_i(r) = re_i, \beta_i(r) = rf_i$ for all $r \in R, i \in \{1, \ldots, m\}$. In particular, $\alpha_i(1) = e_i$ and $\beta_i(1) = f_i$, $1 \leq i \leq n$. By Lemma 5.8, there exists an automorphism θ of $_RR$ such that $\theta^{-1}\beta_i\theta = \alpha_i$, $1 \leq i \leq n$. Since $\theta(r) = r\theta(1)$ for any $r \in R$, $u = \theta(1)$ is a unit of R. Hence

$$
\begin{aligned}
e_i &= \alpha_i(1) = (\theta^{-1}\beta_i\theta)(1) \\
&= \theta^{-1}\beta_i(u) = \theta^{-1}(uf_i) = uf_iu^{-1},
\end{aligned}
$$

as required. ∎

5.10. Corollary. *Let R be a noetherian ring such that every projective indecomposable R-module is strongly indecomposable. Then there exists a set $\{e_i | i = 1, \ldots, m\}$ of orthogonal primitive idempotents of R with $\sum_{i=1}^{m} e_i = 1$. Furthermore, if $\{f_j | j = 1, \ldots, n\}$ is another such set of idempotents of R, then $m = n$ and if the f_j are suitably ordered, there exists a unit u of R such that $u^{-1}e_iu = f_i, i = 1, \ldots, m$.*

Proof. Apply Corollary 2.11 and Proposition 5.9. ∎

6. Matrix rings

Throughout, R denotes an arbitrary ring and, for any integer $n \geq 1$, $M_n(R)$ is the ring of all $n \times n$-matrices over R.

Let $i, j \in \{1, \ldots, n\}$ and let e_{ij} be the matrix with (i, j)-th entry 1 and 0 elsewhere. The elements $e_{ij} \in M_n(R)$ are called *matrix units*. As is customary, we shall identify R with its image in $M_n(R)$ consisting of all scalar matrices $diag(r, r, \ldots, r), r \in R$. As a preliminary to our first result, let us note that the matrix units satisfy the following properties:

(i) $e_{ij}e_{ks} = 0$ if $j \neq k$ and $e_{ij}e_{js} = e_{is}$.

(ii) $1 = e_{11} + \cdots + e_{nn}$.

(iii) The centralizer of $\{e_{ij}\}$ in $M_n(R)$ is R.

(iv) $R \cong e_{11}M_n(R)e_{11}$.

The above motivates the following definition. A finite set $\{v_{ij} | i, j = 1, \ldots, n\}$ of elements of a ring S satisfying (i) and (ii) is called a *set of matrix units* in S.

6.1. Proposition. *Let S be any ring, let $\{v_{ij} | i, j = 1, \ldots, n\}$ be a set of matrix units in S and let R be the centralizer of $\{v_{ij}\}$ in S. Then the map*

$$\begin{cases} M_n(R) & \overset{\psi}{\to} & S \\ (a_{ij}) & \mapsto & \sum_{i,j} a_{ij}v_{ij} \end{cases}$$

is an isomorphism of rings and R-modules. Furthermore, $R \cong v_{11}Sv_{11}$.

Proof. Observe that $M_n(R)$ is a free R-module freely generated by the matrix units e_{ij}. Hence the map $e_{ij} \mapsto v_{ij}$ extends to a homomorphism of R-modules, which clearly coincides with ψ. Applying (i), (ii), and (iii), we immediately deduce that ψ is also a ring homomorphism.

Assume that $\sum_{i,j} a_{ij}v_{ij} = 0$ and fix $k, s \in \{1, \ldots, n\}$. Then, for all $t \in \{1, \ldots, n\}$,

$$0 = v_{tk}(\sum_{i,j} a_{ij}v_{ij})v_{st} = a_{ks}v_{tt}$$

and so

$$0 = \sum_t a_{ks}v_{tt} = a_{ks}(\sum_t v_{tt}) = a_{ks}$$

Thus $(a_{ij}) = 0$ and therefore ψ is injective.

To prove that ψ is also surjective, fix $s \in S$ and, for each i, j, put

$$a_{ij} = \sum_k v_{ki}sv_{jk}$$

Then, for all v_{rt}, we have

$$v_{rt}a_{ij} = \sum_k v_{rt}v_{ki}sv_{jk} = v_{ri}sv_{jt}$$

and

$$a_{ij}v_{rt} = \sum_k v_{ki}sv_{jk}v_{rt} = v_{ri}sv_{jt}$$

Hence a_{ij} commutes with all v_{rt} and so $a_{ij} \in R$. Furthermore, by the foregoing,

$$
\begin{aligned}
\psi((a_{ij})) &= \sum_{i,j} a_{ij}v_{ij} = \sum_{i,j} v_{ii}sv_{jj} \\
&= (\sum_i v_{ii})s(\sum_j v_{jj}) = s,
\end{aligned}
$$

proving that ψ is surjective.

Finally, bearing in mind that

$$\psi(e_{11}M_n(R)e_{11}) = v_{11}Sv_{11} \quad \text{and} \quad e_{11}M_n(R)e_{11} \cong R,$$

the result follows. ∎

For any integer $n \geq 1$ and for any R-module V, we write V^n for the n-th direct power of V.

6.2. Proposition. *Let V be an R-module and let $n \geq 1$. Then*

$$End_R(V^n) \cong M_n(End_R(V))$$

Proof. Fix $i, j \in \{1, \ldots, n\}$ and define $f_{ij} \in End_R(V^n)$ by

$$f_{ij}(v_1, \ldots, v_n) = (0, \ldots, v_j, 0, \ldots, 0)$$

where v_j is in the i-th place. Then clearly

$$f_{ij}f_{ks} = \delta_{jk}f_{is} \quad \text{and} \quad \sum_{i=1}^n f_{ii} = 1$$

If $\psi \in End_R(V^n)$ and $\psi(v_1, \ldots, v_n) = (\psi_1(v_1), \ldots, \psi_n(v_n))$, then ψ obviously centralizes all f_{ij} if and only if $\psi_1 = \psi_2 = \cdots = \psi_n$. Thus the centralizer of $\{f_{ij}\}$ in $End_R(V^n)$ is identifiable with $End_R(V)$. The desired assertion is now a consequence of Proposition 6.1. ■

Let V be an R-module and let $n \geq 1$. Then the R-module V^n can also be viewed as an $M_n(R)$-module in the following natural way. We visualize the elements of V^n as column vectors and, for each $A \in M_n(R)$ and $x \in V^n$, we define Ax as the matrix multiplication. We are now ready to record the following result.

6.3. Proposition. *(i) The map $W \to W^n$ is an isomorphism from the lattice of submodules of V onto the lattice of $M_n(R)$-submodules of V^n.*

(ii) $End_{M_n(R)}(V^n) \cong End_R(V)$.

(iii) The map $V \mapsto V^n$ induces a bijective correspondence between the isomorphism classes of R and $M_n(R)$-modules. The inverse of this correspondence is given by $W \mapsto e_{11}W$.

Proof. (i) The map $W \xrightarrow{f} W^n$ is obviously order-preserving. Thus we need only show that it has an inverse which is also order-preserving.

Consider the projection $\pi_1 : V^n \to V$ on the first factor and, for any submodule X of V^n, put $g(X) = \pi_1(X)$. Then the correspondence $X \mapsto g(X)$ is order-preserving and clearly $(gf)(W) = W$. From the action of $M_n(R)$ on V^n, we see that X can be written in the form $X = V_1^n$ for some submodule V_1 of V. Thus $(fg)(X) = f(V_1) = V_1^n = X$, proving (i).

(ii) Put $S = M_n(R)$ and chose $f \in End_S(V^n)$. Then, by the nature of action of S on V^n, f has the same projections, say λ_f, on all factors. Conversely, any given $\psi \in End_R(V)$ determines an element of $End_S(V^n)$ whose projections on all factors are equal to ψ. Hence the map $\lambda \mapsto \lambda_f$ provides the desired isomorphism.

(iii) It suffices to verify that $e_{11}V^n \cong V$ and $(e_{11}W)^n \cong W$, where V and W are R and $M_n(R)$-modules, respectively. By the definition of e_{11},

$$e_{11}V^n = V \times 0 \times \cdots \times 0 \cong V$$

Note also that

$$W = e_{11}W \oplus \cdots \oplus e_{n1}W \qquad \text{as } R\text{-modules}$$

Since $e_{k1}w = 0$ and only if $e_{11}w = 0$, $1 \leq k \leq n$, $w \in W$, the map

$$e_{11}w_1 + \cdots + e_{n1}w_n \overset{f}{\mapsto} (e_{11}w_1, \ldots, e_{11}w_n)$$

is well defined and is at least an R-isomorphism of W onto $(e_{11}W)^n$. Hence it suffices to show that

$$f(e_{ij}e_{k1}w) = e_{ij}f(e_{k1}w) \qquad (w \in W, i, j, k \in \{1, \ldots, n\})$$

Because the latter is a consequence of the action of $M_n(R)$ on $(e_{11}W)^n$, the result follows. ∎

For any ideal I of R, we put

$$M_n(I) = \{(a_{ij}) \in M_n(R) | a_{ij} \in I \quad \text{with all} \quad i, j \in \{1, \ldots, n\}\}$$

6.4. Proposition. *(i) The map $I \mapsto M_n(I)$ is a bijection between the sets of ideals of R and $M_n(R)$. In particular, R is simple if and only if so is $M_n(R)$.*
(ii) $M_n(R)/M_n(I) \cong M_n(R/I)$.
(iii) If $R = I_1 \oplus \cdots \oplus I_s$ is a two-sided decomposition of R, then

$$M_n(R) = M_n(I_1) \oplus \cdots \oplus M_n(I_s)$$

is a two-sided decomposition of $M_n(R)$. Furthermore, the ideal I_i is indecomposable if and only if so is $M_n(I_i)$.

Proof. (i) It is obvious that $M_n(I)$ is an ideal of $M_n(R)$ and that the given map is injective. Suppose that J is an ideal of R and let $I \subseteq R$ consist of all entries of elements in J. Then I is an ideal of R such that $J = M_n(I)$.

(ii) The natural homomorphism $R \to R/I$ induces a surjective homomorphism $M_n(R) \to M_n(R/I)$ whose kernel is $M_n(I)$.

(iii) This follows from the facts that $Z(M_n(R)) = Z(R)$ and that, for any $r \in R$, $M_n(R)r = M_n(Rr)$. ∎

6.5. Proposition. *For any ring R and any positive integer n,*

$$J(M_n(R)) = M_n(J(R))$$

Proof. Let V be an R-module and let V^n be an $M_n(R)$-module as in Proposition 6.3. Then $ann(V^n) = M_n(ann(V))$, by the definition of V^n. Now apply Proposition 6.3 and Lemma 4.11(iii). ∎

6.6. Corollary. *Let R be a finite direct product of full matrix rings over division rings. Then R is semiprimitive artinian.*

Proof. That R is semiprimitive follows from Proposition 6.5 and Corollary 4.15. To prove that R is artinian, we may assume that $R = M_n(D)$ for some $n \geq 1$ and some division ring D. Now apply Proposition 6.3. ∎

6.7. Lemma. *Let V be an R-module and let e be an idempotent of R. Then $Hom_R(Re, V) \cong eV$ as additive groups. Similarly, if V is a right R-module, then $Hom_R(eR, V) \cong Ve$.*

Proof. If $f \in Hom_R(Re, V)$, then $ef(e) = f(e^2) = f(e) \in eV$. Therefore the map $f \mapsto f(e)$ is a homomorphism from $Hom_R(Re, V)$ to eV. Conversely, if $v \in eV$, then the map $g_v : xe \mapsto xv$ is an R-homomorphism from Re to V. Because the map $v \mapsto g_v$ is an inverse of $f \mapsto f(e)$, the first isomorphism follows. The second isomorphism is proved by a similar argument. ∎

6.8. Lemma. *Let e be an idempotent of a ring R. Then*

$$End_R(Re) \cong (eRe)^\circ \quad and \quad End_R(eR) \cong eRe$$

In particular,

$$End_R(_RR) \cong R^\circ \quad and \quad End_R(R_R) \cong R$$

Proof. Applying Lemma 6.7 for $V = Re$, it follows that the map $f \mapsto f(e)$ is an isomorphism of the additive group of $End_R(Re)$ onto the additive group of eRe. Given $f, g \in End_R(Re)$, write $f(e) = er_1e$ and $g(e) = er_2e$ for some $r_1, r_2 \in R$. Then

$$(fg)(e) = f(er_2e) = er_2er_1e = (er_2e)(er_1e) = g(e)f(e),$$

proving that $f \mapsto f(e)$ reverses multiplication. Because e is the identity element of the ring eRe, the above map preserves identity elements.

This establishes the first isomorphism. The second isomorphism follows from Lemma 6.7 applied to $V = eR$. ∎

6.9. Lemma.   Let e_1 and e_2 be idempotents of a ring R. Then the following conditions are equivalent:

(i) $Re_1 \cong Re_2$.

(ii) There exist elements e_{12} and e_{21} in R such that $e_{12}e_{21} = e_1$ and $e_{21}e_{12} = e_2$.

(iii) There exist elements e_{12} and e_{21} in R such that

$$e_{12}e_{21} = e_1, e_{21}e_{12} = e_2, e_1e_{12}e_2 = e_{12} \quad and \quad e_2e_{21}e_1 = e_{21}$$

In particular, $Re_1 \cong Re_2$ if and only if $e_1R \cong e_2R$.

Proof.   It is clear that (iii) implies (ii). Replacing e_{12} and e_{21} by $e_1e_{12}e_2$ and $e_2e_{21}e_1$, respectively, it follows that (ii) implies (iii). Let $f : Re_1 \to Re_2$ be an isomorphism of R-modules, and let $e_{12} = f(e_1)$ and $e_{21} = f^{-1}(e_2)$. Then $e_{12}e_2 = e_{12}$ and so

$$e_1 = (f^{-1}f)(e_1) = f^{-1}(e_{12}) = f^{-1}(e_{12}e_2) = e_{12}e_{21}$$

Similarly, $e_2 = e_{21}e_{12}$, which proves that (i) implies (ii). Finally, assume that (iii) holds. Then the last two conditions imply that the maps $f : Re_1 \to Re_2, f(re_1) = re_{12}, g : Re_2 \to Re_1, g(re_2) = re_{21}, r \in R$, are well defined homomorphisms of R-modules. Moreover, the first two conditions ensure that g is the inverse of f, hence the result. ∎

6.10. Proposition.   Let $\{e_{ij}|i,j = 1,\ldots,n\}$ be a set of matrix units in a ring A and let R be the centralizer of $\{e_{ij}\}$ in A.

(i) $A = Ae_{11} \oplus \cdots \oplus Ae_{nn}$ and $Ae_{ii} \cong Ae_{jj}$ for all i,j.

(ii) $End_A(Ae_{ii}) \cong R^\circ$, $1 \le i \le n$. In particular, Ae_{ii} is indecomposable (respectively, strongly indecomposable) if and only if R has only trivial idempotents (respectively, if and only if R is local).

Conversely, if $A = I_1 \oplus \cdots \oplus I_n$ is a direct decomposition of $_AA$ and the A-modules I_i and I_j are isomorphic for all i,j, then there exists a set of matrix units $\{e_{ij}|i,j = 1,\ldots,n\}$ such that $I_i = Ae_{ii}$, $1 \le i \le n$.

Proof.   (i) This is a direct consequence of Proposition 2.10 and Lemma 6.9.

(ii) By (i), Proposition 6.1 and Lemma 6.8,

$$End_A(Ae_{ii}) \cong End_A(Ae_{11}) \cong (e_{11}Ae_{11})^\circ \cong R^\circ$$

Conversely, assume that $A = I_1 \oplus \cdots \oplus I_n$ with $I_i \cong I_j$ for all i, j. Then there exist orthogonal idempotents $e_{jj}, j = 1, \ldots, n$ such that $I_j = Ae_{jj}$. Since $I_1 \cong I_j$, $j = 1, \ldots, n$, it follows from Lemma 6.9 that there exist e_{1j}, e_{j1} such that $e_{11}e_{1j}e_{jj} = e_{1j}, e_{jj}e_{j1}e_{11} = e_{j1}, e_{1j}e_{j1} = e_{11}$ and $e_{j1}e_{1j} = e_{jj}$. If we set $e_{ij} = e_{i1}e_{1j}, i, j = 1, \ldots, n$, then a straightforward calculation shows that $\{e_{ij} | i, j = 1, \ldots, n\}$ is a set of matrix units. ∎

6.11. Proposition. *Let A be a ring which possesses two sets of matrix units $\{e_{ij} | i, j = 1, \ldots, n\}$ and $\{f_{ks} | k, s = 1, \ldots, m\}$ such that the centralizers R and S of $\{e_{ij}\}$ and $\{f_{ks}\}$, respectively, are local rings. Then $n = m$ and there exists a unit u of A such that*

$$f_{ij} = u^{-1}e_{ij}u, \quad S = u^{-1}Ru \qquad (1 \leq i, j \leq n)$$

Proof. By Proposition 6.10 amd Theorem 5.4, $n = m$ and

$$A = Ae_{11} \oplus \cdots \oplus Ae_{nn} = Af_{11} \oplus \cdots \oplus Af_{nn}$$

with $Ae_{ii} \cong Af_{ii}, 1 \leq i \leq n$. Hence, by the proof of Proposition 5.9, there exists a unit v of A such that $f_{ii} = v^{-1}e_{ii}v, 1 \leq i \leq n$. Now set

$$u = \sum_{i=1}^{n} e_{i1}vf_{1i}$$

Then u has the inverse $\sum_{i=1}^{n} f_{i1}v^{-1}e_{1i}$ and $f_{ij} = u^{-1}e_{ij}u, i, j = 1, \ldots, n$. Hence $S = u^{-1}Ru$, as required. ∎

6.12. Corollary. *Let R and S be local rings and let n, k be positive integers. Then $M_n(R) \cong M_k(S)$ implies $n = k$ and $R \cong S$.*

Proof. Direct consequence of Proposition 6.11. ∎

7. The Wedderburn-Artin theorem

In this section, we provide the ring-theoretic structure of semiprimitive artinian rings. Throughout, R denotes a ring.

7.1. Lemma. *(Schur's lemma). Let V be a simple R-module. Then $End_R(V)$ is a division ring.*

Proof. Let $f : V \to V$ be a nonzero homomorphism. Since $f(V)$ is a nonzero submodule of V, $f(V) = V$. Because $f \neq 0$, $Ker f \neq V$ and so $Ker f = 0$. Thus f is an isomorphism, as required. ∎

7.2. Theorem. *(Wedderburn-Artin). Let R be a semiprimitive artinian ring.*

(i) There exist only finitely many, say $V_1, \ldots, V_r$ of nonisomorphic simple R-modules and $V_i \cong Re_i$ for some primitive idempotent e_i of R, $1 \leq i \leq r$.

(ii) $_R R \cong n_1 V_1 \oplus \cdots \oplus n_r V_r$ for some positive integers $n_i, 1 \leq i \leq r$, where $n_i V_i$ is a direct sum of n_i copies of V_i.

(iii) $R \cong \prod_{i=1}^{r} M_{n_i}(D_i)$, where $D_i = End_R(V_i)^\circ \cong e_i Re_i$ is a division ring, $1 \leq i \leq r$.

(iv) The integers n_i and r are unique and each D_i is determined up to isomorphism.

Proof. (i) and (ii). Owing to Corollary 4.2, $_R R$ is a finite direct sum of simple submodules. Therefore, by Proposition 2.10, there exist idempotents $e_1, \ldots, e_n$ of R such that $_R R = Re_1 \oplus \cdots \oplus Re_n$ and each Re_i is simple (in particular, e_i is a primitive idempotent of R). We may assume that $Re_1, \ldots, Re_r$, $1 \leq r \leq n$, are all nonisomorphic modules of the set $\{Re_1, \ldots, Re_n\}$. If V is any simple R-module, then by Corollary 5.7 and Lemma 4.11(i), $V \cong Re_i$ for some $i \in \{1, \ldots, r\}$. This establishes (i) and (ii).

(iii) Put $D_i = End_R(V_i)^\circ, 1 \leq i \leq r$. Then we have

$$
\begin{aligned}
R^\circ &\cong End_R(_R R) \quad \text{(by Lemma 6.8)} \\
&\cong \prod_{i=1}^{r} End_R(n_i V_i) \quad \text{(by Corollary 3.8)} \\
&\cong \prod_{i=1}^{r} M_{n_i}(D_i^\circ) \quad \text{(by Proposition 6.2)}
\end{aligned}
$$

and thus

$$R \cong \prod_{i=1}^{r} M_{n_i}(D_i)$$

Since, by Lemma 6.8, $End_R(V_i) \cong (e_i Re_i)^\circ$ we have $D_i \cong e_i Re_i$, proving (iii).

(iv) Assume that $R \cong \prod_{j=1}^{s} M_{k_j}(D_j')$, where each D_j' is a division ring. Then there exist central idempotents $u_1, \ldots, u_s$ of R such that

$$R = Ru_1 \oplus \cdots \oplus Ru_s$$

and $Ru_j \cong M_{k_j}(D_j')$. Hence, by Proposition 6.10, Ru_j is a direct sum of k_j copies of a simple R-module. It is clear that $Ru_i, Ru_j, i \neq j$, have no common composition factors. Thus, by Corollary 5.7, $r = s$ and by renumbering the u_j, we have $k_j = n_j, 1 \leq j \leq r$.

Put $A = M_{k_j}(D_j'), D = D_j'$ and let V be a simple A-module. Then, by Proposition 6.10, $End_A(V) \cong D^\circ$. On the other hand, $V \cong V_j$, hence

$$End_A(V) \cong End_R(V_j) \cong D_j^\circ$$

This proves that $D_j' \cong D_j$ and the result follows. ∎

7.3. Corollary. *For any ring R, the following conditions are equivalent:*

(i) R is semiprimitive artinian.

(ii) R is a finite direct product of full matrix rings over division rings.

(iii) R is semisimple.

(iv) Every R-module is semisimple.

Proof. Direct consequence of Corollaries 4.2 and 6.6 together with Theorem 7.2(iii). ∎

We close by proving the following classical result.

7.4. Theorem. *(Wedderburn). Let A be a finite-dimensional algebra over a field F and let I be an ideal of A which has an F-basis consisting of nilpotent elements. Then I is a nilpotent ideal of A.*

Proof. We may harmlessly assume that F is algebraically closed. Moreover, because $(I + J(A))/J(A)$ is an ideal of $A/J(A)$ having a basis consisting of nilpotent elements, we may assume that $J(A) = 0$. Hence, by Theorem 7.2(iii),

$$A \cong \prod_{i=1}^{r} M_{n_i}(F)$$

for some positive integers $n_1, \ldots, n_r$. Let π_i be the projection of I into $M_{n_i}(F)$, $1 \le i \le r$. It obviously suffices to show that $\pi_i(I) = 0$ for all $i \in \{1, \ldots, r\}$.

Since $\pi_i(I)$ is an ideal of $M_{n_i}(F)$, either $\pi_i(I) = 0$ or $\pi_i(I) = M_{n_i}(F)$. In the latter case, it follows that $M_{n_i}(F)$ has a basis consisting of nilpotent elements. But then the trace of each matrix in $M_{n_i}(F)$ would be zero, a contradiction. Thus $\pi_i(I) = 0$ and the result is established. ∎

8.   Tensor products

Let R be a ring. Given a right R-module V and a left R-module W, the $\mathbb{Z}$-module $V \otimes_R W$, called the *tensor product* of V and W, is defined as follows. Let F be a free $\mathbb{Z}$-module with $V \times W$ as a basis; then each element of F can be uniquely written in the form

$$\sum z_{ij}(v_i, w_j) \qquad (z_{ij} \in \mathbb{Z}, v_i \in V, w_j \in W)$$

with finitely many z_{ij} distinct from 0. Denote by T the subgroup of F generated by all elements of the form

$$(v_1 + v_2, w) \quad -(v_1, w) - (v_2, w)$$
$$(v, w_1 + w_2) \quad -(v, w_1) - (v, w_2)$$
$$(v, rw) \qquad -(vr, w)$$

with $v, v_i \in V$, $w, w_i \in W$ and $r \in R$. Then $V \otimes_R W$ is defined to be the factor group F/T. The image of (v, w) under the natural homomorphism $F \to F/T$ is denoted by $v \otimes w$. Note that the $\mathbb{Z}$-module $V \otimes_R W$ consists of all finite sums $\sum v_i \otimes w_i$ with $v_i \in V, w_i \in W$.

Suppose that V is an (S, R)-bimodule. Then $V \otimes_R W$ can be viewed as a (left) S-module by setting

$$s(v \otimes w) = sv \otimes w \qquad (v \in V, w \in W, s \in S)$$

In particular, if V and W are modules over a commutative ring R, then $V \otimes_R W$ is an R-module.

Let R be a ring, V, V' two right R-modules, W, W' two left R-modules, and $\varphi : V \to V', \psi : W \to W'$ two R-linear maps. Then the map

$$\varphi \otimes \psi : V \otimes_R W \to V' \otimes_R W'$$

defined by

$$(\varphi \otimes \psi)(v \otimes w) = \varphi(v) \otimes \psi(w)$$

is a $\mathbb{Z}$-linear map. Furthermore, if V, V' are (S, R)-bimodules and φ is also S-linear, then $\varphi \otimes \psi$ is S-linear. We now record a number of standard properties of tensor products (see Bourbaki (1972), (1974)).

8.1. Proposition. *Let V, V', V'' be right R-modules, let W, W', W'' be left R-modules and let*

$$V' \xrightarrow{f} V \xrightarrow{g} V'' \to 0$$

$$W' \xrightarrow{s} W \xrightarrow{t} W'' \to 0$$

be exact sequences.

(i) The sequences

$$V' \otimes_R W \xrightarrow{f \otimes 1} V \otimes_R W \xrightarrow{g \otimes 1} V'' \otimes_R W \to 0$$

$$V \otimes_R W' \xrightarrow{1 \otimes s} V \otimes_R W \xrightarrow{1 \otimes t} V \otimes_R W'' \to 0$$

of $\mathbb{Z}$-homomorphisms are exact.

(ii) The homomorphism $g \otimes t : V \otimes_R W \to V'' \otimes_R W''$ is surjective and its kernel is equal to $Im(f \otimes 1) + Im(1 \otimes s)$. In particular, if V' and W' are submodules of V and W, respectively, and f and s are inclusion maps, then the map

$$\begin{cases} (V/V') \otimes_R (W/W') & \to & (V \otimes_R W)/(Im(f \otimes 1) + Im(1 \otimes s)) \\ (v + V') \otimes (w + W') & \mapsto & (v \otimes w) + (Im(f \otimes 1) + Im(1 \otimes s)) \end{cases}$$

is a $\mathbb{Z}$ -isomorphism.

8.2. Proposition.   *Let $(V_i), i \in I$, be a family of right R-modules and $(W_j), j \in J$, a family of left R-modules. Then the map*

$$\begin{cases} (\oplus_{i \in I} V_i) \otimes (\oplus_{j \in J} W_j) & \to & \oplus_{(i,j) \in I \times J}(V_i \otimes_R W_j) \\ (v_i) \otimes (w_j) & \mapsto & (v_i \otimes w_j) \end{cases}$$

is a $\mathbb{Z}$ -isomorphism.

8.3. Proposition.   *Let V be an (S, R)-bimodule. Then*

$$V \otimes_R R \cong V \qquad (as\ S\text{-}modules)$$

8.4.   Proposition.   *If V and W are two free modules over a commutative ring R and $(v_i), (w_j)$ are R-bases of V and W, respectively, then $(v_i \otimes w_j)$ is an R-basis of $V \otimes_R W$.*

8.5. Proposition.   *Let R be a subring of a ring S and let V be a free left R-module with $\{e_1, \ldots, e_n\}$ as an R-basis. Then $S \otimes_R V$ is a free left S-module with $\{1 \otimes e_1, \ldots, 1 \otimes e_n\}$ as an S-basis.*

8.6. Proposition.   *Let V be a right R-module, W a left R-module, V' a submodule of V and W' a submodule of W. If V' is a direct summand of V and W' is a direct summand of W, then the canonical homomorphism*

$$V' \otimes_R W' \to V \otimes_R W$$

is injective and the image of $V' \otimes_R W'$ is a direct summand of the $\mathbb{Z}$ -module $M \otimes_R N$.

Let R be a ring, V a right R-module and W a left R-module. The module V is said to be *flat* if for every injective homomorphism $f : W' \to W$ of left R-modules, the homomorphism

$$1_V \otimes f : V \otimes_R W' \to V \otimes_R W$$

is injective.

8.7. Proposition.   *Every projective right R-module is flat.*

Let V be a right R-module and let A be a left ideal of R. The homomorphism $V \otimes_R A \to V$ which sends $v \otimes a$ to va is called *canonical*; its image is the subgroup VA consisting of all finite sums $\sum v_i a_i$ with $v_i \in V, a_i \in A$. The induced homomorphism $V \otimes_R A \to VA$ is called *canonical*. The canonical map $V \otimes_R R \to V$ is an isomorphism. Observe that if V is an (S, R)-bimodule, then all canonical maps are homomorphisms of S-modules.

8.8. Proposition.   *Let V be a right R-module. Then the following properties are equivalent:*
   (i) V is flat.
   (ii) For any finitely generated submodule W' of a left R-module W, the homomorphism $1 \otimes i : V \otimes_R W' \to V \otimes_R W$ (i is the inclusion map) is injective.
   (iii) For every exact sequence of left R-modules $W' \overset{\varphi}{\to} W \overset{\psi}{\to} W''$ the sequence

$$V \otimes_R W' \overset{1 \otimes \varphi}{\to} V \otimes_R W \overset{1 \otimes \psi}{\to} V \otimes_R W''$$

is exact.
   (iv) For every finitely generated left ideal A of R, the canonical map $V \otimes_R A \to VA$ is an isomorphism.

Let V be a flat right R-module and let W' be a submodule of a left R-module W. Then the canonical injection $V \otimes_R W' \to V \otimes_R W$ allows us to identify $V \otimes_R W'$ with its image in $V \otimes_R W$. In the future, we shall always use this identification.

8.9. Proposition.   *Let V be a flat right R-module.*
   (i) If $f : X \to Y$ is a homomorphism of left R-modules, then

$$Ker(1_V \otimes f) = V \otimes_R (Ker f) \quad and \quad Im(1_V \otimes f) = V \otimes_R (Im f)$$

   (ii) If W', W'' are two submodules of a left R-module W, then

$$V \otimes_R (W' \cap W'') = (V \otimes_R W') \cap (V \otimes_R W'')$$

9.   Group algebras

Let R be a commutative ring and let G be an arbitrary group. The *group algebra* RG of G over R is the free R-module on the elements of G, with multiplication induced by that in G. More explicitly, RG consists of all formal linear combinations $\sum x_g \cdot g, x_g \in R, g \in G$, with finitely many $x_g \neq 0$ subject to

(i) $\sum x_g \cdot g = \sum y_g \cdot g$ if and only if $x_g = y_g$ for all $g \in G$.

(ii) $\sum x_g \cdot g + \sum y_g \cdot g = \sum(x_g + y_g) \cdot g$.

(iii) $(\sum x_g \cdot g)(\sum y_h \cdot h) = \sum z_t \cdot t$, where $t = \sum_{gh=t} x_g y_h$.

(iv) $r(\sum x_g \cdot g) = \sum(rx_g) \cdot g$ for all $r \in R$.

One easily verifies that these operations define RG as an associative R-algebra with $1 = 1_R \cdot 1_G$, where 1_R and 1_G are identity elements of R and G, respectively. Applying the injective homomorphisms

$$\begin{cases} R \to RG \\ r \mapsto r \cdot 1_G \end{cases} \qquad \begin{cases} G \to RG \\ g \mapsto 1_R \cdot g \end{cases}$$

we shall in future identify R and G with their images in RG. With these identifications, the formal sums and products become ordinary sums and products. For this reason, we drop the dot in $x_g \cdot g$. We shall also adopt the convention that $RG \cong RH$ means an isomorphism of R-algebras.

Let $x = \sum x_g g \in RG$. Then the *support* of x, *Supp x*, is defined by

$$Supp\, x = \{g \in G | x_g \neq 0\}$$

It is clear that $Supp\, x$ is a finite subset of G that is empty if and only if $x = 0$.

9.1. Proposition. *Let A be an R-algebra and let*

$$f : G \to U(A)$$

be a homomorphism of G into the unit group $U(A)$ of A. Then the map $f^* : RG \to A$ defined by

$$f^*(\sum x_g g) = \sum x_g f(g)$$

is a homomorphism of R-algebras. In particular, if f is injective and A is R-free with f(G) as a basis, then $RG \cong A$.

Proof. Since RG is R-free freely generated by G, f^* is a homomorphism of R-modules. Let

$$x = \sum x_g g \quad \text{and} \quad y = \sum y_g g$$

be two elements of RG. Then

$$
\begin{aligned}
f^*(xy) &= f^*\left(\sum_{a,b\in G} x_a y_b ab \right) = \sum_{a,b\in G} x_a y_b f(a)f(b) \\
&= \left(\sum_{a\in G} x_a f(a) \right)\left(\sum_{b\in G} y_b f(b) \right) \\
&= f^*(x)f^*(y),
\end{aligned}
$$

as required. ∎

9.2. Corollary. *The map*

$$
\left\{
\begin{array}{ccc}
RG & \to & R \\
\sum x_g g & \mapsto & \sum x_g
\end{array}
\right.
$$

is a homomorphism of R-algebras.

Proof. This is a special case of Proposition 9.1 where $A = R$ and $f(g) = 1$ for all $g \in G$. ∎

The *augmentation ideal* $I(RG)$ of RG is defined to be the kernel of the homomorphism of Corollary 9.2. Thus $I(RG)$ consists of all $x = \sum x_g g \in RG$ for which

$$aug(x) = \sum x_g = 0$$

We shall refer to $aug(x)$ as the *augmentation* of x and to the homomorphism

$$aug : RG \to R$$

as the *augmentation map*.

It follows from the equality

$$\sum x_g g = \sum x_g (g - 1) + \sum x_g$$

that as an R-module, $I(RG)$ is a free module with the elements $g - 1$, $1 \neq g \in G$, as a basis. In the future we shall often suppress refence to R and simply denote the augmentation ideal of RG by $I(G)$.

If X is a subset of RG, we write $RG \cdot X$ and $X \cdot RG$ for the laft and right ideals of RG, respectively, generated by X, i.e.

$$RG \cdot X = \sum_{x \in X} RGx \quad \text{and} \quad X \cdot RG = \sum_{x \in X} xRG$$

Of course, if $RGx = xRG$ for all $x \in X$, then $RG \cdot X = X \cdot RG$ is a two-sided ideal of RG. For example, if N is a normal subgroup of G, then the equalities

$$(n-1)g = g(g^{-1}ng-1) \quad \text{and} \quad g(n-1) = (gng^{-1}-1)g \qquad (n \in N, g \in G)$$

show that $RG \cdot I(N) = I(N) \cdot RG$ is a two-sided ideal of RG. The significance of this ideal comes from the following fact.

9.3. Proposition. *Let $f : G \to H$ be a surjective homomorphism of groups and let $N = Ker f$. Then the map $f^* : RG \to RH$, which is the R-linear extension of f, is a surjective homomorphism of R-algebras with $Ker f^* = RG \cdot I(N)$. In particular,*

$$RG/RG \cdot I(N) \cong R(G/N)$$

Proof. That f^* is a surjective homomorphism of R-algebras is a consequence of Proposition 9.1. It is clear that $RG \cdot I(N) \subseteq Ker f^*$. Consequently, f^* induces a homomorphism $\psi : RG/RG \cdot I(N) \to RH$. The restriction of ψ,

$$\lambda : [G + RG \cdot I(N)]/RG \cdot I(N) \to H$$

is an isomorphism. Thanks to Proposition 9.1, λ^{-1} can be extended to a homomorphism $RH \to RG/RG \cdot I(N)$ which is inverse to ψ. Thus $Ker f^* = RG \cdot I(N)$, as asserted. ∎

Let R be a commutative ring and let G be a group. Given a nonzero R-module V, by a *representation* of G on V, we understand a homomorphism

$$\rho : G \rightarrow Aut_R(V)$$

We say that ρ is *faithful* if $Ker\rho = 1$. By Proposition 9.1, to each representation ρ of G on V, there corresponds an RG-module structure on V given by

$$\left(\sum x_g g\right)v = \sum x_g \rho(g)v \qquad (v \in V, x_g \in R, g \in G)$$

Conversely, any nonzero RG-module V determines a representation ρ of G on V given by $\rho(g)v = gv$ for all $g \in G, v \in V$. Thus there is a bijective correspondence between the class of all nonzero RG-modules and the class of all representations of G on R-modules. We say that the representation ρ of G is *irreducible (indecomposable, completely reducible)* if the corresponding RG-module is simple (indecomposable, semisimple).

Let V be an R-module which is R-free of finite rank. Then

$$Aut_R(V) \cong GL_n(R)$$

where $GL_n(R)$ is the unit group of $M_n(R)$. Thus to each representation ρ of G on V, corresponds a matrix representation $\rho^* : G \rightarrow GL_n(R)$. In particular, if $R = F$ is a field, then all finitely generated F-modules are free of finite rank. In this case, the study of FG-modules, representations of G on F-modules, and matrix representations of G over F are essentially equivalent.

9.4. Proposition. *(Maschke's theorem). Let G be a finite group and let F be a field of characteristic $p \geq 0$. Then FG is semisimple if and only if p does not divide the order of G.*

Proof. If $p > 0$ divides $|G|$, then $x = \sum_{g \in G} g$ satisfies $x \in Z(FG)$ and $x^2 = |G|x = 0$. Hence $FGx \subseteq J(FG) \neq 0$. Conversely, assume that p does not divide $|G|$, and let W be a submodule of an FG-module V. We may write $V = W \oplus W'$ for some F-subspace W'. Let $\theta : V \rightarrow W$ be the projection map. Define $\psi : V \rightarrow W$ by

$$\psi(v) = |G|^{-1} \sum_{x \in G} x \theta x^{-1} v \qquad (v \in V)$$

It will be shown that ψ is a homomorphism of FG-modules such that $\psi(v) = v$ for all $v \in W$. This will obviously complete the proof.

Since for all $v \in V$ and $y \in G$,

$$
\begin{aligned}
\psi(yv) &= |G|^{-1} \sum_{x \in G} x\theta x^{-1} yv = |G|^{-1} \sum_{z \in G} (yz)\theta(yz)^{-1} yv \\
&= |G|^{-1} \sum_{z \in G} yz\theta z^{-1} v = y\psi(v),
\end{aligned}
$$

ψ is an FG-homomorphism. Now fix $v \in W$. Then, for any $x \in G$, $x^{-1}v \in W$, so $\theta(x^{-1}v) = x^{-1}v$. Accordingly, $x\theta x^{-1}v = v$ and $\psi(v) = v$, as required. ∎

Chapter 2

Frobenius and symmetric algebras

In this chapter, we provide an extensive account of the theory of Frobenius and symmetric algebras. As a byproduct of our investigations, we establish a number of important results pertaining to modular representation theory of groups and block theory of group algebras.

After recording some basic facts about Frobenius and symmetric algebras, we provide a large class of Frobenius algebras, namely we show that if A_1 is a Frobenius F-algebra and G a finite group, then any crossed product F-algebra $A_1 * G$ is Frobenius. We then establish an important result due to Harris (1988) which asserts that if A_1 is semisimple, then any crossed product F-algebra $A_1 * G$ is symmetric. As an application, we prove that if N is a normal subgroup of G and $F*G$ is a twisted group algebra of G over F, then $F*G/(F*G)J(F*N)$ is a symmetric F-algebra. Turning to modules over group algebras, we apply the above results to demonstrate that if V is a finitely generated semisimple FN-module, then the F-algebra $End_{FG}(V^G)$ is symmetric.

We then present a number of classical results concerning Frobenius and symmetric algebras, which are applied in the next section to investigate Frobenius uniserial algebras. As an application, we establish Morita's theorem which asserts that if A is a Frobenius uniserial algebra and e_1, e_2 are primitive idempotents of A belonging to the same block of A, then Ae_1 and Ae_2 have the same composition length. This theorem allows us to investigate blocks of uniserial symmetric algebras and their Cartan matrices.

A further section is devoted to various characterizations of Frobenius algebras. Among other characterizations, we show that A is a Frobenius algebra if and only if $_AA$ is a cogenerator and $Soc(_AA) \cong A/J(A)$ and $Soc(A_A) \cong A/J(A)$. We then turn our attention to characters of symmetric algebras and provide a number of applications to projective modular representations of finite groups.

The chapter culminates in a presentation of a number of important results due to Külshammer. These results have many significant applications in block theory, which is demonstrated in the next section. As one of the various applications, we prove that if F is a splitting field for FG, then the number of nonisomorphic simple FG-modules in a block $B = B(e)$ of FG is equal to $dim_F e Rey(FG)$, where $Rey(FG)$ is the Reynolds ideal of $Z(FG)$.

1. Definitions and elementary properties

Our aim is to establish some elementary facts about Frobenius and symmetric algebras. All the information recorded will be frequently applied in our subsequent investigations.

Throughout this section, we put $A^* = Hom_F(A, F)$, where A denotes a finite-dimensional algebra over a field F. The left (right) A-module A means the regular left (right) A-module.

1.1. Lemma. *Let V be a finite-dimensional space over a field F and let $f : V \times V \to F$ be a bilinear form. Then the following conditions are equivalent:*

(i) $f(x, V) = 0$ implies $x = 0$.
(ii) $f(V, x) = 0$ implies $x = 0$.

Proof. Let $e_1, e_2, \ldots, e_n$ be a basis of V and let $b_{ij} = f(e_i, e_j)$. Then it is clear from the bilinearity that $f(x, V) = 0$ if and only if $f(x, e_i) = 0, 1 \le i \le n$. Similarly, $f(V, x) = 0$ if and only if $f(e_i, x) = 0$, $1 \le i \le n$. Now write $x = \sum \lambda_j e_j$ with $\lambda_j \in F$. Then

$$f(x, e_i) = \sum \lambda_j f(e_j, e_i) = \sum b_{ji} \lambda_j$$

Hence $f(x, V) = 0$ if and only if $(\lambda_1, \lambda_2, \ldots, \lambda_n)$ is a solution of the system of homogeneous linear equations

$$b_{1i}\lambda_1 + b_{2i}\lambda_2 + \cdots + b_{ni}\lambda_n = 0 \qquad 1 \le i \le n \qquad (1)$$

Similarly, $f(V, x) = 0$ if and only if the λ's satisfy

$$b_{i1} + b_{i2}\lambda_2 + \cdots + b_{in}\lambda_n = 0 \qquad 1 \le i \le n \qquad (2)$$

We know, from linear algebra, that (1) or (2) has only a zero solution if and only if $det(b_{ij}) \ne 0$. So the lemma is verified. ∎

We say that a bilinear form

$$f : V \times V \to F$$

is *nonsingular* it satisfies the equivalent conditions of Lemma 1.1.

Let $f : A \times A \to F$ be a bilinear form. Then f is said to be *associative* if

$$f(xy, z) = f(x, yz) \quad \text{for all} \quad x, y, z \in A$$

while f is called *symmetric* if

$$f(x, y) = f(y, x) \quad \text{for all} \quad x, y \in A$$

1.2. Corollary. *Let $\psi \in A^*$ and let $f : A \times A \to F$ be defined by $f(x, y) = \psi(xy)$. Then f is an associative bilinear form on A which is symmetric if and only if $\psi(xy) = \psi(yx)$ for all $x, y \in A$. Moreover, the following conditions are equivalent:*
(i) $Ker\psi$ contains no nonzero right ideals of A.
(ii) $Ker\psi$ contains no nonzero left ideals of A.
(iii) f is nonsingular.

Proof. It is clear that f is a bilinear form on A. Given $x, y, z \in A$, we have

$$f(xy, z) = \psi((xy)z) = \psi(x(yz)) = f(x, yz),$$

proving that f is associative. By definition, $f(x, y) = f(y, x)$ if and only if $\psi(xy) = \psi(yx)$. Hence f is symmetric if and only if $\psi(xy) = \psi(yx)$

for all $x, y \in A$.

Finally, observe that $f(x, A) = 0$ if and only if $\psi(xA) = 0$, while $f(A, x) = 0$ if and only if $\psi(Ax) = 0$. The desired conclusion is therefore a consequence of Lemma 1.1. ∎

1.3. Lemma. *The following conditions are equivalent:*

(i) There exists $\psi \in A^*$ such that $Ker\psi$ contains no nonzero right ideals of A.

(ii) There exists $\psi \in A^*$ such that $Ker\psi$ contains no nonzero left ideals of A.

(iii) There exists a bilinear form $f : A \times A \to F$ which is both associative and nonsingular.

Proof. (i) $\Rightarrow$ (ii): Apply Corollary 1.2.

(ii) $\Rightarrow$ (iii): Apply Corollary 1.2.

(iii) $\Rightarrow$ (i): Let $\psi : A \to F$ be defined by $\psi(x) = f(x, 1)$. It is obvious that $\psi \in A^*$. Let $a \in A$ be such that $\psi(a\, A) = 0$. Then $f(ab, 1) = 0$ for all $b \in A$. But $f(ab, 1) = f(a, b)$, so the nonsingularity of f ensures that $a = 0$. ∎

We say that A is a *Frobenius algebra* if it satisfies the equivalent conditions of Lemma 1.3.

1.4. Lemma. *The following conditions are equivalent:*

(i) There exists $\psi \in A^*$ such that $Ker\psi$ contains no nonzero right ideals of A and $\psi(xy) = \psi(yx)$ for all $x, y \in A$.

(ii) There exists $\psi \in A^*$ such that $Ker\psi$ contains no nonzero left ideals of A and $\psi(xy) = \psi(yx)$ for all $x, y \in A$.

(iii) There exists a bilinear form $f : A \times A \to F$ which is associative, nonsingular and symmetric.

Proof. (i) $\Rightarrow$ (ii): Apply Corollary 1.2.

(ii) $\Rightarrow$ (iii): Apply Corollary 1.2.

(iii) $\Rightarrow$ (i): Define $\psi : A \to F$ by $\psi(x) = f(x, 1)$. Then, by Lemma 1.3, $\psi \in A^*$ and $Ker\psi$ contains no nonzero right ideals of A. Moreover, because f is symmetric, we also have

$$\psi(xy) = f(xy, 1) = f(x, y) = f(y, x) = f(yx, 1) = \psi(yx),$$

as asserted. ∎

We say that A is a *symmetric algebra* if it satisfies the equivalent conditions of Lemma 1.4. In order to avoid excessive verbosity, it will be convenient to say that (A, ψ) is a Frobenius (or symmetric) algebra whenever ψ satisfies conditions of Lemma 10.3 (or Lemma 10.4). As a preliminary to the next observation, observe that A^* is an (A, A)-bimodule via

$$\begin{aligned}(af)(x) &= f(xa) \\ (fa)(x) &= f(ax)\end{aligned}$$

for all $x, a \in A$ and $f \in A^*$.

1.5. Lemma. *(i) Assume that (A, ψ) is a Frobenius algebra. Then the map $f : A \to A^*$ given by $f(a)(x) = \psi(xa)$ for all $x, a \in A$ is an isomorphism of left A-modules. Furthermore, if (A, ψ) is a symmetric algebra, then f is an isomorphism of (A, A)-bimodules.*

(ii) Assume that $f : A \to A^*$ is an isomorphism of left A-modules and put $\psi(a) = f(1)(a)$ for all $a \in A$. Then (A, ψ) is a Frobenius algebra. Furthermore, if f is an isomorphism of (A, A)-bimodules, then (A, ψ) is a symmetric algebra.

Proof. (i) It is obvious that for all $a \in A$, $f(a) \in A^*$, and that

$$f(x + y) = f(x) + f(y) \quad \text{for all} \quad x, y \in A$$

Given $r, x \in A$, we also have

$$f(ra)(x) = \psi(xra) = f(a)(xr) = (rf(a))(x),$$

proving that f is a homomorphism of left A-modules. By the definition of ψ, f is injective, and because $dim_F A = dim_F A^*$, f is a bijection. If (A, ψ) is a symmetric algebra, then for all $r, x \in A$, we have

$$f(ar)(x) = \psi(xar) = \psi(rxa) = f(a)(rx) = (f(a)r)(x),$$

as required.

(ii) It is plain that $\psi \in A^*$. Assume that $\psi(a\,A) = 0$. Then, for all $x \in A$,

$$0 = f(1)(ax) = f(x)(a)$$

and hence $a = 0$. Thus (A, ψ) is a Frobenius algebra.

Suppose that f is an isomorphism of (A, A)-bimodules. Then, for all $x, y \in A$, we have

$$\psi(xy) = f(1)(xy) = f(x)(y) = f(1)(yx) = \psi(yx),$$

as desired. ∎

1.6. Corollary. *The following conditions are equivalent:*
(i) A is a Frobenius (respectively, symmetric) algebra.
(ii) $A \cong A^*$ as left A-modules (respectively, $A \cong A^*$ as (A, A)-bimodules).

Proof. Apply Lemma 1.5. ∎

1.7. Lemma. *Let $A_1, \ldots, A_n$ be F-algebras. Then $\prod_{i=1}^n A_i$ is a Frobenius (respectively, symmetric) algebra if and only if each A_i is a Frobenius (respectively, symmetric) algebra.*

Proof. Suppose that $\psi_i \in A^*$ is such that (A_i, ψ_i) is a Frobenius algebra, $1 \leq i \leq n$. Put $A = \prod_{i=1}^n A_i$ and define $\psi : A \to F$ by

$$\psi(a_1, \ldots, a_n) = \psi_1(a_1) + \cdots + \psi_n(a_n)$$

Then obviously $\psi \in A^*$. Put $a = (a_1, \ldots, a_n)$ and assume that $\psi(A\,a) = 0$. Let $x = (x_1, \ldots, x_n) \in A$ be such that $x_j = 0$ for $j \neq i$. Then $\psi(xa) = \psi_i(x_i a_i) = 0$ for all $x_i \in A_i$. Thus each $a_i = 0$ and so $a = 0$, proving that (A, ψ) is a Frobenius algebra.

Conversely, suppose that (A, ψ) is a Frobenius algebra. Let $\lambda_i : A_i \to A$ be the canonical injection and let $\psi_i = \psi \circ \lambda_i$. Then (A_i, ψ_i) is obviously a Frobenius algebra.

If each (A, ψ_i) is a symmetric algebra, then for

$$a = (a_1, \ldots, a_n) \quad \text{and} \quad b = (b_1, \ldots, b_n)$$

we have

$$\psi(ab) = \psi_1(a_1b_1) + \cdots + \psi_n(a_nb_n) = \psi_1(b_1a_1) + \cdots + \psi_n(b_na_n) = \psi(ba),$$

proving that (A, ψ) is a symmetric algebra.

Conversely, if (A, ψ) is a symmetric algebra, then for $a_i, b_i \in A_i$, we have

$$
\begin{aligned}
\psi_i(a_ib_i) &= \psi(\lambda_i(a_ib_i)) = \psi(\lambda_i(a_i)\lambda_i(b_i)) \\
&= \psi(\lambda_i(b_i)\lambda_i(a_i)) = \psi(\lambda_i(b_ia_i)) \\
&= \psi(b_ia_i)
\end{aligned}
$$

Hence each (A_i, ψ_i) is symmetric and the result is established. ∎

1.8. Lemma. *If A is a Frobenius (symmetric) algebra, then for any positive integer n, $M_n(A)$ is a Frobenius (symmetric) algebra.*

Proof. Let $\psi \in A^*$ be such that (A, ψ) is a Frobenius algebra. Define the map $\lambda : M_n(A) \to F$ by

$$\lambda[(a_{ij})] = \sum_{i=1}^n \psi(a_{ii}) \qquad (a_{ij} \in A)$$

Then clearly $\lambda \in M_n(A)^*$. Suppose that $(a_{ij}) \in M_n(A)$ is such that

$$\lambda((a_{ij})M_n(A)) = 0$$

We must show that $(a_{ij}) = 0$. To this end, fix $i, j \in \{1, \ldots, n\}$ and let e_{ij} denote the matrix with (i, j)-th entry 1 and 0 elsewhere. Then, for any $a \in A$

$$\lambda[(a_{ij})ae_{ij}] = \psi(a_{ji}a) = 0$$

Since (A, ψ) is a Frobenius algebra, we conclude that $a_{ji} = 0$. Hence $(a_{ij}) = 0$ and so $M_n(A)$ is a Frobenius algebra.

If (A, ψ) is a symmetric algebra, then

$$
\begin{aligned}
\lambda[(a_{ij})(b_{ij})] &= \sum_{i=1}^n \psi(\sum_{k=1}^n a_{ik}b_{ki}) = \sum_{i=1}^n \sum_{k=1}^n \psi(a_{ik}b_{ki}) \\
&= \sum_{i=1}^n \sum_{k=1}^n \psi(b_{ki}a_{ik}) = \sum_{i=1}^n \sum_{k=1}^n \psi(b_{ik}a_{ki}) \\
&= \lambda[(b_{ij})(a_{ij})],
\end{aligned}
$$

proving that $M_n(A)$ is also symmetric. ∎

1.9. Proposition. *(Eilenberg - Nakayama (1955)). If A is a semisimple algebra, then A is symmetric.*

Proof. By Wedderburn's theorem, A is isomorphic to a finite direct product of full matrix rings over division rings. Hence, by Lemmas 1.7 and 1.8, we may harmlessly assume that A is a division algebra over F. Then, for any nonzero ψ in A^*, (A, ψ) is a Frobenius algebra. Let $[A, A]$ denote the additive subgroup of A generated by all elements of the form $ab - ba$, where $a, b \in A$. It is plain that $[A, A]$ is an F-subspace of A. To prove that (A, ψ) is a symmetric algebra, it suffices to show that $\psi([A, A]) = 0$ for some nonzero ψ in A^*. The latter will follow, provided we demonstrate that $[A, A] \neq A$. This last statement being independent of the ground field, we may assume that F is the centre of A.

Let E/F be a field extension sucht that $A_E = E \otimes_F A$ satisfies $A_E \cong M_n(E)$ for some $n \geq 1$. Since each matrix in $[M_n(E), M_n(E)]$ has trace 0, we have $[A_E, A_E] \neq A_E$. But $[A_E, A_E] = [A, A]_E$, so $[A, A] \neq A$, as asserted. ∎

As a preliminary to the next result, we quote the following standard fact.

1.10. Proposition. *(Noether - Deuring theorem). Let V and W be (left, right or two-sided) A-modules and let E be a field extension of F. Then $V \cong W$ as A-modules if and only if $V_E \cong W_E$ as A_E-modules (here $V_E = E \otimes_F V$ and $A_E = E \otimes_F A$).*

Proof. See Curtis and Reiner (1962). ∎

1.11. Proposition. *(Eilenberg - Nakayama (1955)). (i) If A_1 and A_2 are Frobenius (respectively, symmetric) F-algebras, then so is $A = A_1 \otimes_F A_2$.*

(ii) Let E be a field extension of F and let A be an F-algebra. Then the E-algebra A_E is a Frobenius (respectively, symmetric) algebra if and only if A is so.

Proof. (i) Suppose that A_1 and A_2 are Frobenius F-algebras. Then, by Corollary 1.6, there exists an isomorphism $f_i : A_i \to A_i^*$ of left A_i-modules, $i = 1, 2$. Consider the map $f : A \to A^*$ defined by

$$[f(a_1 \otimes a_2)](b_1 \otimes b_2) = [f_1(a_1)(b_1)][f_1(a_2)(b_2)]$$

for all $a_1, b_1 \in A_1, a_2, b_2 \in A_2$. A routine verification shows that f is an isomorphism of left A-modules. Moreover, if each f_i is an (A_i, A_i)-isomorphism, then f is an (A, A)-isomorphism. The required assertion is therefore a consequence of Corollary 1.6.

(ii) This is a direct consequence of Corollary 1.6 and Proposition 1.10. ∎

We close by demonstrating that any F-algebra is a homomorphic image of a symmetric algebra. As a preliminary, we record the following observation.

1.12. Lemma. *Let R be an arbitrary ring, let V be an (R, R)-bimodule and let $S = R \times V$. Define multiplication on S by*

$$(r_1, v_1)(r_2, v_2) = (r_1 r_2, r_1 v_2 + v_1 r_2) \quad (r_i \in R, v_i \in V)$$

Then S is a ring and R is a homomorphic image of S.

Proof. A routine verification shows that S is a ring. Furthermore, the projection map $S \to R$ is a surjective ring homomorphism. ∎

1.13. Proposition. *(Tachikawa). Every F-algebra is a homomorphic image of a symmetric F-algebra.*

Proof. Let A be an F-algebra and let $B = A \times A^*$ be the F-algebra defined as in Lemma 1.12. By Lemma 1.12, A is a homomorphic image of B. Hence it suffices to verify that B is a symmetric F-algebra. Define $\lambda : B \to F$ by

$$\lambda(a, \varphi) = \varphi(1) \quad (a \in A, \varphi \in A^*)$$

If $x_i = (a_i, \varphi_i), i = 1, 2$, then

$$\lambda(x_1 + x_2) = \lambda(a_1 + a_2, \varphi_1 + \varphi_2) = (\varphi_1 + \varphi_2)(1) = \lambda(x_1) + \lambda(x_2)$$

and obviously $\lambda(\mu(a, \varphi)) = \mu\lambda(a, \varphi)$ for all $\mu \in F$. Thus $\lambda \in B^*$.

Now assume that $b = (a, \varphi)$ is such that $\lambda(bB) = 0$. Then, for all $x \in A$ and $\psi \in A^*$, we have $(a\psi + \varphi x)(1) = 0$. Hence

$$\psi(a) + \varphi(x) = 0 \qquad \text{for all} \quad x \in A, \psi \in A^*$$

In particular, taking $x = 0$, we obtain $\psi(a) = 0$ for all $\psi \in A^*$ and hence $a = 0$. Therefore $\varphi = 0$ also and the result follows. ∎

2.   Frobenius crossed products

In this section, we provide a large class of Frobenius algebras which play an important role in group representation theory.

Let A be an algebra over a commutative ring R. If X, Y are R-submodules of A, then XY denotes the R-submodule of A consisting of all finite sums

$$\sum x_i y_i \qquad (x_i \in X, y_i \in Y)$$

Let G be a multiplicative group. Then A is called a G-graded algebra if there is a family

$$\{A_g | g \in G\}$$

of R-submodules of A indexed by the elements of G such that the following conditions hold:

(a) $A = \oplus_{g \in G} A_g$ (direct sum of R-modules).

(b) $A_x A_y \subseteq A_{xy}$ for all $x, y \in G$.

We shall refer to (a) as a G-grading of A and to A_g as the g-component of A. When (b) is replaced by the stronger condition, namely

(c) $A_x A_y = A_{xy}$ for all $x, y \in G$

we say that A is a strongly G-graded algebra.

From now on, we fix a G-graded algebra A over R and denote by $U(A)$ the unit group of A.

2.1. Lemma.   (i) A_1 is a subalgebra of A with $1 \in A_1$.

(ii) For each $g \in G$, A_g is an (A_1, A_1)-bimodule under left and right multiplication by the elements of A_1.

(iii) A is strongly G-graded if and only if $1 \in A_g A_{g^{-1}}$ for all $g \in G$.

Proof. (i) By definition, A_1 is an R-submodule of A and, by (b), A_1 is multiplicatively closed. Thus we need only verify that $1 \in A_1$. Applying (a), we may write

$$1 = \sum_{g \in G} a_g$$

where $a_g \in A_g$ for all $g \in G$, and all but finitely many of a_g are zero. Now fix $h \in G$ and $a_h' \in A_h$. Since G is a group, it follows from (a) that

$$A = \oplus_{g \in G} A_{gh} \tag{3}$$

Applying (b), we see that the product $a_g a_h'$ lies in A_{gh} for all $g \in G$. Consequently,

$$a_h' = 1 \cdot a_h' = \sum_{g \in G} a_g a_h'$$

is precisely the expansion of a_h' in the decomposition (3). But a_h' already lies in the direct summand $A_{1 \cdot h} = A_h$ in (3). Thus all the $a_g a_h'$ for $g \neq 1$ must be zero and $a_1 a_h'$ must be a_h'. This demonstrates that a_1 acts as a left identity on A_h for all $h \in G$. Invoking (a), it therefore follows that a_1 is a left identity for the algebra A. Hence $a_1 \in A_1$ is the identity element of A, as required.

(ii) This follows by an application of property (b).

(iii) Assume that A is strongly G-graded. Since $1 \in A_1$, it follows from (c) that $1 \in A_g A_{g^{-1}}$ for all $g \in G$.

Conversely, suppose that $1 \in A_g A_{g^{-1}}$ for all $g \in G$. Applying (b), we then obtain

$$\begin{aligned} A_{xy} &= 1 \cdot A_{xy} = A_x A_{x^{-1}} A_{xy} \\ &\subseteq A_x A_{x^{-1} xy} = A_x A_y \end{aligned}$$

for all $x, y \in G$. Hence $A_x A_y = A_{xy}$ for all $x, y \in G$, as required. ∎

We refer to a unit u of A as being *graded* if $u \in A_g$ for some $g \in G$. In this case, we say that g is the *degree* of u and write

$$g = deg(u) = deg_A(u)$$

The set of all graded units of A will be denoted by $GrU(A)$.

2.2. Lemma. *(i) If $u \in Gr\, U(A)$ is of degree g, then u^{-1} is of degree g^{-1}.*

(ii) $Gr\, U(A)$ is a subgroup of $U(A)$ and the map

$$deg : Gr\, U(A) \to G$$

is a group homomorphism with kernel $U(A_1)$.

(iii) the map

$$\begin{cases} Gr\, U(A) & \to & Aut\,(A_1) \\ u & \mapsto & i_u \end{cases}$$

where

$$i_u(x) = uxu^{-1} \qquad \text{for all} \quad x \in A_1$$

is a homomorphism.

(iv) Right multiplication by any $u_g \in A_g \cap U(A)$ is an isomorphism

$$A_1 \to A_1 u_g = A_g$$

of left A_1-modules.

Proof. (i) We may write $u^{-1} = \sum_{x \in G} a_x$ with $a_x \in A_x$ and with finitely many nonzero a_x. Because $u a_x \in A_g A_x \subseteq A_{gx}$ for all $x \in G$, we see that $uu^{-1} = \sum_{x \in G} u a_x$ is a unique expansion for uu^{-1} in the following decomposition:

$$A = \oplus_{x \in G} A_{gx} = \oplus_{x \in G} A_x$$

Owing to Lemma 2.1(i), $uu^{-1} = 1$ lies in $A_1 = A_g A_{g^{-1}}$. Hence the a_x for $x \neq g^{-1}$ must be zero, and 1 must equal $u a_{g^{-1}}$. Thus $u^{-1} = a_{g^{-1}}$ lies in $A_{g^{-1}}$, as required.

(ii) That $Gr\, U(A)$ is a subgroup of $U(A)$ follows from (i) and the fact that $A_x A_y \subseteq A_{xy}$ for all $x, y \in G$. Because $A_x A_y \subseteq A_{xy}$, the given map is a homomorphism with kernel $A_1 \cap U(A)$. By (i), $A_1 \cap U(A_1) \subseteq U(A_1)$ and since the opposite containment is obvious, the desired assertion follows.

(iii) By assumption, $u \in A_g$ for some $g \in G$. Hence, by (i), we have

$$u A_1 u^{-1} \subseteq A_g A_1 A_{g^{-1}} \subseteq A_1$$

Taking into account that $Gr\,U(A)$ is a group, (iii) is established.

(iv) Because u_g is a unit of A, right multiplication by u_g is an A_1-isomorphism of A_1 onto A_1u_g. Since $u_g \in A_g$, we have $A_1u_g \subseteq A_g$. Thanks to (i), we also have

$$A_g = A_g u_g^{-1} u_g \subseteq A_g A_{g^{-1}} u_g \subseteq A_1 u_g$$

This implies that $A_g = A_1 u_g$, as we wished to show. ∎

We now introduce a very important class of group-graded algebras, namely crossed products. Thanks to Lemma 2.2(ii), the sequence of group homomorphisms

$$1 \to U(A_1) \to Gr\,U(A) \overset{deg}{\to} G \to 1 \tag{4}$$

is always exact, except possibly at G. We say that a G-graded algebra A is a *crossed product* of G over A_1, written $A = A_1 * G$, provided the sequence (4) is exact. Hence A is a crossed product of G over A_1 if and only if for any $g \in G$, there exists $\bar{g} \in A_g \cap U(A)$.

In the special case, where (4) is an *exact splitting sequence*, we shall refer to A as a *skew group ring* of G over A_1.

A G-graded algebra A is called a *twisted group ring* of G over A_1 if for all $g \in G$, there exists $\bar{g} \in A_g \cap U(A)$ such that $\bar{g}$ centralizes A_1. In a particular case, where $A_1 \subseteq Z(A)$ and $A_g \cap U(A) \neq \emptyset$ for all $g \in G$, we refer to A as a *twisted group algebra* of G over A_1. Finally, note that if $A_1 * G$ is a twisted group algebra and a skew group ring, $A_1 * G$ is nothing else but the group algebra of G over A_1.

2.3. Lemma. *Let a G-graded algebra A be a crossed product of G over A_1. For any $g \in G$, fix a unit $\bar{g}$ of A in A_g with $\bar{1} = 1$. Then*

(i) A is a strongly G-graded algebra with $A_g = A_1\bar{g} = \bar{g}A_1$.

(ii) A is a free (left and right) A_1-module freely generated by the elements $\bar{g}, g \in G$.

(iii) Define $\alpha : G \times G \to U(A_1)$ by $\alpha(x,y) = \bar{x}\bar{y}\overline{xy}^{-1}$ and put $^g a = \bar{g}a\bar{g}^{-1}$, $g \in G, a \in A_1$. Then

$$(r_1\bar{x})(r_2\bar{y}) = r_1{}^x r_2\alpha(x,y)\overline{xy}$$

for all $x,y \in G, r_1, r_2 \in A_1$.

Proof. (i) Suppose that $u \in A_g \cap U(A)$ for some $g \in G$. By Lemma 2.2(i),

$$u^{-1} \in A_{g^{-1}} \cap U(A)$$

and so $1 = uu^{-1} \in A_g A_{g^{-1}}$. Applying Lemma 2.1(iii), we deduce that A is a strongly G-graded algebra. By Lemma 2.2(iv), $A_g = A_1\bar{g}$ and the argument of that lemma applied to left multiplication shows that $A_g = \bar{g}A_1$, proving (i).

(ii) It follows from (i) and the definition of a G-graded algebra that

$$A = \oplus_{g \in G} A_1\bar{g} = \oplus_{g \in G} \bar{g}A_1$$

which proves the required assertion.

(iii) We have

$$(r_1\bar{x})(r_2\bar{y}) = r_1(\bar{x}r_2\bar{x}^{-1})(\bar{x}\bar{y}) = r_1{}^x r_2 \alpha(x, y)\overline{xy},$$

as required. ∎

In what follows, A_1 denotes a finite-dimensional algebra over a field F.

2.4. Theorem. *Let A_1 be a Frobenius algebra and let G be a finite group. Then any crossed product F-algebra $A_1 * G$ is Frobenius.*

Proof. For any $g \in G$, fix a unit $\bar{g}$ of $A_1 * G$ with $\bar{1} = 1$. Given $a \in A_1$, $g \in G$, put ${}^g a = \bar{g}a\bar{g}^{-1}$. By Lemma 2.3(ii), each element x of $A_1 * G$ can be uniquely written in the form $\sum x_g\bar{g}$ with $x_g \in A_1, g \in G$, while by Lemma 2.3(i), the map $a \mapsto {}^g a$ is an automorphism of A_1.

Let $\psi \in A_1^*$ be such that (A_1, ψ) is a Frobenius algebra. Define the map $\lambda : A_1 * G \to F$ by

$$\lambda\left(\sum x_g\bar{g}\right) = \psi(x_1) \qquad (x_g \in A_1, g \in G)$$

Then it is obvious that λ is an F-linear map. Now assume that $x = \sum x_g\bar{g} \in A_1 * G$ is such that $\lambda(xy) = 0$ for all $y \in A_1 * G$. Since for any $a \in A_1, g \in G$, we have

$$xa\bar{g}^{-1} = \sum_{t \in G} x_t\bar{t}a\bar{g}^{-1} = \sum_{t \in G} x_t{}^t a\bar{t}\bar{g}^{-1}$$

it follows that

$$0 = \lambda(xa\bar{g}^{-1}) = \psi(x_g{}^g a) \quad \text{for all} \quad a \in A_1, g \in G$$

Since $a \mapsto {}^g a$ is an automorphism of A_1, the latter implies that $\psi(x_g A_1) = 0$ for all $g \in G$. But (A_1, ψ) is a Frobenius algebra, hence $x_g = 0$ for all $g \in G$ and therefore $x = 0$. So the proposition is true. ■

3.   Symmetric crossed products

In this section, G denotes a finite group, F a field and A a finite-dimensional F-algebra. Assume that A is a crossed product of G over an F-algebra A_1. It is natural to investigate the following problem.

   Problem. What are necessary and sufficient conditions for A to be a symmetric F-algebra?

Our aim is twofold: first, to demonstrate that the above problem in full generality is extremely complicated; second, to provide a partial solution of this problem, by proving that if A_1 is semisimple, then any crossed product F-algebra $A_1 * G$ is symmetric.

3.1. Lemma.   *Let n be a positive integer and let*

$$A = F[X]/(X^n)$$

Then A is a symmetric F-algebra.

   Proof.   Put $x = X + (X^n)$. Then $\{1, x, \ldots, x^{n-1}\}$ is an F-basis of the commutative F-algebra A. Because any ideal of $F[X]$ containing (X^n) is of the form $(X^i), 0 \le i \le n$, $(X^0) = F[X]$, it follows that $A, Ax, \ldots, Ax^{n-1}$ are all nonzero ideals of A. The map

$$\psi : A \to F$$

given by

$$\psi\left(\sum_{i=0}^{n-1} \lambda_i x^i\right) = \sum_{i=0}^{n-1} \lambda_i$$

is obviously F-linear. Since, for all $i \in \{0, 1, \ldots, n-1\}$, $x^i \notin Ker\psi$, it follows that $Ker\psi$ contains no nonzero ideals of A. Hence (A, ψ) is a symmetric F-algebra, as we wished to show. ∎

We are now ready to prove

3.2. Proposition. *(Dade). There exists a finite-dimensional algebra A over a field F such that A is a crossed product of a finite group G over an F-algebra A_1 such that*
(i) A is a symmetric F-algebra.
(ii) A_1 is not a symmetric F-algebra.

Proof. Put $B = F[X]/(X^2)$, so that $B = F[x]$ with $x = X + (X^2)$ and B is a symmetric F-algebra by Lemma 3.1. Let $A = M_2(B)$ and let $e_{ij}, 1 \le i, j \le 2$, be the usual matrix units of A. Then A is a symmetric F-algebra, by virtue of Lemma 1.8.

The elements $e_{ij}, xe_{ij}, 1 \le i, j \le 2$, form an F-basis of A and

$$xe_{ij} = e_{ij}x, \quad x^2 = 0 \qquad (1 \le i, j \le 2)$$

Let $G = <g>$ be a cyclic group of order 2 and put

$$A_1 = Fe_{11} + Fe_{22} + Fxe_{12} + Fxe_{21}$$
$$A_g = Fe_{12} + Fe_{21} + Fxe_{11} + Fxe_{22}$$

Then an easy verification shows that A becomes a G-graded F-algebra. Because $e_{21} + e_{12} \in A_g$ and

$$(e_{21} + e_{12})^2 = e_{11} + e_{22} = 1$$

it follows that A is a crossed product of G over A_1.

By the foregoing, we are left to verify that A_1 is not a symmetric F-algebra. To this end, note that

$$J(A_1) = Fxe_{21} + Fxe_{12}$$

and

$$A_1 = Ae_{11} \oplus Ae_{22} \qquad \text{(direct sum of } A_1\text{-modules)}$$

Setting $P = A_1 e_{11}$, we have $P = F e_{11} + F x e_{21}$ and $J(P) = J(A_1)P = F x e_{21}$. Hence P is a projective indecomposable A_1-module such that $Soc\, P = J(P)$. Because $e_{11} J(P) = 0$ and $e_{11}(P/J(P)) \neq 0$, we have

$$Soc\, P = J(P) \not\cong P/J(P)$$

Consequently, A_1 is not a symmetric F-algebra, by virtue of Theorem 6.3. ∎

Let A_1 be an arbitrary ring, let G be a finite group and let $A_1 * G$ be a crossed product of G over A_1. As usual, for any $a \in A$, $g \in G$, we put

$$^g a = \bar{g} a \bar{g}^{-1}$$

Then $z \mapsto {}^g z$, $z \in Z(A_1)$ provides an action of G on $Z(A_1)$. Note that if A_1 is an algebra over a field F, then $A_1 * G$ is an F-algebra if and only if G acts trivially on F. From now on, we fix the following data:

E is a G-invariant field contained in $Z(A_1)$.

$K = E^G = \{a \in E |\, {}^g a = a \text{ for all } g \in G\}$.

F is a subfield of K such that K/F is a finite field extension and $dim_E A_1 < \infty$.

$t = tr_{E/K} : E \to K$.

Observe that

$$t(E) = K \tag{1}$$

and

$$t(^g a) = t(a) \quad \text{for all} \quad a \in E, g \in G \tag{2}$$

(see Lang (1984, Chapter VIII)).

Let $\varphi \in Hom_E(A_1, E)$. We say that φ is G-invariant if

$$\varphi(^g a) = {}^g \varphi(a) \quad \text{for all} \quad g \in G, a \in A_1 \tag{3}$$

3.3. Lemma. *If (A_1, φ) is a Frobenius E-algebra, $0 \neq \lambda \in Hom_F(K, F)$ and $f : A_1 * G \to F$ is defined by*

$$f(\sum x_g \bar{g}) = \lambda(t(\varphi(x_1))), \qquad (x_g \in A_1, g \in G)$$

then

(i) $(A_1 * G, f)$ is a Frobenius F-algebra.

(ii) If (A_1, φ) is a symmetric E-algebra and φ is G-invariant, then $(A_1 * G, f)$ is a symmetric F-algebra.

Proof. (i) If $\psi = \lambda \circ t \circ \varphi$, then $\psi \in Hom_F(A_1)$ and, by (1), $(\lambda \circ t)(E) = F$. Suppose that $\psi(aA_1) = 0$ for some $a \in A_1$. If $\varphi(aA_1) \neq 0$, then $\varphi(aA_1) = E$ and $\psi(aA_1) = (\lambda \circ t)(E) = F = 0$, a contradiction. Thus $\varphi(aA_1) = 0$ and so $a = 0$, proving that (A_1, ψ) is a Frobenius F-algebra. The required assertion is now a consequence of Theorem 2.4.

(ii) Assume the additional hypotheses of (ii) and let $x = \sum x_g \bar{g}, y = \sum y_g \bar{g}$ be two elements of $A_1 * G$, $x_g, y_g \in A_1$. Then we have

$$
\begin{aligned}
f(xy) &= \lambda(t(\varphi(\sum_{g \in G} x_g \bar{g} y_{g^{-1}} \overline{g^{-1}}))) \\
&= \sum_{g \in G} \lambda(t(\varphi(x_g(\bar{g} y_{g^{-1}} \overline{g^{-1}})))) \\
&= \sum_{g \in G} \lambda(t(\varphi(\bar{g} y_{g^{-1}} \overline{g^{-1}} x_g)))
\end{aligned}
$$

because $\bar{g} y_{g^{-1}} \overline{g^{-1}} \in A_1$ and $\varphi(ab) = \varphi(ba)$ for all $a, b \in A_1$. Thus

$$
\begin{aligned}
f(xy) &= \sum_{g \in G} \lambda(t(\varphi[^g(y_{g^{-1}} \overline{g^{-1}} x_g \bar{g})]))) \\
&= \sum_{g \in G} \lambda(t(^g\varphi(y_{g^{-1}} \overline{g^{-1}} x_g \bar{g}))) \quad \text{(by (3))} \\
&= \sum_{g \in G} \lambda(t(\varphi(y_{g^{-1}} \overline{g^{-1}} x_g \bar{g}))) \quad \text{(by (2))} \\
&= \lambda(t\varphi(\sum_{g \in G} y_{g^{-1}} \overline{g^{-1}} x_g \bar{g})) \\
&= f(yx),
\end{aligned}
$$

as desired. ■

For future use, we next record the following elementary observation.

3.4. Lemma. *Let R be a commutative ring and let a G-graded R-algebra A be a crossed product of G over A_1. Let $e_1, \ldots, e_n$ be orthogonal idempotents of $Z(A) \cap A_1$ with $1 = e_1 + \cdots + e_n$. Then*

$$
A = \oplus_{i=1}^n e_i A
$$

is a direct sum decomposition of A into ideals $e_i A$, where each $e_i A$ is a G-graded R-algebra which is a crossed product of G over $e_i A_1$ such that

    (i) e_i is the identity element of $e_i A$.

    (ii) $(e_i A)_g = e_i A_g$ for all $g \in G$.

    (iii) $e_i \bar{g} \in U(e_i A) \cap ((e_i A)_g)$ for all $g \in G$.

    (iv) $U(A) = \oplus_{i=1}^{n} U(e_i A)$.

Proof. The verification is straightforward and therefore will be omitted. ∎

We next introduce reduced characteristic polynomials which will play an important role in the proof of the main result.

Let A be a finite-dimensional algebra over a field F. Then A is said to be *separable* if for every field extension E of F, $A_E = E \otimes_F A$ is a semisimple E-algebra.

For any given $a \in A$, let $char.pol._{A/F}(a)$ be the characteristic polynomial of the F-linear map $x \mapsto ax$, $x \in A$. The *trace map* $T_{A/F}$ and *norm map* $N_{A/F}$ are defined by

$$char.pol._{A/F}(a) = X^m - T_{A/F}(a)X^{m-1} + \cdots + (-1)^m N_{A/F}(a),$$

where $m = dim_F A$.

Suppose that A is a separable F-algebra. Then there exists a field extension E/F and an isomorphism

$$f : A_E \to \prod_{i=1}^{s} M_{n_i}(E)$$

of E-algebras, for some positive integers s and n_i, $1 \le i \le s$. For any given $a \in A$, let us put

$$f(1 \otimes a) = (\varphi_1(a), \ldots, \varphi_s(a)) \quad (\varphi_i(a) \in M_{n_i}(E))$$

Then the *reduced characteristic polynomial* of $a \in A$ is defined by

$$red.char.pol._{A/F}(a) = \prod_{i=1}^{s} char.pol.\varphi_i(a),$$

where $char.pol.\varphi_i(a)$ is the characteristic polynomial of the matrix $\varphi_i(a)$. It can be shown (see Curtis and Reiner (1981, p.159)) that this reduced characteristic polynomial is independent of the choice of E and of the isomorphism f. Note also that its coefficients all lie in the ground field F.

Let us put $m = \sum n_i$ and write

$$red.\,char.\,pol._{A/F}(a) = X^m - (tr_{A/F}(a))X^{m-1} + \cdots + (-1)^m nr_{A/F}(a)$$

for $a \in A$. Then $tr_{A/F}$ is called the *reduced trace*, and $nr_{A/F}$ is called the *reduced norm*. One immediately verifies that

$$tr_{A/F} : A \to F$$

is an F-linear map such that

$$tr_{A/F}(ab) = tr_{A/F}(ba) \quad \text{for all} \quad a, b \in A \tag{4}$$

For convenience of reference, we next quote the following two standard facts.

3.5. Proposition. *Let A be a separable F-algebra. Then the reduced trace $tr_{A/F}$ gives rise to a symmetric bilinear associative nonsingular form, namely*

$$(a, b) \mapsto tr_{A/F}(ab) \qquad (a, b \in A)$$

Proof.   See Curtis and Reiner (1981, p. 165). ∎

3.6. Proposition. *Let A be a semisimple F-algebra. Then A is separable if and only if the centres of the division algebras associated with the Wedderburn components of A are separable field extensions of F.*

Proof.   See Curtis and Reiner (1981, p.145). ∎

Let A, B be isomorphic rings and let $\sigma : A \to B$ be a ring isomorphism. Assume further that F is a field contained in $Z(A)$ and such that A is a finite-dimensional separable F-algebra. Setting $L = \sigma(F)$, it follows that L is a field contained in $Z(B)$, B is a finite-dimensional separable L-algebra and

$$dim_F A = dim_L B$$

For any given $a \in A$, we put

$$red.\,char.\,pol._{A/F}(a) = X^n + \lambda_1 X^{n-1} + \cdots + \lambda_{n-1} X + \lambda_n$$

where $n \geq 1$ is an integer and $\lambda_i \in F$ for all $i \in \{1, \ldots, n\}$.

3.7. Lemma. *With the notation above, we have*

$$red.\,char.\,pol._{B/L}(\sigma(a)) = X^n + \sigma(\lambda_1) X^{n-1} + \cdots + \sigma(\lambda_{n-1}) X + \sigma(\lambda_n)$$

Proof. Denote by $\bar{F}$ and $\bar{L}$ the algebraic closures of F and L, respectively. It is a standard fact of field theory (see Lang (1984, VII, Theorem 2.8)) that $\sigma : F \to L$ can be extended to a field isomorphism $\bar{\sigma} : \bar{F} \to \bar{L}$. Hence there exists a ring isomorphism

$$\alpha : \bar{F} \otimes_F A \to \bar{L} \otimes_L B$$

such that

$$\alpha(\lambda \otimes a) = \bar{\sigma}(\lambda) \otimes \sigma(a)$$

for all $\lambda \in \bar{F}$ and all $a \in A$. Bearing in mind that $\bar{F} \otimes_F A$ is a finite-dimensional semisimple $\bar{F}$-algebra, we may find an $\bar{F}$-algebra isomorphism

$$\beta : \bar{F} \otimes_F A \to \prod_{i=1}^{m} M_{r_i}(\bar{F})$$

for some positive integer m and some positive integers r_i for all $i \in \{1, \ldots, m\}$. The isomorphism $\bar{\sigma} : \bar{F} \to \bar{L}$ induces the ring isomorphism

$$\rho : \prod_{i=1}^{m} M_{r_i}(\bar{F}) \to \prod_{i=1}^{m} M_{r_i}(\bar{L})$$

Setting $\gamma = \rho \circ \beta \circ \alpha^{-1}$, we thus obtain an $\bar{L}$-algebra isomorphism

$$\gamma : \bar{L} \otimes_L B \to \prod_{i=1}^{m} M_{r_i}(\bar{L})$$

such that $\gamma \circ \alpha = \rho \circ \beta$.

Write $\beta(1 \otimes a) = (\varphi_1(a), \varphi_2(a), \ldots, \varphi_m(a))$ for $\varphi_i(a) \in M_{r_i}(\bar{F})$, $1 \leq i \leq m$, so that

$$red.\,char.\,pol._{A/F}(a) = \prod_{i=1}^{m} char.\,pol.(\varphi_i(a))$$

Also put

$$\gamma(1 \otimes \sigma(a)) = (\psi_1(\sigma(a)), \psi_2(\sigma(a)), \ldots, \psi_m(\sigma(a)))$$

for $\psi_i(\sigma(a)) \in M_{r_i}(\bar{L})$, $1 \leq i \leq m$, so that

$$red. \, char. \, pol._{B/L}(\sigma(a)) = \prod_{i=1}^{m} char. \, pol. \, \psi_i(\sigma(a))$$

Then we have

$$
\begin{aligned}
\rho(\beta(1 \otimes a)) &= (\rho(\varphi_1(a)), \ldots, \rho(\varphi_m(a))) \\
&= \gamma(\alpha(1 \otimes a)) \\
&= (\psi_1(\sigma(a)), \ldots, \psi_m(\sigma(a))
\end{aligned}
$$

Hence $\rho(\varphi_i(a)) = \psi_i(\sigma(a))$ for all $i \in \{1, \ldots, m\}$ and

$$\bar{\sigma}[char. \, pol. \, (\varphi_i(a))] = char. \, pol. \, \psi_i(\sigma(a)) \quad (1 \leq i \leq m)$$

thus completing the proof. ∎

We are now ready to prove our main result.

3.8. Theorem. *(Harris (1988)). Let G be a finite group and let A_1 be a finite-dimensional semisimple algebra over a field F. Then any F-algebra A which is a crossed product of G over A_1 is symmetric.*

Proof. Choose primitive orthogonal idempotents $e_1, \ldots, e_n$ of $Z(A_1)$ such that $1 = e_1 + \cdots + e_n$. Setting $B_i = e_i A_1$, we then have

$$A_1 = B_1 \oplus \cdots \oplus B_n$$

where each B_i is a simple F-algebra. The map $e_i \mapsto {}^g e_i = \bar{g} e_i \bar{g}^{-1}, g \in G$, provides an action of G on $\{e_1, \ldots, e_n\}$. If $f_1, \ldots, f_k$, $k \leq n$, are the G-orbit sums, then $f_1, \ldots, f_k$ are mutually orthogonal idempotents of $Z(A) \cap A_1$. Bearing in mind that a finite direct product of symmetric F-algebras is a symmetric F-algebra (Lemma 1.7), it follows from Lemma 3.4 that it suffices to assume that G acts transitively on $\{e_1, \ldots, e_n\}$.

Put $e = e_1, f = 1 - e, B = B_1 = eA_1, E = Z(B)$ and let H denote the stabilizer of e. Also put $A_H = \{\sum x_h \bar{h} | x_h \in A_1, h \in H\}$. Then E is a field, $F \cong Fe = eF \subseteq E$ and A_H is an F-algebra which is a crossed product of H over A_1. It is also clear that e, f are orthogonal idempotents of $Z(A_H) \cap A_1$ with $e + f = 1$. Owing to Lemma 3.4, eA_H is a crossed product of H over B.

For each $h \in H$, put $\tilde{h} = e\bar{h}$. Then $\tilde{h}$ is a graded unit of eA_H of degree h and the map $\mu \mapsto \tilde{h}\mu\tilde{h}^{-1}, \mu \in E, h \in H$, provides an action of H as a group of automorphisms of E. Let $K = E^H$ be the H-fixed subfield of E, so that

$$F = Fe = eF \subseteq K$$

Put $t = tr_{E/K} : E \to K$ and fix $0 \neq \lambda \in Hom_F(K, F)$. Observe that B is a finite-dimensional simple E-algebra, hence the reduced trace $tr_{B/E} \in Hom_E(B, E)$ is defined by Proposition 3.6. Moreover, by (4), we have

$$tr_{B/E}(ab) = tr_{B/E}(ba) \quad \text{for all} \quad a, b \in B$$

Note also that, by Proposition 3.5, the kernel of $tr_{B/E}$ contains no nonzero right ideals of B and, by Lemma 3.7, $tr_{B/E}$ is H-invariant.

Now define $f : eA_H \to F$ as in Lemma 3.3 with $tr_{B/E}$ playing the role of φ. Then, by Lemma 3.3(ii), (eA_H, f) is a symmetric F-algebra. Therefore, for any graded unit u of A_H and any $b \in B$,

$$f(eubeu^{-1}) = f(b) \tag{5}$$

Choose $g_1, \ldots, g_n$ in G such that $e_i = \bar{g}_i e_1 \bar{g}_i^{-1}$. Then $B_i = \bar{g}_i B_1 \bar{g}_i^{-1}$ for all $i \in \{1, \ldots, n\}$. Next define

$$\varphi : A_1 = \oplus_{i=1}^n B_i \to F$$

as follows: if $y = \sum_{i=1}^n y_i \in A_1$ for all $y_i \in B_i$, put

$$\varphi(y) = \sum_{i=1}^n f(\bar{g}_i^{-1} y_i \bar{g}_i)$$

The clearly $\varphi \in Hom_F(A_1, F)$, $\varphi(ab) = \varphi(ba)$ for all $a, b \in A_1$ and $Ker\varphi$ contains no nonzero right ideals of A_1. Invoking Lemma 3.3(ii), with F playing the role of E, we are left to verify that

$$\varphi(^g x) = \varphi(x) \quad \text{for all} \quad x \in A_1, g \in G$$

The latter, of course, will follow provided we show that for any fixed $j \in \{1, \ldots, n\}$, $g \in G$ and $z \in B_j$, we have

$$\varphi(^g z) = \varphi(z)$$

We have $\bar{g} B_j \bar{g}^{-1} = \bar{g} \bar{g}_j B_1 \bar{g}_j^{-1} \bar{g}^{-1} = B_k$ for some $k \in \{1, \ldots, n\}$. Thus $\bar{g} \bar{g}_j = u \bar{g}_k$ where u is a graded unit of A_H. Bearing in mind that $\bar{g}_j^{-1} z \bar{g}_j \in B_1 = B$, we have

$$u \bar{g}_j^{-1} z \bar{g}_j u^{-1} = e u \bar{g}_j^{-1} z \bar{g}_j e u^{-1}$$

and hence, by (5), $f(u \bar{g}_j^{-1} z \bar{g}_j u^{-1}) = f(\bar{g}_j^{-1} z \bar{g}_j)$. But then

$$
\begin{aligned}
\varphi(^g z) &= \varphi(\bar{g} z \bar{g}^{-1}) = f(\bar{g}_k^{-1} \bar{g} z \bar{g}^{-1} \bar{g}_k) \\
&= f(u \bar{g}_j^{-1} z \bar{g}_j u^{-1}) = f(\bar{g}_j^{-1} z \bar{g}_j) \\
&= \varphi(z),
\end{aligned}
$$

thus completing the proof. ∎

As a preliminary to the next result, we record the following simple observation.

3.9. Lemma.   *Let R be a commutaitve ring, let A_1 be an R-algebra and let an R-algebra A be a crossed product of G over A_1. Then the R-algebra $A/A \cdot J(A_1)$ is a crossed product of G over $A_1/J(A_1)$.*

Proof.   For any $g \in G$, choose a unit $\bar{g}$ of A in A_g with $\bar{1} = 1$. It is clear that for all $g \in G$,

$$\bar{g} J(A_1) \bar{g}^{-1} = J(A_1)$$

which implies that $\bar{g} J(A_1) = J(A_1) \bar{g}$. Since $A = \oplus_{g \in G} A_1 \bar{g}$, it follows that

$$A \cdot J(A_1) = \oplus_{g \in G} J(A_1) \bar{g}$$

Hence $A/A \cdot J(A_1)$ is a G-graded R-algebra by setting

$$(A/A \cdot J(A_1))_g = (A_g + A \cdot J(A_1))/A \cdot J(A_1)$$

for all $g \in G$. Since $\bar{g} + A \cdot J(A_1)$ is a graded unit of $A/A \cdot J(A_1)$ of degree g and since

$$
\begin{aligned}
(A/A \cdot J(A_1))_1 &= (A_1 + A \cdot J(A_1))/A \cdot J(A_1) \\
&\cong A_1/(A_1 \cap A \cdot J(A_1)) \\
&= A_1/J(A_1),
\end{aligned}
$$

the result follows. ■

3.10. Corollary. *Let G be a finite group and let A_1 be a finite-dimensional algebra over a field F. If an F-algebra A is a crossed product of G over A_1, then $A/A \cdot J(A_1)$ is a symmetric F-algebra.*

Proof. Owing to Lemma 3.9, $A/A \cdot J(A_1)$ is a crossed product of G over the finite-dimensional semisimple F-algebra $A_1/J(A_1)$. Now apply Theorem 3.8. ■

As a preliminary to another application of Theorem 3.8, we record the following property.

3.11. Lemma. *Let R be a commutative ring and let A be a G-graded R-algebra. If N is a normal subgroup of G, then A can be regarded as an G/N-graded R-algebra by setting*

$$
A_{gN} = \oplus_{x \in gN} A_x
$$

Moreover, if A is a crossed product of G over A_1, then A can also be regarded as a crossed product of G/N over

$$
A_N = \oplus_{x \in N} A_x
$$

Proof. Let T be a transversal for N in G. It is obvious that

$$
A = \oplus_{t \in T} A_{tN}
$$

Given $t_1, t_2 \in T$, we have

$$
\begin{aligned}
A_{t_1 N} A_{t_2 N} &= (\oplus_{x \in t_1 N} A_x)(\oplus_{y \in t_2 N} A_y) \\
&\subseteq \oplus_{x \in T_1 N, y \in t_2 N} A_{xy} = \oplus_{z \in t_1 t_2 N} A_z \\
&= A_{t_1 t_2 N},
\end{aligned}
$$

proving the first assertion.

Assume that A is a crossed product of G over A_1. Then, for all $g \in G$, there exists a unit $\bar{g}$ of A in A_g, in which case $\bar{g} \in A_{gN}$, as required. ∎

3.12. Corollary. *Let $F * G$ be a crossed product of a finite group G over a field F. If N is a normal subgroup of G, then*

$$F * G / (F * G) J(F * N)$$

is a symmetric F^G-algebra, where F^G is the fixed field of G.

Proof. We first observe that $A = F * G$ is an F^G-algebra. Since G is finite, F/F^G is a finite Galois extension. In particular, F is a finite-dimensional algebra over the field F^G. Hence $A_N = F * N$ is a finite-dimensional algebra over F^G. By Lemma 3.11, A can be regarded as a crossed product of G/N over A_N. Hence, by Corollary 3.10, $A/A \cdot J(A_N)$ is a symmetric F^G-algebra, as required. ∎

The following result for ordinary group algebras is due to Harris (1988).

3.13. Corollary. *Let $F * G$ be a twisted group algebra of a finite group G over a field F. If N is a normal subgroup of G, then*

$$F * G / (F * G) J(F * N)$$

is a symmetric F-algebra.

Proof. Since G acts trivially on F, we have $F = F^G$. Now apply Corollary 3.12. ∎

4. Symmetric endomorphism algebras

In this section, we are concerned with the problem of finding circumstances under which the endomorphism algebra of a given module is symmetric. First, we focus our attention on endomorphism rings of

projective modules over symmetric algebras. The following two preliminary observations will clear our path.

4.1. Lemma. *Let S be an arbitrary ring and let V be an S-module. If W is a direct summand of V and $e : V \to W$ is an idempotent projection, then*

$$End_S(W) \cong eEnd_S(V)e$$

Proof. We have $W = eV$ and $V = W \oplus U$, where $U = (1 - e)V$. It is clear that the subring K of $End_S(V)$ defined by

$$K = \{\psi \in End_S(V) | \psi(W) \subseteq W, \psi(U) = 0\}$$

is isomorphic to $End_S(W)$. Now $\psi(U) = 0$ if and only if $\psi = \psi e$. Because

$$\psi(ev) = e\psi e(v) + (1 - e)\psi e(v) \qquad \text{for all} \quad v \in V$$

we have $\psi(W) \subseteq W$ if and only if $(1 - e)\psi e = 0$, which happens if and only if $\psi e = e\psi e$. Hence $K = eEnd_S(V)e$ and the result follows. ∎

4.2. Lemma. *Let A be a symmetric algebra over a field F. Then, for any idempotent e of A, eAe is a symmetric F-algebra.*

Proof. Let $\psi \in Hom_F(A, F)$ be such that (A, ψ) is a symmetric algebra. Denote by λ the restriction of ψ to eAe. Then obviously $\lambda \in Hom(eAe, F)$. Now assume that $\lambda(eAex) = 0$ for some $x \in eAe$. Then $x = exe = xe$ and

$$0 = \psi(eAex) = \psi(eAexe) = \psi(Aexe) = \psi(Ax),$$

which implies $x = 0$. ∎

4.3. Proposition. *Let A be a finite-dimensional algebra over a field F. Then A is symmetric if and only if for any finitely generated projective A-module P, the F-algebra $End_A(P)$ is symmetric.*

Proof.   Assume that A is symmetric and let P be a finitely generated projective A-module. Then P is a direct summand of a free A-module Q of rank $n \geq 1$. Since

$$End_A(Q) \cong M_n(End_A({}_A A)) \cong M_n(A°)$$

it follows from Lemma 1.8 that $End_A(Q)$ is a symmetric algebra. By Lemma 4.1, $End_A(P) \cong eEnd_A(Q)e$ for some idempotent e of $End_A(Q)$. Since $End_A(Q)$ is a symmetric algebra, it follows from Lemma 4.2 that $End_A(P)$ is also a symmetric algebra.

Conversely assume that for any finitely generated projective A-module P, the F-algebra $End_A(P)$ is symmetric. Then, choosing $P = {}_A A$, it follows that

$$End_A(P) = End_A({}_A A) \cong A°$$

is symmetric. But then A is also symmetric, hence the result. ∎

The rest of the section will be devoted to the study of the endomorphism algebra of induced modules. Throughout, N denotes a normal subgroup of a finite group G, R an arbitrary commutative ring and RG the group algebra of G over R.

Let V be an RN-module. Then the induced module V^G is defined by

$$V^G = RG \otimes_{RN} V$$

It is clear that $g \otimes V$ is an RN-submodule of V^G, $g \in G$. The module V is said to be *G-invariant* if $g \otimes V \cong V$ for all $g \in G$.

If W is another RN-module, we say that V *weakly divides* W if there exists a positive integer k and an injective homomorphism

$$f : V \to kW = W \oplus \cdots \oplus W \quad (k \text{ copies})$$

such that $f(V)$ is a direct summand of kW. We shall say that V and W are *weakly isomorphic* if each weakly divides the other. Of course, this is an equivalence relation among RN-modules. Finally, we say that V is *weakly G-invariant* if it is weakly isomorphic to each $g \otimes V, g \in G$.

4.4. Lemma. *Let V be an RN-module.*
(i) The map

$$\begin{cases} End_{RN}(V) & \to & End_{RG}(V^G) \\ f & \mapsto & 1 \otimes f \end{cases}$$

is an injective homomorphism of R-algebras whose image consists of all $\psi \in End_{RG}(V^G)$ for which $\psi(1 \otimes V) \subseteq 1 \otimes V$.
(ii) If $f : 1 \otimes V \to g \otimes V$ is an RN-isomorphism, then f can be uniquely extended to an RG-automorphism of V^G.

Proof. (i) It is clear that the given map is an injective homomorphism of R-algebras. Furthermore, for all $f \in End_{RN}(V)$, $1 \otimes f$ sends $1 \otimes V$ into $1 \otimes V$. Assume that $\psi \in End_{RG}(V^G)$ is such that $\psi(1 \otimes V) \subseteq 1 \otimes V$. Then there exists $f \in End_{RN}(V)$ such that

$$\psi(1 \otimes v) = 1 \otimes f(v) \quad \text{for all} \quad v \in V$$

Since ψ is uniquely determined by its restriction to $1 \otimes V$, we have $\psi = 1 \otimes f$, as required.

(ii) It is easy to see that f can be uniquely extended to an RG-homomorphism $\psi : V^G \to V^G$. Since, for any $x \in G$,

$$\psi(x \otimes V) = x\psi(1 \otimes V) = xf(1 \otimes V) = xg \otimes V,$$

it follows that ψ is surjective.

To prove that ψ is injective, fix a transversal T for N in G. Then obviously

$$V^G = \oplus_{t \in T} t \otimes V$$

and, therefore, any given $w \in V^G$ can be uniquely written in the form

$$w = \sum_{t \in T} t \otimes v_t \quad (v_t \in V)$$

If $\psi(w) = 0$, then

$$\sum_{t \in T} tf(1 \otimes v_t) = 0$$

Since $tf(1 \otimes v_t) \in tg \otimes V$ and Tg is a new transversal for N in G, we have $tf(1 \otimes v_t) = 0$. Hence $1 \otimes v_t = 0$ for all $t \in T$ and so $w = 0$, as required. ∎

4.5. Proposition. *(Dade). Let V be an RN-module, let $E = End_{RG}(V^G), S = End_{RN}(V)$, and, for any $g \in G$, define E_{gN} by*

$$E_{gN} = \{f \in E | f(1 \otimes V) \subseteq g^{-1} \otimes V\}$$

(i) E is a G/N-graded R-algebra with E_{gN} as its gN-component, $g \in G$ and the identity component of E is identifiable with S.

(ii) E is strongly G/N-graded if and only if V is weakly G-invariant, while E is a crossed product of G/N over S if and only if V is G-invariant.

Proof. (i) Note that if $xN = yN$, then $x \otimes V = y \otimes V$ which shows that E_{gN} is well-defined. Denote by T a transversal for N in G. Then

$$V^G = \oplus_{t \in T} t \otimes V$$

where each $t \otimes V$ is an RN-module. Let $\pi_t : V^G \to t \otimes V$ be the projection map. Then we obviously have

$$\sum_{t \in T} \pi_t = 1 \tag{1}$$

Now fix $f \in E$ and let $g_t : 1 \otimes V \to t \otimes V$ be the RN-homomorphism which is the restriction of $\pi_t \circ f$ to $1 \otimes V$. It follows from the definition of induced modules that there exists $f_t \in E$ such that

$$f_t(1 \otimes v) = g_t(1 \otimes v) \in t \otimes V \quad \text{for all} \quad v \in V \tag{2}$$

This shows that $f_t \in E_{t^{-1}N}$ and, invoking (1) and (2), we also have

$$\left(\sum_{t \in T} f_t\right)(1 \otimes v) = \sum_{t \in T} g_t(1 \otimes v) = \sum_{t \in T}(\pi_t f)(1 \otimes v)$$

$$= \left(\sum_{t \in T} \pi_t\right) f(1 \otimes v) = f(1 \otimes v),$$

whence $f = \sum_{t \in T} f_t$.

Next assume that $\sum_{t \in T} f_t = 0$ with $f_t \in E_{t^{-1}N}$. For any $v \in V$, we then have

$$v_t = f_t(1 \otimes v) \in t \otimes V$$

and $\sum_{t \in T} v_t = 0$. Hence $v_t = 0$ and $f_t = 0$ for all $t \in T$, proving that

$$E = \oplus_{t \in T} E_{tN}$$

If $f \in E_{xN}, \varphi \in E_{yN}$, then

$$f(1 \otimes V) \subseteq x^{-1} \otimes V, \varphi(1 \otimes V) \subseteq y^{-1} \otimes V$$

and therefore

$$(f\varphi)(1 \otimes v) \subseteq f(y^{-1} \otimes V) \subseteq y^{-1}f(1 \otimes V) \subseteq y^{-1}x^{-1} \otimes V$$

Thus $f\varphi \in E_{xyN}$ and so E is a G/N-graded R-algebra with E_{gN} as its gN-component. Finally, by Lemma 4.4(i), the identity component of E is identifiable with S, proving (i).

(ii) Owing to Lemma 2.1(iii), E is strongly G/N-graded if and only if for any given $g \in G$, there exists $\varphi_i \in E_{gN}, \psi_i \in E_{g^{-1}N}, 1 \le i \le n = n(g)$, such that

$$\varphi_1\psi_1 + \cdots + \varphi_n\psi_n = 1 \qquad (3)$$

It will now be demonstrated that (3) is equivalent to V and $g \otimes V$ to be weakly isomorphic, which will imply the first assertion.

If (3) holds, then any map

$$f : 1 \otimes V \to n(g \otimes V), v \mapsto (\psi_1(v), \ldots, \psi_n(v))$$

is an injective RN-homomorphism. Furthermore, the map

$$\begin{cases} n(g \otimes V) & \to & 1 \otimes V \\ (v_1, \ldots, v_n) & \mapsto & \varphi_1(v_1) + \cdots + \varphi_n(v_n) \end{cases}$$

is a homomorphism which is a left inverse to f. This shows that V weakly divides $g \otimes V$. By symmetry, V weakly divides $g^{-1} \otimes V$ and so $g \otimes V$ weakly divides V, proving that V and $g \otimes V$ are weakly isomorphic.

Conversely, assume that V and $g \otimes V$ are weakly isomorphic. Then V weakly divides $g \otimes V$, so there is a positive integer n and an injective RN-homomorphism $\psi : 1 \otimes V \to n(g \otimes V)$ such that $\psi(1 \otimes V)$ is a direct summand of $n(g \otimes V)$. We may write $\psi(x) = (\psi_1(x), \ldots, \psi_n(x))$, where $\psi_i : 1 \otimes V \to g \otimes V$ is an RN-homomorphism, in which case ψ_i

can be regarded as an element of $E_{g^{-1}N}$. Let $f : n(g \otimes V) \to 1 \otimes V$ be an RN-homomorphism which is a left inverse to ψ. Then f determines RN-homomorphisms

$$\varphi_i : g \otimes V \to 1 \otimes V \qquad (1 \le i \le n)$$

(which can be viewed as elements of E_{gN}) such that (3) holds.

If V is G-invariant, then E is a crossed product of G/N over S, by virtue of Lemma 4.4(ii). Conversely, assume that for any $g \in G$ there exists $f_g \in U(E) \cap E_{gN}$. Then $f_{g^{-1}} : 1 \otimes V \to g \otimes V$ is an injective RN-homomorphism such that $f_{g^{-1}} f_g$ is an automorphism of $1 \otimes V$ and hence of $g \otimes V$. Thus $f_{g^{-1}}$ is an RN-isomorphism, as required. ∎

4.6. Lemma. *Let H be a subgroup of G, let V be an RH-module, and let W be an RG-module. Then the induced module $V^G = RG \otimes_{RH} V$ satisfies the following property.*

$$Hom_{RG}(V^G, W) \cong Hom_{RH}(V, W_H) \quad \text{as } R\text{-modules}$$

Proof. Given $\lambda \in Hom_{RH}(V, W_H)$, we may define $\lambda^* \in Hom_{RG}(V^G, W)$ by

$$\lambda^*(x \otimes v) = x(\lambda(v)) \quad v \in V, x \in RG$$

The map $\lambda \mapsto \lambda^*$ is obviously R-linear. Furthermore, if $\lambda^* = 0$, then

$$\lambda^*(1 \otimes v) = \lambda(v) = 0 \quad \text{for all} \quad v \in V$$

and so $\lambda = 0$. Finally, given $\psi \in Hom_{RG}(V^G, W)$, define $\varphi : V \to W$ by $\varphi(v) = \psi(1 \otimes v)$, for all $v \in V$. Then $\varphi \in Hom_{RH}(V, W_H)$ and $\psi = \varphi^*$ as required. ∎

For the rest of this section, N denotes a normal subgroup of a finite group G, F an arbitrary field and V an FN-module. Our aim is to apply the results so far obtained in order to prove that if V is semisimple, then the F-algebra $End_{FG}(V^G)$ is symmetric. In what follows all modules are assumed to be finitely generated over their ground rings. The following two preliminary results will clear our path.

4.7. Lemma. *Let V be a simple FN-module and let H be the inertia group of V, i.e. $H = \{g \in G | g \otimes V \cong V\}$. Then any FH-homomorphism $\theta : V^H \to V^H$ extends to a unique FG-homomorphism $\theta' : V^G \to V^G$ and the map*

$$\left\{ \begin{array}{ccc} End_{FH}(V^H) & \to & End_{FG}(V^G) \\ \theta & \mapsto & \theta' \end{array} \right.$$

is an isomorphism of F-algebras.

Proof. Let $g_1, g_2, \ldots, g_s$ be a transversal for N in H and $g_1, \ldots, g_k, k \geq s$, a tranversal for N in G. Then

$$V^H = \oplus_{i=1}^s g_i \otimes V \quad \text{and} \quad V^G = \oplus_{i=1}^k g_i \otimes V$$

If $\theta \in End_{FH}(V^H)$, define $\theta' \in End_{FG}(V^G)$ by

$$\theta'(g_i \otimes v) = g_i\theta(1 \otimes v) \qquad (v \in V)$$

Then it is obvious that θ' is a unique element of $End_{FG}(V^G)$ extending θ. Hence the map $\theta \mapsto \theta'$ is an injective homomorphism of F-algebras.

Now $(V^H)_N$ is the sum of all submodules of $(V^G)_N$ isomorphic to V. Since V is assumed to be simple, we deduce that for any $\psi \in End_{FG}(V^G)$, we have $\psi(V^H) \subseteq V^H$. This proves that the given map is surjective and the result is established. ∎

Two FN-modules V_1 and V_2 are said to be *conjugate* if $V_2 \cong g \otimes V_1$ for some $g \in G$.

4.8. Lemma. *Let V be a semisimple FN-module. Then there exist a positive integer s, simple nonconjugate FN-modules $V_1, \ldots, V_s$ and positive integers $n_1, \ldots, n_s$ such that*

$$End_{FG}(V^G) \cong \prod_{i=1}^s M_{n_i}\left(End_{FH_i}(V_i^{H_i})\right)$$

where H_i is the inertia group of V_i.

Proof. Write $V = \oplus_{i=1}^k m_i V_i$ where each V_i is simple and $m_i \geq 1$ is an integer. We may assume that $V_1, \ldots, V_s$ ($s \leq k$) are such that

$V_1, \ldots, V_s$ are mutually nonconjugate and each $V_j, j \in \{1, \ldots, k\}$, is conjugate to some $V_i, i \in \{1, \ldots, s\}$ (hence $V_j^G \cong V_i^G$). Hence we may find positive integers $n_1, \ldots, n_s$ such that

$$V^G \cong \oplus_{i=1}^s n_i V_i^G \cong (\oplus_{i=1}^s n_i V_i)^G$$

Thus we may harmlessly assume that $V = \oplus_{i=1}^s n_i V_i$.

Let T denote a transversal for N in G containing 1. Suppose that $i, j \in \{1, \ldots, s\}$ and $1 \neq j$. Then, by Lemma 4.6,

$$Hom_{FG}((n_i V_i)^G, (n_j V_j)^G) \cong Hom_{FN}(n_i V_i, (n_j V_j)_N^G)$$

Consequently,

$$Hom_{FG}((n_i V_i)^G, (n_j V_j)^G) \cong n_i n_j (\oplus_{t \in T} Hom_{FN}(V_i, t \otimes V_j)) = 0$$

since V_i and $t \otimes V_j$ are simple and nonisomorphic FN-modules for all $t \in G$. This fact immediately implies that

$$
\begin{aligned}
End_{FG}(V^G) & \cong \prod_{i=1}^s End_{FG}(n_i V_i^G) \\
& \cong \prod_{i=1}^s M_{n_i}(End_{FG}(V_i^G)) \\
& \cong \prod_{i=1}^s M_{n_i}(End_{FH_i}(V_i^{H_i}))
\end{aligned}
$$

where the last isomorphism follows by virtue of Lemma 4.7. ∎

We have now accumulated all the information necessary to provide the following class of symmetric endomorphism algebras.

4.9. Theorem. *(Harris (1988)). Let F be an arbitrary field and let V be a finitely generated semisimple FN-module. Then the F-algebra $End_{FG}(V^G)$ is symmetric.*

Proof. Applying Lemmas 1.7 and 1.8 together with Lemma 4.8, we may harmlessly assume that V is simple and G-invariant. Owing to Proposition 4.5(ii), the F-algebra $End_{FG}(V^G)$ is then a crossed product of G/N over $S = End_{FN}(V)$. SInce V is simple, S is a division ring by Schur's lemma. Moreover, since V is finitely generated, S is of finite F-dimension. The desired conclusion is therefore a consequence of Theorem 3.8. ■

We close by proving the following result.

4.10. Theorem. *Let F be an arbitrary field and let V be a finitely generated G-invariant FN-module. Then the F-algebra*

$$End_{FG}(V^G)/End_{FG}(V^G)J(End_{FN}(V))$$

is symmetric.

Proof. Put $E = End_{FG}(V^G)$ and $S = End_{FN}(V)$. Since V is G-invariant, it follows from Proposition 4.5 that E is a crossed product of G/N over S. Hence, by Lemma 3.9, the F-algebra $E/E \cdot J(S)$ is a crossed product of G/N over $S/J(S)$. Taking into account that $S/J(S)$ is a finite-dimensional semisimple F-algebra, the result follows by applying Theorem 3.8. ■

5. Projective covers and injective hulls

In this section, we provide some background information which will be required for subsequent investigations of Frobenius and symmetric algebras. In what follows, R denotes an arbitrary ring and all modules are assumed to be *finitely generated left modules*.

We know that each R-module is a homomorphic image of a projective module. For some modules V a stronger assertion is possible. A homomorphism

$$f : V \to W$$

of R-modules is said to be *essential* if for every proper submodule V' of V, $f(V') \neq f(V)$.

We say that an R-module P is a *projective cover* of V in case P is projective and there is an essential epimorphism

$$P \to V$$

5.1. Lemma. *(i) A homomorphism $f : V \to W$ of R-modules is essential if and only if $Ker f$ is superfluous.*

(ii) If $V_1, \ldots, V_n$ are superfluous submodules of V, then so is $V_1 + \cdots + V_n$.

(iii) If P_i is a projective cover of V_i, then $P_1 \oplus \cdots \oplus P_n$ is a projective cover of $V_1 \oplus \cdots \oplus V_n$.

(iv) A projective cover of a simple module is indecomposable.

Proof. (i) Suppose that $f : V \to W$ is an essential homomorphism. If W is a submodule of V such that $V = W + Ker f$, then $f(V) = f(W)$ and thus W cannot be a proper submodule of V. This proves that $V = W$ and so $Ker f$ is indeed a superfluous submodule of V.

Conversely, suppose that $Ker f$ is superfluous. If V' is a submodule of V with $f(V) = f(V')$, then $V = V' + Ker f$. Since $Ker f$ is superfluous, we must have $V = V'$ and so f is essential.

(ii) Assume that W is a submodule of V such that

$$(V_1 + \cdots + V_n) + W = V$$

Then $V_1 + (V_2 + \cdots + V_n + W) = V$ and, since V_1 is superfluous, we have $V_2 + \cdots + V_n + W = V$. Now apply induction on n.

(iii) If $f_i : P_i \to V_i$ is an essential epimorphism, then

$$\oplus f_i : \oplus P_i \to \oplus V_i$$

is an epimorphism whose kernel $\oplus_{i=1}^n Ker f_i$ is superfluous, by virtue of (ii). Thus, by (i), $\oplus f_i$ is essential, which ensures that $\oplus P_i$ is a projective cover of $\oplus V_i$.

(iv) Assume that $\psi : P \to V$ is an essential epimorphism, where V is a simple R-module. If $P = P' \oplus P''$, for some proper submodules P' and P'', then $\psi(P') = \psi(P'') = 0$ since 0 is the only proper submodule of V. But then $V = \psi(P') + \psi(P'') = 0$, a contradiction. Thus P is indecomposable, as desired. ∎

One of the consequences of the lemma below is that if a module does have a projective cover, then it is unique up to isomorphism.

5.2. Lemma.   *Suppose that an R-module V has a projective cover P and let M be an R-module such that V is a homomorphism image of M.*

(i) If M is projective, then P is isomorphic to a direct summand of M.

(ii) If M has a projective cover L, then P is isomorphic to a direct summand of L.

(iii) If M is another projective cover of V, then $M \cong P$.

Proof.   (i) Fix an epimorphism $\varphi : M \to V$ and an essential epimorphism $\psi : P \to V$. Because M is projective, there exists a homomorphism $\theta : M \to P$ such that $\varphi = \psi \circ \theta$. Since ψ is an essential epimorphism and $\psi \circ \theta$ is an epimorphism, θ is also an epimorphism. But P is projective, hence there is a splitting homomorphism $f : P \to M$ and so $M = Im f \oplus Ker \theta$. Since f is injective, (i) is established.

(ii) Suppose that L is a projective cover of M and let $f : L \to M$ be an epimorphism. Then $\varphi f : L \to V$ is an epimorphism, hence by (i) P is isomorphic to a direct summand of L.

(iii) Assume that M is another projective cover of V and choose φ in (i) to be essential. It will be shown that θ is also essential, from which will follow that $M = Im f \cong P$.

Suppose that M' is a proper submodule of M and put $P' = \theta(M')$. Then

$$\varphi(M') = (\psi\theta)(M') = \psi(M') \neq V$$

and thus $P' \neq P$. Hence θ is essential, as required. ∎

From now on, we write $P(V)$ for a projective cover of V. Owing to Lemma 5.2(iii), if $P(V)$ exists, then it is unique up to isomorphism.

5.3. Theorem.   *Let R be an artinian ring and let V be an R-module.*

(i) V has a projective cover.

(ii) A projective R-module W is a projective cover of V if and only if $W/J(R)W \cong V/J(R)V$.

(iii) If V is projective and if $V/J(R)V = \oplus_{i=1}^{n} V_i$ is a direct sum decomposition into simple submodules, then $V \cong \oplus_{i=1}^{n} P(V_i)$ is a de-

composition as a direct sum of indecomposable R-modules.

(iv) If V and W are projective R-modules, then $V \cong W$ if and only if

$$V/J(R)V \cong W/J(R)W$$

Proof. (i) Let L be a projective module such that $V = L/S$ for some submodule S of L. For any submodule W of S, let $f_W : L/W \to L/S$ be the canonical homomorphism. Choose W to be minimal in S such that f_W is essential; such a submodule exists, since f_S is essential and S is artinian. We claim that W is a direct summand of L; if sustained, it will follow that L/W is a projective module and hence that L/W is a projective cover of V.

Let W' be a submodule of L minimal among those whose projection $\varphi : W' \to L/W$ is surjective, and let $\pi : L \to L/W$ be the projection map. Because L is projective, there is a homomorphism $\pi' : L \to W'$ such that $\pi = \varphi \circ \pi'$, and the minimality of W' forces $\pi'(L) = W'$. Let L' denote the kernel of π'. The projection f_L factors into

$$L/L' \to L/W \to L/S$$

and the two factors are essential. Since L' is contained in W, the minimality of W implies that $W = L'$, i.e. that φ is an isomorphism. It follows that $L = W \oplus W'$, as required.

(ii) Let W be a projective module, let $f : W \to V$ be an essential epimorphism and let $\pi : V \to V/J(R)V$ be the projection map. Since π is essential, so is πf and hence $Ker\pi f \subseteq J(R)W$ (Lemma 5.1(i) and Proposition 1.4.9). On the other hand

$$f(J(R)W) \subseteq J(R)V$$

and thus $J(R)W \subseteq Ker\pi f$. Hence $Ker\pi f = J(R)W$ and so

$$W/J(R)W \cong V/J(R)V$$

Conversely, suppose that W is a projective R-module such that the above isomorphism holds. Let $f : V \to V/J(R)V$ be the projection map and let $g : W \to V/J(R)V$ be the homomorphism induced by the given isomorphism. Because f is an essential epimorphism and W

is projective, there exists an epimorphism $g' : W \to V$ with $fg' = g$. Thus

$$Ker g' \subseteq Ker g = J(R)W$$

is a superfulous submodule of W. Hence W is a projective cover of V, as required.

(iii) By Propositions 1.4.9 and 1.4.12, $J(R)V$ is a superfluous submodule of V. Because V is projective, it follows that $P(V/J(R)V) = V$. Applying Lemma 5.1(iii), (iv) we conclude that

$$V \cong P(V_1) \oplus \cdots \oplus P(V_n)$$

where each $P(V_i)$ is indecomposable.

(iv) If $V \cong W$, then obviously $V/J(R)V \cong W/J(R)W$. Conversely, suppose that V and W are projective R-modules such that $V/J(R)V \cong W/J(R)W$. By the Krull-Schmidt theorem, we may write

$$V/J(R)V = V_1 \oplus \cdots \oplus V_n$$

and

$$W/J(R)W = W_1 \oplus \cdots \oplus W_n$$

where each V_i, W_i is simple and $V_i \cong W_i$, $1 \leq i \leq n$. Then, by (iii), we have

$$W \cong P(W_1) \oplus \cdots \oplus P(W_n) \cong P(V_1) \oplus \cdots \oplus P(V_n) \cong V,$$

as required. ∎

Suppose that R is an artinian ring and let $n \geq 1$ be such that

$$R = U_1 \oplus \cdots \oplus U_n$$

where the U_i are principal indecomposable R-modules. By the Krull-Schmidt theorem, the U_i are uniquely determined up to isomorphism and the order in which they appear. Moreover, the above decomposition determines a complete set $\{e_1, \ldots, e_n\}$ of primitive idempotents in R such that $U_i = Re_i$, $1 \leq i \leq n$. Except when $J(R) = 0$, the principal indecomposable R-modules form only a small subclass of all indecomposable R-modules. However, it is the class which plays a very important role due to the following result.

5.4. Theorem. *Let R be an artinian ring.*
(i) The following conditions are equivalent:
(a) V is a projective cover of a simple R-module.
(b) V is a projective indecomposable R-module.
(c) V is a principal indecomposable R-module.
(ii) If $Re_1, \ldots, Re_m$ are all nonisomorphic principal indecomposable R-modules, then $Re_1/J(R)e_1, \ldots, Re_m/J(R)e_m$ are all nonisomorphic simple R-modules. Furthermore, for each $i \in \{1, \ldots, m\}$, Re_i is a projective cover of $Re_i/J(R)e_i$.

Proof. (i) The implications (a)$\Rightarrow$ (b) and (c) $\Rightarrow$ (a) are consequences of Lemma 5.1(iv) and Theorem 5.3(iii), respectively. Now suppose that (b) holds. Since V is projective, there exists $n \geq 1$ such that $R^n = V \oplus V'$, where R^n denotes a direct sum of n copies of R. Note that the indecomposable components of R^n are principal indecomposable R-modules. Since V is indecomposable, (c) follows by applying the Krull-Schmidt theorem.

(ii) By (i), there exist simple R-modules $V_1, \ldots, V_m$ such that $Re_i = P(V_i)$, in which case $V_i \cong Re_i/J(R)e_i$ by Theorem 5.3(ii). If $V_i \cong V_j$, then $Re_i \cong Re_j$ and hence $i = j$. If V is any simple R-module, then by (i), $P(V) \cong P(V_i)$ for some $i \in \{1, \ldots, m\}$. Hence, by Theorem 5.3(ii), $V \cong V_i$ as required. ∎

For the rest of this section, we fix a finite-dimensional algebra A over a field F. Unless explicitly stated otherwise, all A-modules are assumed to be left and finitely generated.

Let V be an A-module. Then V is said to be *injective* if for every injective A-homomorphism $Y \xrightarrow{\alpha} X$ and every A-homomorphism $Y \xrightarrow{\beta} V$ there exists an A-homomorphism $X \xrightarrow{\gamma} V$ such that $\beta = \gamma \circ \alpha$. Thus V is injective if and only if V is a direct summand of any module containing V. The reader may immediately verify that if $V_1, \ldots, V_n$ are A-modules, then $\oplus_{i=1}^n V_i$ is injective if and only if each V_i is injective.

5.5. Lemma.   *Let V be an A-module. Then the following conditions are equivalent:*

(i) V is injective.

(ii) Every homomorphism $W \to V$ of A-modules can be extended to a homomorphism from any module containing W to V.

(iii) If I is a left ideal of A and $\varphi \in Hom_A(I, V)$, then φ can be extended to an element of $Hom_A(A, V)$.

Proof.   (i) $\Rightarrow$ (ii): Suppose that V is injective and that $\varphi : W \to V$ is a homomorphism of A-modules. If X is an A-module containing W and $\alpha : W \to X$ is the natural injection, then by definition there exists an A-homomorphism $\varphi^* : X \to V$ such that $\varphi = \varphi^* \circ \alpha$. Hence φ^* extends φ.

(ii) $\Rightarrow$ (iii): Obvious.

(iii) $\Rightarrow$ (i): Suppose that V satisfies (iii), let $X \xrightarrow{\alpha} Y$ be an injective A-homomorphism and let $X \xrightarrow{\beta} V$ be an A-homomorphism. Put $\alpha(X) = X_0$ and note that, since α is injective, we can define $\gamma_0 : X_0 \to V$ by $\gamma_0 = \beta \circ \alpha^{-1}$. Now let S be the set of all ordered pairs (U, γ) such that U is a submodule of $Y, U \supseteq X_0, \gamma : U \to V$, and γ extends γ_0. If (U_1, γ_1) and (U_2, γ_2) are in S, we write $(U_1, \gamma_1) \leq (U_2, \gamma_2)$ if and only if $U_1 \subseteq U_2$ and γ_2 extends γ_1. Since $(X_0, \gamma_0) \in S$, it follows easily from Zorn's Lemma that S contains a maximal element, say, (U, γ). We show that $U = Y$ which will obviously complete the proof.

Let $y \in Y$ and let $I = \{r \in R | ry \in U\}$. Then I is a left ideal of A and we have an A-homomorphism $\varphi : I \to V$ given by $\varphi(r) = \gamma(ry)$. By assumption φ extends to an A-homomorphism $\varphi^* : A \to V$. With this map we can now define $\gamma^* : U + Ay \to V$ by $\gamma^*(u + ry) = \gamma(u) + \varphi^*(r)$. This is easily seen to be well defined and $(U, \gamma) \leq (U + Ry, \gamma^*)$. Hence, by the maximality of (U, γ) we have $y \in U$, and since y was arbitrary we have $U = Y$, as required. ■

5.6. Lemma.   *Let V be an A-module. Then there exists a (finitely generated) injective A-module that contains V as a submodule.*

Proof.   Set $M = Hom_F(A, V)$ and observe that M is an A-module via

$$(a\varphi)(b) = \varphi(ba) \qquad (a, b \in A, \varphi \in M)$$

We claim that M is injective. Indeed, let I be a left ideal of A and let $\psi : I \to M$ be an A-homomorphism. Define $\mu : I \to V$ by

$$\mu(x) = \psi(x)(1) \qquad \text{for all} \quad x \in I$$

Then μ is clearly an F-homomorphism and hence we may extend μ to an F-homomorphism $\lambda : A \to V$. Now define $\theta : A \to M$ by

$$\theta(a)(b) = \lambda(ba) \qquad \text{for all} \quad a, b \in A$$

Then θ is an A-homomorphism extending ψ and so M is injective, by Lemma 5.5. Since the map $V \to M$ given by $v \mapsto f_v$ with $f_v(a) = av$, $a \in A, v \in V$, is an injective A-homomorphism, the result follows. ∎

Let V be an A-module. An A-module M containing V is called an *injective hull* of V if M is injective and, for any injective A-module M_1 with $V \subseteq M_1 \subseteq M$ we have $M_1 = M$. By Lemma 5.6, such an M always exists.

To investigate some properties of injective hulls, we need the following concept. A submodule W of an A-module V is called *essential* if W has nonzero intersection with any nonzero submodule of V.

5.7. Lemma. *Let V be a submodule of an injective A-module M and let W be a submodule of M maximal with respect to V being an essential submodule of W. Then $M = X \oplus W$ for some submodule X of M.*

Proof. Assume that X is a submodule of M such that $X \cap W = 0$ and $X' \cap W \neq 0$ if $X' \subseteq M$ and X is a proper submodule of X'. Denote by $\bar{W}$ the image of W in M/X. Then $\bar{W}$ is an essential submodule of M/X. Because M is injective, the injective homomorphism $\bar{W} \to M, w+X \mapsto w, w \in W$, can be extended to an injective homomorphism $\psi : M/X \to M$ (Lemma 5.5). Setting $K = \psi(M/X)$, we have $V \subseteq W \subseteq K$. Let Y be a nonzero submodule of K, say $Y = \psi(Z)$, where Z is a nonzero submodule of M/X. Then $Z \cap \bar{W} \neq 0$ and hence

$0 \neq \psi(Z) \cap \psi(\bar{W}) = Y \cap W$. It follows that $(Y \cap W) \cap V \neq 0$, since V is an essential submodule of W. Thus $Y \cap V \neq 0$, so V is an essential submodule of K, therefore $K = W = \psi(\bar{W})$. It follows that $\bar{W} = M/X$ and hence that $M = X \oplus W$, as required. ∎

We are now ready to prove our main result on injective hulls.

5.8. Theorem. *Let V be a submodule of an injective A-module M.*

(i) M is an injective hull of V if and only if V is an essential submodule of V.

(ii) If M is indecomposable, then M is an injective hull of V. The converse is true if $Soc(V)$ is simple.

(iii) Any two injective hulls of V are isomorphic.

Proof. (i) Assume that M is an injective hull of V and let W be as in Lemma 5.7. Then W is an injective A-module with $V \subseteq W \subseteq M$. Hence $W = M$ and so V is an essential submodule of M.

Conversely, suppose that V is an essential submodule of M and let W be an injective A-module with $V \subseteq W \subseteq M$. By Lemma 5.5, the inclusion map $V \to W$ extends to an A-homomorphism $M \to W$. Because V is an essential submodule of M, the kernel of this homomorphism is zero. Hence $M = W$ and so M is an injective hull of V.

(ii) If M is indecomposable, then V is an essential submodule of M, by Lemma 5.7. Thus, by (i), M must be an injective hull of V.

Conversely, suppose that M is an injective hull of V and that $Soc(V)$ is simple. Then, by (i), V is an essential submodule of M and, by hypothesis $Soc(V)$ is contained in any nonzero submodule of M. Hence, if $M = X \oplus Y$ for some nonzero submodules X, Y of M, then $Soc(V) \subseteq X \cap Y = 0$, a contradiction. Thus M is indecomposable.

(iii) Let M_1 and M_2 be two injective hulls of V. Since V is an essential submodule of M_1, it follows from Lemma 5.5 that there exists an injective A-homomorphism $M_2 \to M_1$. A similar argument shows the existence of an injective A-homomorphism $M_1 \to M_2$ and thus $M_1 \cong M_2$. ∎

We close by providing some further general information which will be used in the next section.

5.9. Lemma. *Let e be an idempotent of an arbitrary ring R. Then*

(i) For any ideal I of R, $Re \cap I = Ie$.

(ii) $Re/J(R)e \cong \bar{R}\bar{e}$ where $\bar{R} = R/J(R)$ and $\bar{e} = e + J(R)$.

Proof. (i) Assume that $x \in Re \cap I$. Then $x = ae$ for some $a \in R$. Therefore

$$x = ae = ae^2 = xe \in Ie,$$

proving that $Re \cap I \subseteq Ie$. The opposite containment being trivial, (i) follows.

(ii) Applying (i) for $I = J(R)$, we have

$$
\begin{aligned}
Re/J(R)e &= Re/(Re \cap J(R)) \cong (Re + J(R))/J(R) \\
&= (R/J(R))(e + J(R)) = \bar{R}\bar{e}
\end{aligned}
$$

as required. ∎

Again, assume that A is a finite-dimensional algebra over a field F. Note that if Ae is a principal indecomposable A-module, then by Lemma 5.9 and Theorem 5.4, $\bar{A}\bar{e}$ is a simple A-module.

5.10. Lemma. *Let $A = \oplus_{i=1}^{n} Ae_i$ be a decomposition of A into principal indecomposable A-modules and let $\bar{e}_i$ be the image of the primitive idempotent e_i of A in $\bar{A} = A/J(A)$. Then, for any A-module V,*

$$dim_F e_i V = dim_F Hom_A(Ae_i, V) = m\, dim_F End_A(\bar{A}\bar{e}_i)$$

where m is the multiplicity of the simple A-module $\bar{A}\bar{e}_i$ as a composition factor of V.

Proof. Owing to Lemma 1.6.7, the map $Hom_A(Ae_i, V) \mapsto e_i V, f \mapsto f(e)$ is an isomorphism of additive groups. Since this map is obviously F-linear, we have

$$Hom_A(Ae_i, V) \cong e_i V \qquad \text{as } F\text{-spaces}$$

The equality $dim_F Hom_A(Ae_i, V) = m dim_F End_A(\bar{A}\bar{e}_i)$ is trivial if V is simple. Therefore we may assume that V has a maximal nonzero submodule X. Since Ae_i is projective, the exact sequence

$$0 \to X \to V \to V/X \to 0$$

induces an exact sequence

$$0 \to Hom_A(Ae_i, X) \to Hom_A(Ae_i, V) \to Hom_A(Ae_i, V/X) \to 0$$

Thus we have

$$dim_F Hom_A(Ae_i, V) = dim_F Hom_A(Ae_i, X) + dim_F Hom_A(Ae_i, V/X)$$

If m_1 and m_2 are the multiplicities of $\bar{A}\bar{e}_i$ as a composition factor of X and V/X, respectively, then applying induction on $dim\, V$, we have

$$dim_F Hom_A(Ae_i, X) = m_1 dim_F End_A(\bar{A}\bar{e}_i)$$

and

$$dim_F Hom_A(Ae_i, V/X) = m_2 dim_F End_A(\bar{A}\bar{e}_i)$$

Adding up to these equalities yields

$$
\begin{aligned}
dim_F Hom_A(Ae_i, V) &= (m_1 + m_2) dim_F End_A(\bar{A}\bar{e}_i) \\
&= m dim_F End_A(\bar{A}\bar{e}_i),
\end{aligned}
$$

as desired. ∎

5.11. Corollary. *Let e be a primitive idempotent of A, let $\bar{e}$ be the image of e in $\bar{A} = A/J(A)$ and let V be an A-module. Then the following conditions are equivalent:*
(i) $\bar{A}\bar{e}$ is a composition factor of V.
(ii) $Hom_A(Ae, V) \neq 0$.
(iii) $eV \neq 0$.

Proof. Apply Lemma 5.10. ∎

Let $e_1, \ldots, e_n$ be a complete set of primitive idempotents of A. Then

$$A = Ae_1 \oplus \cdots \oplus Ae_n$$

is a decomposition of A as a direct sum of principal indecomposable modules. Let $\bar{e}_i$ be the image of e_i in $\bar{A} = A/J(A)$. We know, from Lemma 5.9(ii), that

$$Ae_i/J(A)e_i \cong \bar{A}\bar{e}_i \qquad (1 \le i \le n)$$

Suppose that the numbering is so chosen that $Ae_1, \dots, Ae_m$ are all nonisomorphic among the $Ae_j, 1 \le j \le n$. Then, by Theorem 5.4(ii),

$$\bar{A}\bar{e}_1, \bar{A}\bar{e}_2, \dots, \bar{A}\bar{e}_m$$

are all nonisomorphic simple A-modules. Given $i, j \in \{1, \dots, m\}$, let c_{ij} be the multiplicity of the simple A-module $\bar{A}\bar{e}_j$ as a composition factor of Ae_i. The nonnegative integers c_{ij} are called the *Cartan invariants* of A, and the $m \times m$ matrix $C = (c_{ij})$ is called the *Cartan matrix* of A. Note that the Cartan matrix C is only determined up to a permutation of its rows and columns (depending on the numbering of $Ae_1, \dots, Ae_m$).

Let A be an algebra over a field F. Then F is called a *splitting field* for A if for any simple A-module V, $dim_F End_A(V) = 1$.

5.12. Theorem.   *Let F be a splitting field for A, let*

$$A = Ae_1 \oplus \cdots \oplus Ae_n$$

be a decomposition of A into direct sum of principal indecomposable modules, and let $C = (c_{ij})$ be the Cartan matrix of A. Then

(i) $c_{ij} = dim_F e_j Ae_i$.

(ii) The multiplicity of Ae_i as a direct summand of A in the sense of the Krull-Schmidt theorem is $dim_F \bar{A}\bar{e}_i$.

(iii) The multiplicity of $\bar{A}\bar{e}_i$ as a composition factor of A is $dim_F e_i A$.

Proof.   (i) Since F is a splitting field for A, $dim_F End_A(\bar{A}\bar{e}_i) = 1$. Applying Lemma 5.10, we conclude that the multiplicity of $\bar{A}\bar{e}_j$ as a composition factor of Ae_i is equal to $dim_F e_j Ae_i$, as required.

(ii) Since F is a splitting field for A, $\bar{A}$ is a direct product of full matrix rings over F. Hence the multiplicity of $\bar{A}\bar{e}_i$ in $\bar{A}$ is equal to $dim_F \bar{A}\bar{e}_i$. Owing to Theorem 5.4, Ae_i is determined, up to isomorphism, by $\bar{A}\bar{e}_i$ and therefore $dim_F \bar{A}\bar{e}_i$ also equals the multiplicity of Ae_i as a direct summand of A.

(iii) Direct consequence of Lemma 5.10. ∎

6.   Classical results

In this section, we provide a number of classical results concerning Frobenius and symmetric algebras. In what follows, A denotes a finite-dimensional algebra over a field F and all A-modules are assumed to be finitely generated.

For each subset X of A, let $l(X)$ and $r(X)$ denote the left and right annihilators of X defined by

$$l(X) = \{a \in A | aX = 0\}$$
$$r(X) = \{a \in A | Xa = 0\}$$

It is clear that $l(X)$ and $r(X)$ are left and right ideals of A, respectively. Moreover, if X is a left (respectively, right) ideal of A, then $l(X)$ (respectively, $r(X)$) is a two-sided ideal of A.

Suppose that (A, ψ) is a Frobenius algebra and that X is a subset of A. Define the subsets ${}^{\perp}X$ and $X^{\perp}$ of A by

$$^{\perp}X = \{a \in A | \psi(aX) = 0\}$$
$$X^{\perp} = \{a \in A | \psi(Xa) = 0\}$$

It is plain that if X is a subspace of A, then so are ${}^{\perp}X$ and $X^{\perp}$.

6.1. Theorem.   *Let (A, ψ) be a Frobenius algebra and let X be an F-subspace of A.*

(i) $({}^{\perp}X)^{\perp} = {}^{\perp}(X^{\perp}) = X$ and $dim_F X^{\perp} = dim_{\frac{1}{F}} X = dim_F A - dim_F X$.

(ii) If X is a left (respectively, right) ideal of A, then $r(X) = X^{\perp}$ (respectively, $l(X) = {}^{\perp}X$). In particular, for every idempotent e of A

$$(Ae)^{\perp} = (1 - e)A$$
$$A/(Ae)^{\perp} \cong eA \quad \text{as right A-modules}$$
$$dim_F eA = dim_F Ae$$

(iii) If X is a left ideal of A, then

$$dim_F X + dim_F r(X) = dim_F A \quad \text{and} \quad l(r(X)) = X$$

If X is a right ideal of A, then

$$dim_F X + dim_F l(X) = dim_F A \quad and \quad r(l(X)) = X$$

(iv) The mapping $X \mapsto r(X)$ is a duality of the lattice of left ideals of A onto the lattice of right ideals of A, that is $X \mapsto r(X)$ is a bijection with

$$r(X_1 + X_2) = r(X_1) \cap r(X_2) \quad and \quad r(X_1 \cap X_2) = r(X_1) + r(X_2)$$

for all left ideals X_1, X_2 of A.

Similarly $X \mapsto l(X)$ is a duality of the lattice of right ideals of A onto the lattice of left ideals of A.

(v) If (A, ψ) is symmetric, then

(a) $r(X) = l(X)$ for any two-sided ideal X of A.

(b) The Cartan matrix of A is symmetric, provided that F is a splitting field for A.

Proof. (i) Let $\{a_1, \ldots, a_n\}$ be an F-basis of A such that $\{a_1, \ldots, a_m\}$ $m \leq n$ is an F-basis of X. The map

$$\begin{cases} A & \to & F \times F \times \cdots \times F \quad (n \text{ times}) \\ a & \mapsto & (\psi(a_1 a), \psi(a_2 a), \ldots, \psi(a_n a)) \end{cases}$$

is an isomorphism of F-spaces such that the image of $X^\perp$ consists of all $(\lambda_1, \ldots, \lambda_n)$ with $\lambda_1 = \cdots = \lambda_m = 0$. Thus

$$dim_F X^\perp = dim_F A - dim_F X$$

and, by a similar argument,

$$dim_F^\perp X = dim_F A - dim_F X$$

It follows that

$$dim_F(^\perp X)^\perp = dim_F X$$

Since $(^\perp X)^\perp \supseteq X$, we deduce that $(^\perp X)^\perp = X$. A similar argument shows that $^\perp(X^\perp) = X$, as required.

(ii) Suppose that X is a right ideal of A. Then $a \in {}^\perp X$ if and only if $\psi(aX) = 0$ or if and only if $\psi(aXA) = 0$. Because the latter equality is equivalent to $aX = 0$, we have $l(X) = {}^\perp X$. The equality $r(X) = X^\perp$ for a left ideal X of A follows by the same argument.

Now assume that e is an idempotent of A. Then, for all $x \in A$,

$$x \in (Ae)^\perp \Longleftrightarrow Aex = 0 \Longleftrightarrow ex = 0 \Longleftrightarrow x \in (1 - e)A$$

which implies

$$(Ae)^\perp = (1 - e)A$$

and

$$A/(Ae)^\perp \cong eA \tag{1}$$

Finally, $dim_F Ae = dim_F eA$, by applying (1) and (i).

(iii) This is a direct consequence of (i) and (ii).

(iv) It is a consequence of (iii) that the given maps are bijections. We obviously have

$$r(X_1 + X_2) = r(X_1) \cap r(X_2) \quad \text{and} \quad l(X_1 + X_2) = l(X_1) \cap l(X_2)$$

Applying (iii), it follows that

$$
\begin{aligned}
l(X_1 \cap X_2) &= l(r(l(X_1)) \cap r(l(X_2))) \\
&= l(r(l(X_1) + l(X_2))) \\
&= l(X_1) + l(X_2)
\end{aligned}
$$

The corresponding assertion for left ideals follows similarly.

(v) Suppose that (A, ψ) is a symmetric algebra. Then ${}^\perp X = X^\perp$ for any subset X of A. Thus (a) follows by virtue of (ii).

Let $e_1, \ldots, e_n$ be a complete set of orthogonal primitive idempotents of A and let $C = (c_{ij})$ be the Cartan matrix of A. Then

$$A = \oplus_{i=1}^n Ae_i = \oplus_{i=1}^n e_i A$$

and hence

$$A = \oplus_{i,j=1}^n e_j Ae_i \quad (\text{direct sum of } F\text{-spaces})$$

Now fix $i, j, k, s \in \{1, \ldots, n\}$ such that $(k, s) \neq (j, i)$. Then

$$0 = \psi(e_i Ae_j e_k Ae_s) = \psi(e_k Ae_s e_i Ae_j)$$

and hence $e_k A e_s \subseteq (e_i A e_j)^\perp$. Thus

$$\oplus_{(k,s)\neq(j,i)} e_k A e_s \subseteq (e_i A e_j)^\perp$$

and therefore, by (i) and Theorem 5.12,

$$
\begin{aligned}
c_{ji} &= dim_F e_i A e_j = dim_F A - dim_F (e_i A e_j)^\perp \\
&\leq dim_F A - \sum_{(k,s)\neq(j,i)} dim_F e_k A e_s \\
&= dim_F e_j A e_i = c_{ij}
\end{aligned}
$$

A similar argument shows that $c_{ij} \leq c_{ji}$, proving (b). ∎

Turning to factor algebras, we now prove the following useful results.

6.2. Theorem. *(Nakayama (1939)). Let X be an ideal of the F-algebra A.*

(i) Assume that (A, ψ) is a Frobenius algebra. Then A/X is a Frobenius algebra if and only if there exists $a \in A$ such that $r(X) = aA = Aa$.

(ii) Assume that (A, ψ) is a symmetric algebra. Then A/X is a symmetric algebra if and only if there exists $a \in Z(A)$ such that $r(X) = aA$.

Proof. (i) Suppose that $(A/X, \mu)$ is a Frobenius algebra, and let $\pi : A \to A/X$ be the natural map. Then, by Lemma 1.5(i), there exists $a \in A$ such that $\mu\pi = \psi_a$, where $\psi_a(b) = \psi(ba)$ for all $b \in A$. It will be shown that $r(X) = aA = Aa$. Since $\psi_a(X) = \psi(Xa) = 0$, we have $a \in X^\perp$. Hence $aA \subseteq X^\perp$ and, by Theorem 6.1(ii), $aA \subseteq r(X)$.

Assume that $b \in l(aA)$ or $b \in l(Aa)$. Then $ba = 0$ and thus

$$\mu((A/X)(b + X)) = (\mu\pi)(Ab) = \psi_a(Ab) = \psi(Aba) = 0$$

Thus $b \in X$ and therefore

$$l(aA) \subseteq X \quad \text{and} \quad l(Aa) \subseteq X$$

Applying Theorem 6.1(iii), we conclude that

$$aA = r(l(aA)) \supseteq r(X),$$

proving that $r(X) = aA$. It follows that $Aa \subseteq r(X)$ and hence $l(Aa) \supseteq X$, again by Theorem 6.1(iii). Thus $X = l(Aa)$ and therefore $r(X) = rl(Aa) = Aa$.

Conversely, suppose that $r(X) = aA = Aa$ for some $a \in A$. Consider the map

$$\begin{cases} A/X \xrightarrow{\mu} F \\ b + X \mapsto \psi(ba) \end{cases} \quad (b \in A)$$

Since $a \in r(X)$, μ is well defined and obviously an F-linear map. Assume that $b \in A$ is such that $\mu((A/X)(b+X)) = 0$. Then $\psi(Aba) = 0$, so $ba = 0$ and hence $b \in l(aA)$. However, $r(X) = aA$ implies that $X = l(r(X)) = l(aA)$; hence $b \in X$. Thus A/X is a Frobenius algebra, as required.

(ii) Suppose that A/X is symmetric. By (i), $r(X) = aA$ for some $a \in A$. On the other hand

$$\psi(xya) = \psi_a(xy) = \psi_a(yx) = \psi(yxa) = \psi(xay)$$

for all $x, y \in A$. Thus $ay = ya$ for all $y \in A$ and $a \in Z(A)$.

Conversely, assume that $r(X) = aA$ with $a \in Z(A)$. Then, by (i), A/X is obviously symmetric, thus completing the proof. ∎

The right socle of A is defined to be the socle of A_A.

6.3. Theorem. *Suppose that A is a Frobenius algebra.*
(i) The left and right socles of A coincide. Furthermore,

$$dim_F Soc(A) = dim_F(A/J(A))$$

(ii) For each primitive idempotent $e \in A$, $Soc(A)e$ is the unique simple submodule of Ae. In particular,

$$Soc(Ae) = Soc(A)e$$

(iii) If A is symmetric, then
(a) There exists $z \in Z(A)$ such that $Soc(A) = Az$ and $J(A) = l(z)$.
(b) For any primitive idempotent $e \in A$,

$$Soc(Ae) \cong Ae/J(A)e$$

Proof. (i) and (ii): Setting $J = J(A)$, it follows that $r(J)$ and $l(J)$ are the left and right socles of A, respectively. Our plan is to show that $r(J) \subseteq l(J)$ and that $er(J)$ is a unique simple submodule of eA. The same argument (reversing left and right) will prove that $l(J) \subseteq r(J)$ and $l(J)e$ is the unique simple submodule of Ae. Since $r(J) = J^\perp$ and $dim_F J^\perp = dim_F A - dim_F J = dim_F A/J$ (Theorem 6.1(i), (ii)), properties (i) and (ii) will be established.

Let e be a primitive idempotent of A. Then, by Theorem 5.4(ii), Ae/Je is a simple module and hence Je is a unique maximal submodule of Ae. Owing to Theorem 6.1(i), (iv), the map $X \mapsto X^\perp$ is an inclusion reversing bijection between the set of left ideals X of A contained in Ae and the set of right ideals of A containing $(Ae)^\perp$. We thus obtain an inclusion-reversing bijection $X \mapsto X^\perp/(Ae)^\perp$ between the set of submodules of Ae and the set of submodules of $A/(Ae)^\perp$. But, by Theorem 6.1(ii),

$$(Ae)^\perp = (1 - e)A \quad \text{and} \quad A/(Ae)^\perp \cong eA$$

Thus $X \mapsto eX^\perp$ is an inclusion-reversing bijection between the set of submodules of Ae and submodules of eA. In particular, $e(Je)^\perp = er(J)$ is a unique simple submodule of eA.

Since J annihilates all simple A-modules, we have $er(J)J = 0$. Since e is an arbitrary primitive idempotent of A, we conclude that $r(J)J = 0$ and therefore $r(J) \subseteq l(J)$, as required.

(iii) By hypothesis, A is symmetric and, by Proposition 1.9, $A/J(A)$ is also symmetric. Hence, by Theorem 6.2(ii), $r(J(A)) = Az$ for some $z \in Z(A)$. Since $Soc(A) = r(J(A))$, it follows that $Soc(A) = Az$. By Theorem 6.1(iii), we also have

$$J(A) = l(r(J(A)) = l(Az) = l(z),$$

proving (a).

To prove (b), put $S = Soc(A)$ and let $\psi \in A^*$ be such that (A, ψ) is a symmetric algebra. By (ii), $Soc(Ae) = Se$ is a simple submodule of Ae. Since Je is a unique maximal submodule of Ae, it suffices to show that

$$Hom_A(Ae, Se) \neq 0$$

By Lemma 1.6.7, the latter is equivalent to $eSe \neq 0$. Assume by way of contradiction that $eSe = 0$. Then

$$0 = \psi(eSe) = \psi(Se) = \psi(ASe)$$

and hence $Se = 0$, which is impossible. This completes the proof of the theorem. ∎

6.4. Corollary. *Let A be a symmetric algebra and let $e_1, \ldots, e_m$ be primitive idempotents of A such that $Ae_1, \ldots, Ae_m$ are all nonisomorphic principal indecomposable A-modules. If $C = (c_{ij}), 1 \leq i, j \leq m$ is the Cartan matrix of A, then for any given $i \in \{1, \ldots, m\}$, either $c_{ii} \geq 2$ or Ae_i is simple, $c_{ii} = 1$ and $c_{ij} = 0$ for all $j \neq i$.*

Proof. Fix $i \in \{1, \ldots, n\}$ and assume that $c_{ii} < 2$. By Theorem 6.3(iii) and Lemma 5.9(ii),

$$Soc(Ae_i) \cong Ae_i / J(A)e_i \cong \bar{A}\bar{e}_i$$

where $\bar{e}_i$ is the image of e_i in $\bar{A} = A/J(A)$. Hence $c_{ii} = 1$ and Ae_i is simple. Since Ae_i is simple, we obviously have $c_{ij} = 0$ for all $j \neq i$, as asserted. ∎

6.5. Theorem. *Let A be a Frobenius algebra.*
(i) A is an injective left A-module.
(ii) An A-module V is injective if and only if V is projective.
(iii) If P is a projective indecomposable A-module, then P is an injective hull of $Soc(P)$. Moreover, if A is symmetric, then P is a projective cover of $Soc(P)$.

Proof. (i) Let I be a left ideal of A and let $f \in Hom_A(I, A)$. By Lemma 5.5, it suffices to show that f can be extended to an element of $End_A(A)$. By hypothesis, there exists $\psi \in A^*$ such that (A, ψ) is a Frobenius algebra. Given $a \in A$, define $\lambda_a : I \to A$ by $\lambda_a(x) = xa$. Then each λ_a lies in $Hom_A(I, A)$ and clearly λ_a can be extended to an element λ'_a of $End_A(A)$ by setting $\lambda'_a(x) = xa$ for all $x \in A$. Thus it suffices to verify that for $M = \{\lambda_a | a \in A\}$

$$dim_F(Hom_A(I, A) \leq dim_F M$$

To this end, observe that the mapping $a \mapsto \lambda_a$ is an F-homomorphism of A onto M with kernel $r(I)$. Therefore, by Theorem 6.1(iii),

$$dim_F M = dim_F A - dim_F r(I) = dim_F I = dim_F(Hom_F(I, F))$$

and thus we need only verify that the F-linear map

$$\begin{cases} Hom_A(I, A) & \rightarrow & Hom_F(I, F) \\ f & \mapsto & \psi f \end{cases}$$

is injective. But if $(\psi f)(I) = 0$, then the left ideal $f(I)$ is in the kernel of ψ. Thus $f(I) = 0$ and therefore $f = 0$, as required.

(ii) By (i), the regular module $_A A$ is injective. Hence every free A-module is injective. If V is projective, then V is a direct summand of a free A-module. Hence V is a direct summand of an injective A-module and so is injective.

Conversely, assume that V is injective. Since V is a finite direct sum of indecomposable A-modules, we may assume that V is indecomposable. Let V_1 be a simple submodule of V. Owing to Theorem 5.8(ii), V is an injective hull of V_1. By Theorems 6.3(iii) and 5.4(ii), there exists a primitive idempotent e of A such that $V_1 \cong Soc(Ae)$. But, by the foregoing, Ae is an indecomposable injective A-module. Thus, by Theorem 5.8(ii), Ae is an injective hull of $Soc(Ae)$. Hence $V \cong Ae$ by Theorem 5.8(iii), proving that V is projective.

(iii) Let P be a projective indecomposable A-module. Then, by Theorem 5.4, $P \cong Ae$ for some primitive idempotent e of A. Since $Soc(Ae)$ is a submodule of an injective indecomposable A-module Ae, it follows from Theorem 5.8(ii), that Ae is an injective hull of $Soc(Ae)$. Hence P is an injective hull of $Soc(P)$. If A is symmetric, then by Theorem 6.3(iii), $Soc(Ae) \cong Ae/J(A)e$. Since, by Theorem 5.4(ii), Ae is a projective cover of $Ae/J(A)e$, the result follows. ∎

Let V be an A-module. Then the *dual module* $V^* = Hom_F(V, F)$ is a right A-module with the action of A on V^* given by

$$(\psi a)(v) = \psi(av) \qquad (a \in A, v \in V, \psi \in V^*)$$

If V is a right A-module, then a similar definition gives a left A-module V^*. We now record the following properties of left (or right) A-modules V and W.

6.6. Lemma. *(i) V is indecomposable if and only if V^* is indecomposable.*

(ii) V is projective if and only if V^* is injective.

(iii) $(V^*)^* \cong V$.

(iv) $(V \oplus W)^* \cong V^* \oplus W^*$.

(v) The map $W \mapsto W^\perp$, where $W^\perp = \{f \in V^*| f(W) = 0\}$ yields an inclusion reversing bijection between the submodules of V and the submodules of V^*. Moreover, for any submodule W of V,

(a) $dim_F W + dim_F W^\perp = dim_F V$.

(b) $(W^\perp)^\perp = W$.

(c) $(V/W)^* \cong W^\perp$ and $V^*/W^\perp \cong W^*$.

Proof. The verification is straightforward and therefore will be omitted. ∎

6.7. Theorem. *Let A be a Frobenius algebra, let $Ae_1, \ldots, Ae_k$ be all nonisomorphic principal indecomposable A-modules and write*

$$A \cong \oplus_{i=1}^k n_i A_i e_i \tag{2}$$

for some positive integers $n_i, 1 \leq i \leq k$. Then there exists a permutation π of $\{1, \ldots, k\}$ such that

(i) $n_i = n_{\pi(i)}$ and $Ae_i \cong (e_{\pi(i)}A)^*$ for all $i \in \{1, \ldots, k\}$.

(ii) $Soc(Ae_i) \cong \bar{A}\bar{e}_{\pi(i)}$ for all $i \in \{1, \ldots, k\}$

where $\bar{e}_{\pi(i)}$ is the image of $e_{\pi(i)}$ in $\bar{A} = A/J(A)$.

Proof. (i) By Lemma 1.6.9, $e_1 A, \ldots, e_k A$ are all nonisomorphic principal indecomposable right A-modules. Since

$$A \cong \oplus_{i=1}^k n_i e_i A \quad \text{(as right A-modules)}$$

we also have, by Lemma 6.6,

$$A^* \cong \oplus_{i=1}^k n_i (e_i A)^* \quad \text{(as left A-modules)} \tag{3}$$

where $(e_1 A)^*, \ldots, (e_k A)^*$ are pairwise nonisomorphic. By Corollary 1.6, the left A-modules A and A^* are isomorphic. Hence, by applying the Krull-Schmidt theorem to the decompositions (2) and (3), we find a permutation π of $\{1, \ldots, k\}$ such that (i) holds.

(ii) Note that $e_{\pi(i)}A$ has a unique maximal A-submodule $e_{\pi(i)}J(A)$. Hence, by Lemma 6.6(v), $(e_{\pi(i)}A)^*$ has a unique minimal A-submodule $(e_{\pi(i)}J(A))^{\perp}$, which by Lemma 6.6(v) is isomorphic to $(\bar{e}_{\pi(i)}\bar{A})^*$. On the other hand, $\bar{A}\bar{e}_{\pi(i)}$ and $(\bar{e}_{\pi(i)}\bar{A})^*$ belong to the same simple component of $\bar{A}$ and hence are isomorphic, proving (ii). ■

6.8. Corollary. *Let A be a symmetric algebra. Then, for any primitive idempotent e of A,*

$$Ae \cong (eA)^*$$

Proof. We preserve the notation of Theorem 6.7. By Theorem 6.3(iii), $Soc(Ae_i) \cong \bar{A}\bar{e}_i$. Hence, by Theorem 6.7(ii), $\pi(i) = i$ for all $i \in \{1, \ldots, k\}$. The desired conclusion is therefore a consequence of Theorem 6.7(i). ■

6.9. Theorem. *Let (A, ψ) be a Frobenius algebra. Then there exists an F-algebra automorphism σ of A such that*

(i) $\psi(xa) = \psi(\sigma(a)x)$ for all $a, x \in A$.

(ii) If I is a left ideal of A, then there are A-isomorphisms

$$I \cong [A/\sigma(r(I))]^*$$

and

$$A/I \cong [\sigma(r(I))]^*$$

(iii) If π is a permutation given by Theorem 6.7, then

$$Ae_{\pi(i)} \cong A\sigma(e_i)$$

(iv) If A is a symmetric algebra, then σ is the identity automorphism.

Proof. (i) Owing to Lemma 1.5(i), the map $f : A \to A^*$, where $f(a)(x) = \psi(xa)$ for all $x, a \in A$, is an isomorphism of left A-modules. Similarly, we have an isomorphism $g : A \to A^*$ such that $g(a)(x) = \psi(ax)$ for all $x, a \in A$. Thus, for any given $a \in A$ there exists a unique $\sigma(a) \in A$ such that

$$f(a) = g(\sigma(a))$$

and hence such that (i) holds.

By the foregoing, we are left to verify that σ is an F-algebra automorphism of A. It is clear that σ is an F-automorphism of A. Since

$$\psi(\sigma(ab)x) = \psi((xa)b) = \psi(\sigma(b)xa) = \psi(\sigma(a)\sigma(b)x),$$

we must also have $\sigma(ab) = \sigma(a)\sigma(b)$ for all $a, b \in A$, as required.

(ii) Consider the A-isomorphism $f : A \to A^*$ given in (i) and keep the notation of Lemma 6.6(v). We identify A^{**} with A by means of

$$a(\chi) = \chi(a) \quad \text{for all} \quad a \in A, \chi \in A^*$$

Then we have

$$\begin{aligned}
f(I)^{\perp} &= \{a \in A | a(f(I)) = 0\} \\
&= \{a \in A | f(I)(a) = 0\} \\
&= \{a \in A | \psi(aI) = 0\} \\
&= {}^{\perp}I
\end{aligned}$$

We now claim that ${}^{\perp}I = \sigma(r(I))$; if sustained, the required isomorphisms will follow by virtue of Lemma 6.6(v) (b), (c).

To substantiate our claim, we first note that $\sigma(r(I)) = r(\sigma(I))$. Hence, by (i) and Theorem 6.1(ii),

$$a \in \sigma(r(I)) \iff \psi(\sigma(I)a) = 0 \iff \psi(aI) = 0 \iff a \in {}^{\perp}I$$

as required.

(iii) Consider the left ideal $I = Ae_i$. Then $r(I) = (1 - e_i)A$ and $\sigma(r(I)) = (1 - \sigma(e_i))A$. Hence

$$A/\sigma(r(I)) \cong \sigma(e_i)A$$

and therefore, by (ii), $Ae_i \cong (\sigma(e_i)A)^*$. On the other hand, by Theorem 6.7(ii), $Ae_i \cong (e_{\pi(i)}A)^*$, which proves that $\sigma(e_i)A \cong e_{\pi(i)}A$ by Lemma 6.6(iii). Thus, by Lemma 6.9, $A\sigma(e_i) \cong Ae_{\pi(i)}$, as asserted.

(iv) Assume that A is symmetric. Then, for all $a, x \in A$, we have

$$0 = \psi(xa) - \psi(ax) = \psi((\sigma(a) - a)x)$$

Thus $\sigma(a) = a$ for all $a \in A$, as asserted. $\blacksquare$

The automorphism σ in Theorem 6.9 is called the *Nakayama auto-morphism* of the Frobenius algebra A. Our next aim is to provide an application of Theorem 6.5(iii). As a preliminary, we record the following piece of information.

Let V be an A-module and let $P(V)$ be a projective cover of V. Then, by definition of $P(V)$, there is an exact sequence

$$0 \to \Omega(V) \to P(V) \to V \to 0$$

where $\Omega(V)$ is a superfluous submodule of $P(V)$ and thus

$$\Omega(V) \subseteq J(A)P(V)$$

We shall refer to $\Omega(V)$ as the *Heller module* of V. As we shall see below, V determines the isomorphism class of $\Omega(V)$.

6.10. Lemma.   *(Schanuel's lemma). Let V, W_1, W_2 be A-modules and let P_1, P_2 be projective A-modules such that the sequences*

$$0 \to W_1 \to P_1 \xrightarrow{f_1} V \to 0$$

$$0 \to W_2 \to P_2 \xrightarrow{f_2} V \to 0$$

are exact. Then

$$W_1 \oplus P_2 \cong W_2 \oplus P_1$$

Proof.   Let X be the submodule of $P_1 \oplus P_2$ defined by

$$X = \{(x_1, x_2) | f_1(x_1) = f_2(x_2)\}$$

If $\pi : P_1 \oplus P_2 \to P_1$ is the projection map, then $\pi(P_1) = P_1$ and

$$X \cap Ker\pi = \{(0, x_2) | x_2 \in Ker f_2\} \cong W_2$$

Thus we have an exact sequence

$$0 \to W_2 \to X \to P_1 \to 0$$

Because P_1 is projective, we conclude that $X \cong W_2 \oplus P_1$. By the same argument, $X \cong W_1 \oplus P_2$ and the result follows. ∎

6.11. Corollary. *Let V be an A-module. Then V determines the isomorphism class of $\Omega(V)$.*

Proof. Let P_1 and P_2 be two projective covers of V and let

$$0 \to \Omega_i(V) \to P_i \to V \to 0 \quad (i = 1, 2)$$

be the corresponding exact sequences. By Lemma 6.10,

$$\Omega_1(V) \oplus P_2 \cong \Omega_2(V) \oplus P_1$$

while by Lemma 5.2(iii), $P_1 \cong P_2$. Hence, by the Krull-Schmidt theorem,

$$\Omega_1(V) \cong \Omega_2(V),$$

as required. ∎

An A-module V is said to be *projective-free* if V has no projective module as a direct summand.

6.12. Lemma. *Let A be a Frobenius algebra and let V be a projective-free A-module. Then*

$$Soc(\Omega(V)) = Soc(P(V))$$

Proof. Set $P = P(V)$ and write $P = \oplus_{i=1}^{n} P_i$ where each P_i is an indecomposable projective module. We know, from Theorem 6.3(ii), that $Soc(P_i)$ is simple, $1 \le i \le n$. Since $P/\Omega(V) \cong V$ is projective-free, and, by Theorem 6.5(ii), every projective A-module is injective, it follows easily that $P_i \cap \Omega(V) \ne 0$, $1 \le i \le n$. Hence $Soc(P_i) \subseteq \Omega(V)$ for all i. Thus $Soc(P) \subseteq \Omega(V)$ and since obviously $Soc(\Omega(V)) \subseteq Soc(P)$, the result follows. ∎

We are now ready to record the following application of Theorem 6.5(iii) and Lemma 6.12.

6.13. Corollary.   *If A is a symmetric algebra and V is a projective-free A-module, then*

$$Soc(\Omega(V)) \cong V/J(A)V$$

Proof.   It is a consequence of Theorem 6.5(iii), that for $P = P(V)$, $Soc(P) \cong P/J(A)P$. Since $\Omega(V) \subseteq J(A)P$, the isomorphism $V \cong P/\Omega(V)$ induces an isomorphism $V/J(A)V \cong P/J(A)P$. Hence, by Lemma 6.12,

$$Soc(\Omega(V)) = Soc(P) \cong V/J(A)V,$$

as asserted. ∎

Let A be a Frobenius algebra and let V be an A-module. Then, by Theorem 6.5(ii), V is injective if and only if V is projective. We close by providing another characterization of injective A-modules, due to Ikeda.

Again, assume that (A, ψ) is a Frobenius algebra. By Lemma 1.5(i), the map $f : A \to A^*$ given by

$$f(a)(x) = \psi(xa) \qquad \text{for all} \quad x, a \in A$$

is an isomorphism of left A-modules. Let $a_1, \ldots, a_n$ be an F-basis of A and let $a_1^*, \ldots, a_n^*$ be the dual basis of A^*, i.e.

$$a_i^*(a_j) = \delta_{ij} \quad \text{for all} \quad i, j \in \{1, \ldots, n\}$$

Setting $b_j = f^{-1}(a_j^*)$, it then follows that $b_1, \ldots, b_n$ is an F-basis of A such that

$$\psi(a_i b_j) = \delta_{ij} \quad \text{for all} \quad i, j \in \{1, \ldots, n\} \tag{4}$$

We shall refer to the basis $b_1, \ldots, b_n$ of A as the ψ-dual of $a_1, \ldots, a_n$.

6.14. Lemma.   *Let (A, ψ) be a Frobenius algebra, let $a_1, \ldots, a_n$ be an F-basis of A and let the basis $b_1, \ldots, b_n$ of A be the ψ-dual of $a_1, \ldots, a_n$. Then, for any A-module V and any $\lambda \in End_F(V)$,*

$$\sum_{i=1}^{n} b_i \lambda a_i \in End_A(V)$$

Proof. For any given $a \in A$, write

$$a_i a = \sum_{j=1}^{n} \mu_{ij}(a)a_j \quad \mu_{ij}(a) \in F, \quad 1 \leq i \leq n \tag{5}$$

We claim that

$$ab_i = \sum_{j=1}^{n} \mu_{ji}(a)b_j \quad 1 \leq i \leq n \tag{6}$$

Indeed, write $ab_i = \sum_{j=1}^{n} \alpha_{ji}(a)b_j$ with $\alpha_{ji} \in F$. Then, for any $k \in \{1, \ldots, n\}$,

$$a_i ab_k = \sum_{j=1}^{n} \mu_{ij}(a)a_j b_k = \sum_{j=1}^{n} \alpha_{jk}a_i b_j$$

Hence, applying (4), we obtain

$$\psi(a_i ab_k) = \mu_{ik}(a) = \alpha_{ik}(a),$$

proving (6).

Now fix an A-module V and $\lambda \in End_F(V)$. Then, invoking (5) and (6), we obtain

$$a\left(\sum_{i=1}^{n} b_i \lambda a_i\right) = \sum_{i=1}^{n}\sum_{j=1}^{n} \mu_{ji}(a)b_j \lambda a_i \quad a \in A$$

and

$$\left(\sum_{i=1}^{n} b_i \lambda a_i\right)a = \sum_{i=1}^{n}\sum_{j=1}^{n} \mu_{ij}(a)b_i \lambda a_j \quad a \in A$$

Because the two double summands are the same, the result is established. ∎

6.15. Theorem. *Let (A, ψ) be a Frobenius algebra, let $a_1, \ldots, a_n$ be an F-basis of A and let the basis $b_1, \ldots, b_n$ of A be the ψ-dual of $a_1, \ldots, a_n$. Then, for any given A-module V, the following conditions are equivalent:*
(i) V is projective.
(ii) V is injective.
(iii) There exists $\lambda \in End_F(V)$ such that

$$\sum_{i=1}^{n} b_i \lambda a_i = 1_V$$

where 1_V is the identity map on V.

Proof. By Theorem 6.5(ii), (i) and (ii) are equivalent. In what follows we prove that (ii) is equivalent to (iii).

(ii)$\Rightarrow$(iii): Assume that V is injective and put $W = A \otimes_F V$. Then the F-space W may be viewed as a left A-module in a natural way. Note that each element of W can be uniquely written in the form

$$\sum b_i \otimes v_i \qquad (v_i \in V)$$

Applying (5) and (6), it follows that the map

$$\begin{cases} V \xrightarrow{\alpha} W \\ v \mapsto \sum b_i \otimes a_i v \end{cases}$$

is an injective A-homomorphism. Because V is injective, we see that $\alpha(V)$ is a direct summand of W and there exists $\beta \in End_A(W)$ which projects W onto $\alpha(V)$.

Now write $1 = \sum \gamma_i a_i$ with $\gamma_i \in F$. Define $\delta \in End_F(W)$ by setting

$$\delta(\sum b_i \otimes v_i) = \sum 1 \otimes \gamma_i v_i \qquad (v_i \in V)$$

We claim that

$$\sum b_i \delta a_i = 1 \tag{7}$$

If sustained, then setting $\lambda = \alpha^{-1} \beta \delta \beta \alpha$ one immediately verifies that λ satisfies (iii).

To prove (7), fix $v \in V$ and note that, by (5) and (6),

$$\begin{aligned}
\sum_i b_i \delta a_i (b_j \otimes v) &= \sum_i b_i \delta (\sum_k b_k \mu_{kj}(a_i) \otimes v) \\
&= \sum_i b_i (\sum_k 1 \otimes \gamma_k \mu_{kj}(a_i) v) \\
&= \sum_i b_i \otimes (\sum_k \gamma_k \mu_{kj}(a_i) v)
\end{aligned}$$

But $\psi(a_i b_j) = \delta_{ij}$, hence by (5),

$$\mu_{kj}(a_i) = \psi(a_k a_i b_j)$$

and therefore

$$\sum_k \gamma_k \mu_{kj}(a_i) = \sum_k \gamma_k \psi(a_k a_i b_j)$$
$$= \psi(\sum_k \gamma_k a_k a_i b_j)$$
$$= \psi(a_i b_j) = \delta_{ij}$$

Thus we must have

$$(\sum b_i \delta a_i)(b_j \otimes v) = b_j \otimes v,$$

as desired.

(iii) $\Rightarrow$ (ii): Suppose that $\lambda \in End_F(V)$ satisfies (iii). To prove that V is injective, assume that V is a submodule of W and let $\alpha \in End_F(W)$ be a projection of W upon V. Define $\beta \in End_F(W)$ by

$$\beta = \sum b_i \lambda \alpha a_i$$

Owing to Lemma 6.14, $\beta \in End_A(V)$ and one easily verifies that β projects W upon V. It follows that V is a direct summand of W, as we wished to demonstrate. ∎

7. Frobenius uniserial algebras

Throughout this section, A denotes a finite-dimensional algebra over a field F and all A-modules are assumed to be finitely generated.

Let V be an A-module. Then the descending chain

$$V \supseteq J(A)V \supseteq J(A)^2 V \supseteq \cdots$$

of submodules of V is called the (lower) *Loewy series* of V. Since $J(A)$ is nilpotent, there is an integer k, called the *Loewy length* of V such that

$$J(A)^{k-1}V \neq o \quad \text{but} \quad J(A)^k V = 0$$

Noye that if $J(A)^i V = J(A)^{i+1}V$ for $i < k$, then

$$J(A)^{k-1}V = J(A)^{k-i-1}J(A)^i V = J(A)^{k-i-1}J(A)^{i+1}V = J(A)^k V = 0,$$

a contradiction. Thus, if k is the Loewy length of V, then

$$V \supset J(A)V \supset J(A)^2 V \supset \cdots \supset J(A)^{k-1} V \supset 0$$

The module V is said to be *uniserial* if the above chain is a composition series of V. Thus V is uniserial if and only if it has a unique composition series. Since a uniserial module has a unique simple submodule, it must be *indecomposable*. We shall refer to the algebra A as being *uniserial* if every indecomposable A-module is uniserial. Our first aim is to provide a characterization of Frobenius uniserial algebras. We begin by recording the following preliminary result.

7.1. Lemma.   *Let V be a uniserial A-module. Then V is isomorphic to a factor module of a projective indecomposable A-module.*

Proof.   There exists a projective A-module P and a surjective A-homomorphism $f : P \to V$. Write $P = \oplus_{i=1}^{n} P_i$, where each P_i is indecomposable. If $f(P_i) \subset V$ for all i, then $f(P_i) \subseteq J(A)V$ for all i, since $J(A)V$ is a unique maximal submodule of the uniserial A-module V. But then

$$f(P) \subseteq J(A)V \neq V,$$

a contradiction. Thus $f(P_i) = V$ for some $i \in \{1, \ldots, n\}$, as asserted. ∎

Note that the dual of a uniserial module is a uniserial right module. Hence, if A is uniserial, then every indecomposable right A-module is uniserial.

7.2.   Theorem.   *(Nakayama (1939)).   The following conditions are equivalent:*

(i) The algebra A is uniserial.

(ii) All indecomposable projective and indecomposable injective A-modules are uniserial.

(iii) For any primitive idempotent e of A, the modules Ae and eA are uniserial.

Proof.   It is a consequence of Lemma 6.6, that (ii) is equivalent to (iii). Because (i) obviously implies (ii), it suffices to verify that (ii) implies (i).

Assume that all indecomposable projective and indecomposable injective A-modules are uniserial. Let V be an arbitrary A-module. Our aim is to show, by induction on $dim_F V$, that V is a direct sum of uniserial modules. This will clearly prove the result.

If $dim_F V = 1$, then there is nothing to prove. So assume that $dim_F V > 1$. If V is simple, then V is obviously uniserial. Hence we may harmlessly assume that V has a proper simple submodule V', in which case V' is uniserial. Denote by W a uniserial submodule of V with $dim_F W$ as large as possible. Then we must have $W \neq 0$.

Let X be a submodule of V, maximal with respect to the property that $W \cap X = 0$. Suppose that X_1/X and X_2/X are simple submodules of V/X. Then $X_1 \cap W \neq 0$ and $X_2 \cap W \neq 0$ and hence $X_1 \cap X_2 \cap W \neq 0$ since W is uniserial. Hence $X_1 \cap X_2 \neq X$ and so $X_1 = X_2$, proving that $Soc(V/X)$ is simple.

Since $Soc(V/X)$ is simple, it follows from Theorem 5.8(ii) that V/X is a submodule of an indecomposable injective A-module. Thus, by hypothesis, V/X must be uniserial. Applying Lemma 7.1, we deduce that there is a surjective homomorphism.

$$f : P \to V/X$$

for a suitable indecomposable projective A-module P. Since P is projective, there exists a homomorphism

$$g : P \to V$$

such that $\varphi \circ g = f$, where $\varphi : V \to V/X$ is the natural surjection. Since, by hypothesis, P is uniserial, it follows that $g(P)$ is a uniserial submodule of V.

Now $W \cap X = 0$ implies that W is isomorphic to a submodule of V/X. Therefore, by the maximality of $dim_F W$,

$$dim_F g(P) \leq dim_F W \leq dim_F(V/X)$$

On the other hand, since $f = \varphi \circ g$ and f is surjective,

$$dim_F g(P) \geq dim_F(V/X)$$

The conclusion is that $g(P) \cong V/X \cong W$. It follows that

$$V = g(P) + X \quad \text{and} \quad g(P) \cap X = 0$$

Hence $V = g(P) \oplus X$ and the required assertion follows by induction hypothesis. ∎

7.3. Corollary. *Let A be a Frobenius algebra. Then A is uniserial if and only if each projective indecomposable A-module is uniserial.*

Proof. Apply Theorems 7.2 and 6.5(ii). ∎

Our next aim is to demonstrate that if e_1, e_2 are primitive idempotents of a uniserial Frobenius algebra A belonging to the same block of A, then Ae_1 and Ae_2 have the same composition length. We need a number of preliminary results, some of which are of independent interest.

7.4. Lemma. *Suppose that an A-module V has a unique maximal submodule V_1, and let e be a primitive idempotent of A such that $V/V_1 \cong Ae/J(A)e$. Then V is a homomorphic image of Ae.*

Proof. By Corollary 5.11, $e(V/V_1) \neq 0$ so that $eV \not\subseteq V_1$. Choose $v \in V$ such that $ev \notin V_1$. Then $Aev \not\subseteq V_1$ and hence $V = Aev$. It follows that the map

$$\begin{cases} Ae & \to & V \\ ae & \mapsto & aev \end{cases}$$

is a surjective A-homomorphism, as desired. ∎

In what follows, $c(V)$ denotes the composition length of an A-module V.

7.5. Lemma. *Let A be a uniserial algebra, let e_1, e_2 be two primitive idempotents of A and let m, n be positive integers such that*

$$J(A)^m e_1 \neq 0 \quad \text{and} \quad J(A)^n e_2 \neq 0$$

(i) If $J(A)^m e_1/J(A)^{m+1} e_1 \cong Ae_2/J(A)e_2$, then

$$J(A)^m e_1 \cong Ae_2/J(A)^s e_2$$

where $s = c(J(A)^m e_1)$.
   (ii) If $J(A)^m e_1/J(A)^{m+1} e_1 \cong J(A)^n e_2/J(A)^{n+1} e_2$, then

$$J(A)^{m-1} e_1/J(A)^m e_1 \cong J(A)^{n-1} e_2/J(A)^n e_2$$

where, by convention, $J(A)^0 = A$

Proof. (i) Put $V = J(A)^m e_1$. Because A is uniserial, V has a unique maximal submodule $J(A)^{m+1} e_1$. Therefore, by Lemma 7.4,

$$V \cong Ae_2/J(A)^s e_2 \quad \text{for some} \quad s \geq 1$$

Taking into account that s is the composition length of $Ae_2/J(A)^s e_2$, the required property is established.
   (ii) Let e_3, e_4 be primitive idempotents of A such that

$$J(A)^{m-1} e_1/J(A)^m e_1 \cong Ae_3/J(A)e_3$$

and

$$J(A)^{n-1} e_2/J(A)^n e_2 \cong Ae_4/J(A)e_4 \tag{1}$$

Owing to (i), we have

$$J(A)^{m-1} e_1 \cong Ae_3/J(A)^s e_3, \ J(A)^{n-1} e_2 \cong Ae_4/J(A)^t e_4 \tag{2}$$

for some positive integers s, t. Moreover, $s \geq 2$ and $t \geq 2$, because if, say $s = 1$, then $J(A)^m e_1 = 0$, a contradiction. Applying (2), we conclude that

$$J(A)^m e_1/J(A)^{m+1} e_1 \cong J(A)e_3/J(A)^2 e_3 \tag{3}$$

and

$$J(A)^n e_2/J(A)^{n+1} e_2 \cong J(A)e_4/J(A)^2 e_4 \tag{4}$$

Let e denote a primitive idempotent of A such that

$$J(A)^m e_1/J(A)^{m+1} e_1 \cong Ae/J(A)e$$

It follows then from (1), (3) and (4) that

$$J(A)e_3/J(A)^2e_3 \cong Ae/J(A)e \cong J(A)e_4/J(A)^2e_4 \qquad (5)$$

Therefore, by Corollary 5.11 and (5), we have

$$eJ(A)e_3/eJ(A)^2e_3 \neq 0$$

and

$$eJ(A)e_4/eJ(A)^2e_4 \neq 0$$

Again, applying the right analogue of Corollary 5.11, we deduce that $eJ(A)/eJ(A)^2$ has composition factors isomorphic to $e_3A/e_3J(A)$ and $e_4A/e_4J(A)$. Moreover, $eJ(A)/eJ(A)^2$ is simple, so

$$e_3A/e_3J(A) \cong e_4A/e_4J(A)$$

and thus

$$Ae_3/J(A)e_3 \cong Ae_4/J(A)e_4 \qquad (6)$$

On the other hand, invoking (2), we also have

$$J(A)^{m-1}e_1/J(A)^m e_1 \cong Ae_3/J(A)e_3$$

and

$$J(A)^{n-1}e_2/J(A)^n e_2 \cong Ae_4/J(A)e_4$$

Therefore, applying (6), the required assertion is established. ∎

We are now ready to prove the following important result.

7.6. Theorem. *(Kupisch (1959)). Let A be an indecomposable uniserial algebra. Then there is a numbering $Ae_1, Ae_2, \ldots, Ae_n$ of the nonisomorphic principal indecomposable A-modules such that*
(i) $c(Ae_i) \geq 2$ for $1 \leq i \leq n$.
(ii) $J(A)e_i/J(A)^2e_i \cong Ae_{i+1}/J(A)e_{i+1}$ for $1 \leq i \leq n$.

$$J(A)e_n/J(A)^2e_n \cong Ae_1/J(A)e_1 \quad if \quad J(A)e_n \neq 0$$

(iii) $c(Ae_{i+1}) \geq c(Ae_i) - 1$, where the indices are taken modulo n.

Proof. Let $Ae_1, Ae_2, \ldots, Ae_n$ be all nonisomorphic principal indecomposable A-modules. For any given $i \in \{1, \ldots, n\}$, define the
sequence F_i by

$$F_i = (i_1, \ldots, i_m, i_{m+1}, \ldots,)$$

where $i_1 = i$ and

$$J(A)e_{i_m}/J(A)^2 e_{i_m} \cong Ae_{i_{m+1}}/J(A)e_{i_{m+1}} \tag{7}$$

and, by convention, if $J(A)e_{i_m} = 0$, then the sequence F_i terminates. If
$J(A)^k e_{i_m} \neq 0$, then by (7) and Lemma 7.5(i), the induction on k gives

$$J(A)^k e_{i_m}/J(A)^{k+1} e_{i_m} \cong Ae_{i_{m+k}}/J(A)e_{i_{m+k}} \tag{8}$$

Similarly, it follows from (7) that

$$\text{if} \quad i_m = j_s \in F_i \cap F_j, \quad \text{then} \quad i_{m+k} = j_{s+k} \tag{9}$$

Thus if $i_1 \in F_j$, then F_i is a subsequence of F_j. In what follows, we
write $F_i \subseteq F_k$ to indicate that all elements in F_i belong to F_k.

Our next aim is to show that

$$\text{if} \quad F_i \cap F_k \neq \emptyset, \quad \text{then either} \quad F_i \subseteq F_k \quad \text{or} \quad F_k \subseteq F_i \tag{10}$$

To this end, among the common elements of F_i and F_k, let us choose
$i_s = k_t$ with the smallest index s. In case $s = 1$, we have $i_1 \in F_k$ and so
F_i is a subsequence of F_k. In case $s > 1$, we must have $t = 1$, Indeed,
assume by way of contradiction that $s > 1$ and $t > 1$. Then, by (7), we
have

$$J(A)e_{i_{s-1}}/J(A)^2 e_{i_{s-1}} \cong J(A)e_{k_{t-1}}/J(A)^2 e_{k_{t-1}},$$

which implies that $i_{s-1} = k_{t-1}$, contrary to our choise of i_s. This proves
the validity of (10).

Now choose $Ae_i \in \{Ae_1, Ae_2, \ldots, Ae_n\}$ such that F_i contains the
maximum number of distinct elements. If Ae_k has a common composition factor, say $Ae_s/J(A)e_s$, with Ae_t, where $t \in F_i$, then

$$J(A)^j e_k/J(A)^{j+1} e_k \cong J(A)^r e_t/J(A)^{r+1} e_t \cong Ae_s/J(A)e_s$$

for some $j, r \geq 1$. We conclude that s belongs to F_k as well as F_t
and hence to F_i. It follows from (10) and the maximality of F_i that

$F_k \subseteq F_i$, i.e. $k \in F_i$. Because A is indecomposable, we conclude that the sequence F_i contains all the elements $j = 1, \ldots, n$. By (9), it follows that these elements occur as the first n terms of F_i. This gives us the numbering of n left ideals $Ae_1, \ldots, Ae_n$ for which

$$J(A)e_i/J(A)^2 e_i \cong Ae_{i+1}/J(A)e_{i+1} \quad (1 \leq i \leq n) \tag{11}$$

In particular, this demonstrates that

$$J(A)e_i \neq 0 \qquad (1 \leq i \leq n)$$

Now assume that $J(A)e_n \neq 0$. Then

$$J(A)e_n/J(A)^2 e_n \cong Ae_t/J(A)e_t$$

which implies $t = 1$. If $1 < t \leq n$, then by (ii) we have

$$J(A)e_{t-1}/J(A)^2 e_{t-1} \cong J(A)e_n/J(A)^2 e_n$$

and thus, by Lemma 7.5(ii),

$$Ae_{t-1} = Ae_n$$

It therefore follows that

$$J(A)e_n/J(A)^2 e_n \cong Ae_1/J(A)e_1,$$

proving (ii). Property (iii) is a consequence of (11) and Lemma 7.5(i). Finally, bearing in mind that (i) is a consequence of (iii), the result is established. ∎

In order to apply the results above, we need to introduce the concept of a block. While on the subject, we shall present certain properties of blocks which will be required for subsequent investigations.

7.7. Proposition. *Let A be an F-algebra.*
(i) There exists a decomposition

$$A = B_1 \oplus \cdots \oplus B_n$$

of A into indecomposable two-sided ideals $B_i \neq 0$ with $B_iB_j = 0$ for $i \neq j$.

(ii) Write $1 = e_1 + \cdots + e_n$ with $e_i \in B_i$. Then the e_i are mutually orthogonal centrally primitive idempotents and $B_i = Ae_i = e_iA$.

(iii) $Z(A) = Z(A)e_1 \oplus \cdots \oplus Z(A)e_n$ is a direct decomposition of $Z(A)$ into indecomposable ideals and $Z(A)e_i = B_i \cap Z(A)$.

(iv) $J(Z(A)e_i) = J(Z(A))e_i$ and $Z(A)e_i/J(Z(A)e_i)$ is a finite field extension of F.

Proof. (i) and (ii). The verification is straightforward and therefore will be omitted.

(iii) The first statement and the inclusion $Z(A)e_i \subseteq B_i \cap Z(A)$ follow from (ii). Assume that $z \in B_i \cap Z(A)$. Because e_i is the identity element of B_i, we have $z = ze_i \in Z(A)e_i$, proving that $B_i \cap Z(A) \subseteq Z(A)e_i$.

(iv) It is clear that $J(Z(A)e_i) = J(Z(A))e_i$. Since each e_i is primitive, it follows from Theorem 5.4(ii) that the $Z(A)$-module $Z(A)e_i/J(Z(A)e_i)$ is simple. The latter imples that the commutative algebra $Z(A)e_i/J(Z(A)e_i)$ is simple and hence is a field. That this field is a finite extension of F follows from the assumption that A is finite-dimensional over F. ∎

We now introduce an important concept of blocks. Let

$$A = \oplus_{i=1}^{n} B_i = \oplus_{i=1}^{n} Ae_i$$

be the decomposition defined in Proposition 7.7. We shall refer to B_i as a *block* and to e_i as a *block idempotent* of A. We shall also write $B = B(e)$ to indicate that B is a block containing the block idempotent e.

7.8. Proposition. Let $A = B(e_1) \oplus \cdots \oplus B(e_n)$ be the direct decomposition of A into sum of blocks.

(i) For every A-module V, $V = \oplus_{i=1}^{n} e_i V$ is a direct decomposition of V into A-modules. In particular, if V is indecomposable, then $V = e_i V$ for a unique $i \in \{1, \ldots, n\}$.

(ii) For any left ideal I of A, we have

$$I = \oplus_{i=1}^{n}(I \cap B(e_i))$$

In particular, each indecomposable left ideal lies in exactly one of the $B(e_i)$.

(iii) If A_1 and A_2 are ideals of A and $A = A_1 \oplus A_2$, then A_1 and A_2 are direct sums of blocks. In particular, the blocks are the only indecomposable ideals of A which are direct summands of A.

Proof.   (i) Since e_i is central, $e_i V$ is an A-module. Because the e_i are orthogonal, we also have

$$\sum_{i=1}^{n} e_i V = \oplus_{i=1}^{n} e_i V$$

Bearing in mind that

$$V = 1 \cdot V = (\sum_{i=1}^{n} e_i) V \subseteq \sum_{i=1}^{n} e_i V \subseteq V$$

the assertion is proved.

(ii) Owing to (i), $I = \oplus_{i=1}^{n} e_i A$ and thus

$$e_i \subseteq I \cap B(e_i) = I \cap e_i A \subseteq e_i I$$

Thus we must have

$$I = \oplus_{i=1}^{n} (I \cap B(e_i)),$$

proving (ii).

(iii) By (ii), it follows that

$$A = A_1 \oplus A_2 = \oplus_{i=1}^{n}(A_1 \cap B(e_i)) \oplus \oplus_{i=1}^{n}(A_2 \cap B(e_i))$$

Therefore for all i,

$$B(e_i) = (A_1 \cap B(e_i)) \oplus (A_2 \cap B(e_i))$$

and since $B(e_i)$ is indecomposable we have $B(e_i) \subseteq A_1$ or $B(e_i) \subseteq A_2$. Thus

$$A_k = \oplus B(e_i) \qquad (k = 1, 2)$$

$$B(e_i) \subseteq A_k$$

as required. ■

Let $e_1, \ldots, e_n$ be the block idempotents of A and let V be an indecomposable A-module. It follows from Proposition 7.8 that there exists $i \in \{1, \ldots, n\}$ such that

$$e_i V = V \quad \text{and} \quad e_j V = 0 \quad \text{for all} \quad j \neq i$$

In this case we say that V lies in the block $B(e_i)$. Note that if V lies in the block $B(e_i)$, then

(a) All A-modules isomorphic to V lie in $B(e_i)$.

(b) $v = 1 \cdot v = e_i v$ for all $v \in V$.

The above provides a classification of all indecomposable (and, in particular, all simple) A-modules into blocks. Our next aim is to tie together blocks and principal indecomposable modules.

Let U and V be principal indecomposable A-modules. We say that U and V are *linked*, written $U \sim V$, provided there exists a sequence

$$U_1 = U, \ldots, U_n = V$$

of principal indecomposable modules such that any two neighbouring left ideals in this sequence have a common composition factor. It is clear that $\sim$ is an equivalence relation on the set of all principal indecomposable A-modules. In what follows, we denote by

$$X_1, X_2, \ldots, X_m$$

the equivalence classes of $\sim$.

7.9. Proposition. *Let $A = \oplus_{i=1}^{n} Ae_i$ be a decomposition of A into principal indecomposable modules, and for each $j \in \{1, \ldots, m\}$, put*

$$I_j = \oplus_{Ae_i \in X_j} Ae_i$$

(i) $I_1, \ldots, I_m$ are all blocks of A.

(ii) Two principal indecomposable A-modules are linked if and only if they belong th the same block.

Proof. It is clear that (ii) is a direct consequence of (i). To prove (i), note that we obviously have $A = \oplus_{j=1}^{m} I_j$. By Proposition 7.8(iii), it suffices to show that each I_j is a two-sided ideal contained in a block.

Suppose that $Ae_i \subseteq I_j$ and $Ae_k \subseteq I_l$ with $j \neq l$. Then Ae_i and Ae_k have no composition factors in common. In particular, Ae_k has no composition factor isomorphic to $\bar{A}e_i$, where $\bar{e}_i$ is the image of e_i in $\bar{A} = A/J(A)$. Hence, by Corollary 5.11, $e_i Ae_k = 0$. This shows that $I_j I_l = 0$ for $j \neq l$. Thus

$$I_j A = I_j (\oplus_{l=1}^m I_l) = I_j^2 \subseteq I_j$$

which proves that I_j is a two-sided ideal of A.

If Ae_i and Ae_j have the composition factor $\bar{A}\bar{e}_k$ in common, then by Corollary 5.11, $e_k Ae_i \neq 0$ and $e_k Ae_j \neq 0$. By Proposition 7.8(ii), there exist blocks B, B' and B'' such that $Ae_i \subseteq B, Ae_j \subseteq B'$ and $Ae_k \subseteq B''$. It follows that $0 \neq e_k Ae_i \subseteq B''B$ and hence $B = B''$. Similarly, $0 \neq e_k Ae_j \subseteq B''B'$ and so $B' = B''$. Thus $Ae_i + Ae_j \subseteq B$ and repeated application of this argument demonstrates that I_j is contained in a block of A. This proves the required assertion. ∎

Turning to Frobenius uniserial algebras, we now prove the following result.

7.10. Theorem. *(Morita (1951)). Let A be a Frobenius uniserial algebra. If e_1, e_2 are primitive idempotents of A belonging to the same block of A, then Ae_1 and Ae_2 have the same composition length.*

Proof. Owing to Theorem 7.6(ii) and Proposition 7.8(ii), we may harmlessly assume that

$$J(A)e_1/J(A)^2 e_1 \cong Ae_2/J(A)e_2$$

By Corollary 5.11,

$$e_2 J(A)e_1 \nsubseteq J(A)^2$$

Let b denote an element in $e_2 J(A)e_1$ not contained in $J(A)^2$. Then $Ab \subseteq J(A)e_1$, and $Ab \nsubseteq J(A)^2 e_1$. Since $J(A)e_1$ is a uniserial module, the latter implies $Ab = J(A)e_1$. Similarly, $bA \subseteq e_2 J(A)$ and $bA \nsubseteq e_2 J(A)^2$, and therefore $bA = e_2 J(A)$, since $e_2 J(A)$ is also uniserial. Since $e_2 b = b$, the map

$$\left\{ \begin{array}{l} Ae_2 \overset{\psi}{\to} Ab = J(A)e_1 \\ ae_2 \mapsto ae_2 b \end{array} \right.$$

is a surjective homomorphism.

We claim that

$$Ker\psi = Ae_2 \cap l(J(A))$$

Indeed, $Ker\psi = J(A)^k e_2$ for some $k \geq 1$ and $J(A)^k e_2 b = 0$. Hemce

$$
\begin{aligned}
J(A)^k e_2 (bA) &= J(A)^k e_2 e_2 J(A) \\
&= J(A)^k e_2 J(A) \\
&= 0
\end{aligned}
$$

and therefore $Ker\psi \subseteq Ae_2 \cap l(J(A))$. The opposite containment being obvious, the claim is established.

It follows from the claim above that

$$Ae_2/(Ae_2 \cap l(J(A))) \cong J(A)e_1$$

Because A is Frobenius, $l(J(A)) = r(J(A))$ (Theorem 6.3(i)) and thus

$$Ae_2 \cap l(J(A)) = Soc(Ae_2)$$

proving that

$$Ae_2/Soc(Ae_2) \cong J(A)e_1$$

The desired conclusion now follows by virtue of Theorem 6.3(ii). ■

7.11. Corollary. *(Morita (1951)). Let B be a block of a uniserial symmetric algebra A. Then there is a numbering $Ae_1, Ae_2, \ldots, Ae_n$ of the nonisomorphic principal indecomposable modules in B such that they have the following composition factors:*

$$
\begin{aligned}
Ae_1 &: \quad V_1, V_2, \ldots, V_n, V_1, V_2, \ldots, V_n, \ldots, V_1, V_2, \ldots, V_n, V_1 \\
Ae_2 &: \quad V_2, V_3, \ldots, V_1, V_2, V_3, \ldots, V_1, \ldots, V_2, V_3, \ldots, V_1, V_2 \\
Ae_3 &: \quad V_3, V_4, \ldots, V_2, V_3, V_4, \ldots, V_2, \ldots, V_3, V_4, \ldots, V_2, V_3 \\
&\quad - \quad - \quad - \quad - - - \quad - - - \quad - - - \quad - - - \\
Ae_n &: \quad V_n, V_1, \ldots, V_{n-1}, V_n, V_1, \ldots, V_{n-1}, \ldots, V_n, V_1, \ldots, V_{n-1}, V_n
\end{aligned}
$$

where

$$V_i = Ae_i/J(A)e_i, \quad 1 \leq i \leq n$$

In particular, the Cartan matrix of B is of the form

$$
\begin{bmatrix}
s+1 & s & \cdots s & s \\
s & s+1 & \cdots s & s \\
\vdots & & & \\
s & s & \cdots s & s+1
\end{bmatrix}
$$

Proof. The last column in the table for the composition factors comes from the fact that $Soc(Ae_i) \cong V_i$ by Theorem 6.3. The rest follows from Theorems 7.6(ii) and 7.10. ∎

8. Characterizations of Frobenius algebras

In this section, we fix a finite-dimensional algebra A over a field F. As usual, all A-modules below are assumed to be finitely generated. We also assume (unless explicitly stated otherwise) that all A-modules are left A-modules. Our aim is to provide a number of important characterizations of Frobenius algebras.

Let V be a (left or right) A-module. Then V is called a *cogenerator* if for any A-module U,

$$
\bigcap_{f \in Hom_A(U,V)} Ker f = 0
$$

8.1. Lemma. *Let V be a (left or right) A-module. Then V is a cogenerator if and only if for any simple A-module W, the injective hull of W is isomorphic to a submodule of V.*

Proof. We shall treat the case where V is a left A-module. A similar argument will establish the case where V is a right A-module. Assume that V is a cogenerator and let W be a simple A-module with injective hull $I(W)$. Since V is a cogenerator,

$$
\bigcap_{f \in Hom_A(I(W),V)} Ker f = 0
$$

and hence there exists $f \in Hom_A(I(W), V)$ with $W \not\subseteq Ker f$. Because W is simple, we must have $W \cap Ker f = 0$. But W is an essential submodule of $I(W)$, hence $Ker f = 0$ and so $I(W) \cong f(I(W))$.

Conversely, assume that for any simple A-module W, the injective hull of W is isomorphic to a submodule of V. Let $U \neq 0$ be an arbitrary A-module and choose $0 \neq u \in U$. Let M be a maximal submodule of Au and let $W = Au/M$. If $I(W)$ is an injective hull of W, then by hypothesis there exists an injective A-homomorphism $\psi : I(W) \to V$. Let

$$\pi : Au \to Au/M$$

be the natural homomorphism. Then $\pi(u) = u + M \neq 0$. Since $I(W)$ is an injective module, the homomorphism $\pi : Au \to I(W)$ can be extended to a homomorphism $\lambda : U \to I(W)$ (Lemma 5.5). In particular, $\lambda(u) = \pi(u) \neq 0$. Setting $f = \psi \circ \lambda$, it follows that $f \in Hom_A(U, V)$ and $u \notin Ker f$. Thus

$$\bigcap_{f \in Hom_A(U,V)} Ker f = 0$$

proving that V is a cogenerator. ∎

8.2. Lemma. *Let $V_1, \ldots, V_r$ be all nonisomorphic simple A-modules and let $I(V_i)$ be the injective hull of V_i, $1 \leq i \leq r$. Then $I(V_i), \ldots, I(V_r)$ are all nonisomorphic injective indecomposable A-modules.*

Proof. By the definition and Theorem 5.8(ii), each $I(V_i)$ is injective and indecomposable. Let U and W be two simple essential submodules of an injective A-module V. To prove that $I(V_1), \ldots, I(V_r)$ are nonisomorphic it suffices by Theorem 5.8(i), to show that $U = W$. But, by definition of an essential submodule, $U \cap W \neq 0$. Hence $U = W$, as required.

Finally, let V be an arbitrary injective indecomposable A-module. Choose $i \in \{1, \ldots, r\}$ such that V_i is isomorphic to a submodule L of V. Then, by Theorem 5.8(ii), V is an injective hull of L and thus $V \cong I(V_i)$. ∎

8.3. Lemma. *Let $V_1, \ldots, V_r$ be all nonisomorphic simple A-modules and let $I(V_i)$ be the injective hull of V_i, $1 \leq i \leq r$.*

(i) If an A-module V is a cogenerator, then each $I(V_i)$ is isomorphic to an indecomposable direct summand of V.

(ii) If an A-module V is an injective cogenerator, then $I(V_i), \ldots, I(V_r)$

are nonisomorphic indecomposable A-modules such that

$$V \cong \oplus_{i=1}^{r} n_i I(V_i)$$

for some positive integers $n_i, 1 \le i \le r$.

Proof. (i) By Lemma 8.1, V contains a copy of each $I(V_i)$ and any such copy being an injective submodule of V is a direct summand of V. Since, by Lemma 8.2, each $I(V_i)$ is indecomposable, the required assertion follows.

(ii) Apply (i) and Lemma 8.2. ∎

8.4. Lemma. *We have*

$$Soc(A^*) = \{\psi \in A^* | J(A) \subseteq Ker\psi\}$$

Proof. Note that $Soc(A^*) = \{\psi \in A^* | J(A)\psi = 0\}$. Hence it suffices to show that $J(A)\psi = 0$ if and only if $J(A) \subseteq Ker\psi$. If $J(A)\psi = 0$, then for all $a \in A$, $(a\psi)(1) = \psi(a) = 0$ and so $J(A) \subseteq Ker\psi$.

Conversely, assume that $J(A) \subseteq Ker\psi$. Then, for all $a \in J(A), x \in A$,

$$0 = \psi(xa) = (a\psi)(x)$$

and therefore $J(A)\psi = 0$, as required. ∎

8.5. Lemma. *(i) A^* is an injective hull of $Soc(A^*)$ and $A/J(A) \cong Soc(A^*)$.*

(ii) The left A-module A^* is an injective cogenerator.

Proof. (i) We claim that A^* is injective. Since $Soc(A^*)$ is an essential submodule of A^*, it will follow from Theorem 5.8 that A^* is an injective hull of $Soc(A^*)$.

Let W be an A-module, let V be a submodule of W and let $g : V \to A^*$ be an A-homomorphism. By Lemma 5.5, it suffices to show that g can be extended to a homomorphism $h : W \to A^*$. Consider the F-homomorphism $\alpha : A^* \to F$ given by $\psi \mapsto \psi(1)$. Then we may choose an F-homomorphism $\beta : W \to F$ such that $\alpha \circ g = \beta$. Define $h : W \to A^*$ by

$$h(w)(x) = \beta(xw) \qquad (w \in W, x \in A)$$

It is clear that h preserves addition. Furthermore, for any $a, x \in A$ and $w \in W$,

$$h(aw)(x) = \beta(xaw) = h(w)(xa) = ah(w)(x),$$

proving that h is an A-homomorphism. Since, for all $v \in V$ and $a \in A$,

$$
\begin{aligned}
h(v)(a) &= \beta(av) = (\alpha \circ g)(av) \\
&= g(av)(1) = ag(v)(1) \\
&= g(v)(a),
\end{aligned}
$$

it follows that A^* is injective.

By the foregoing, we are left to verify that

$$A/J(A) \cong Soc(A^*)$$

By Lemma 8.4, the map

$$
\left\{
\begin{array}{ccc}
Soc(A^*) & \xrightarrow{f} & (A/J(A))^* \\
\psi & \mapsto & \bar{\psi}
\end{array}
\right.
$$

where $\bar{\psi}(a + J(A)) = \psi(a), a \in A$, is well defined. It is clear that f is an injective A-homomorphism. Furthermore, for any $\mu \in (A/J(A))^*$ and the natural map $\pi : A \to A/J(A)$, we have $\mu \circ \pi \in Soc(A^*)$ and $\overline{\mu \circ \pi} = \mu$. Thus f is an A-isomorphism.

Now $A/J(A)$ is emisimple, hence by Proposition 1.9 and Corollary 1.6, there is an $A/J(A)$-isomorphism

$$g : (A/J(A))^* \to A/J(A)$$

We may clearly view g as an A-isomorphism, in which case

$$g \circ f : Soc(A^*) \to A/J(A)$$

is a required A-isomorphism.

(ii) By (i), $Soc(A^*) \cong A/J(A)$ and hence A^* contains a copy of any simple A-module. Since A^* is injective, it must also contain a copy of an injective hull of any simple A-module. The desired conclusion is therefore a consequence of Lemma 8.1. ■

8.6. Lemma. *If A is an injective A-module, then for any left ideals X, Y of A,*

$$r(X \cap Y) = r(X) + r(Y)$$

Proof. It is clear that $r(X) + r(Y) \subseteq r(X \cap Y)$. Conversely, assume that $a \in r(X \cap Y)$. Then the map

$$\begin{cases} X + Y \xrightarrow{f} A \\ \quad x + y \mapsto ya \end{cases}$$

is well defined: for $x, x_1 \in X, y, y_1 \in Y$,

$$x + y = x_1 + y_1 \quad \Rightarrow \quad x - x_1 = y_1 - y \in X \cap Y$$
$$\Rightarrow \quad y_1 a = (y_1 - y)a + ya = ya$$

and is obviously an A-homomorphism. Since A is injective, it follows from Lemma 5.5(ii) that there exists $z \in A$ such that

$$f(x + y) = (x + y)z = ya \quad \text{for all} \quad x \in X, y \in Y$$

In particular, taking $y = 0$, we obtain $z \in r(X)$, and taking $x = 0$ we obtain $f(y) = yz = ya$, for all $y \in Y$. The latter implies $a - z \in r(Y)$ and hence $a = z + (a - z) \in r(X) + r(Y)$, as required. ■

8.7. Lemma. *Assume that A satisfies the following two conditions:*

(a) For any left ideals X, Y of A,

$$r(X \cap Y) = r(X) + r(Y)$$

(b) For any right ideal X of A,

$$r(l(X)) = X$$

Then $_A A$ is injective.

Proof. Let X be a left ideal of A and let $f : X \to A$ be an A-homomorphism. Owing to Lemma 5.5(ii), it suffices to verify that f can be extended to an A-homomorphism $A \to A$, or equivalently that there exists $a \in A$ such that $f(x) = xa$ for all $x \in X$.

We argue by induction on the number n of generators of X. First assume that $n = 1$, say $X = Ax_0$ for some $x_0 \in X$. Since $yx_0 = 0$ for $y \in A$ implies $f(yx_0) = yf(x_0) = 0$, we have $l(x_0) \subseteq l(f(x_0))$. Thus

$$l(x_0 A) \subseteq l(f(x_0)A)$$

and, by (b),

$$x_0 A = r(l(x_0 A)) \supseteq r(l(f(x_0)A)) = f(x_0)A$$

Hence there exists $a \in A$ with $f(x_0) = x_0 a$ and so

$$f(yx_0) = (yx_0)a \qquad \text{for all} \quad y \in A$$

proving the case $n = 1$.

Let $X = \sum_{i=1}^{n+1} Ax_i$ for some $x_i \in X$. Then, by induction hypothesis, there exist $z_1, z_2 \in A$ such that

$$f\left(\sum_{i=1}^{n} y_i x_i\right) = \left(\sum_{i=1}^{n} y_i x_i\right) z_1$$

and

$$f(y_{n+1} x_{n+1}) = y_{n+1} x_{n+1} z_2$$

for all $y_1, \ldots, y_{n+1}$ in A. Applying (a), we have

$$z_1 - z_2 \in r\left(\sum_{i=1}^{n} Ax_i \cap Ax_{n+1}\right) = r(\sum_{i=1}^{n} Ax_i) + r(Ax_{n+1})$$

and hence

$$z_1 - z_2 = c - d \quad \text{with} \quad c \in r(\sum_{i=1}^{n} Ax_i), d \in r(Ax_{n+1})$$

Setting $a = z_1 - c = z_2 - d$, we obtain

$$
\begin{aligned}
f(\sum_{i=1}^{n+1} y_i x_i) &= f(\sum_{i=1}^{n} y_i x_i) + f(y_{n+1} x_{n+1}) \\
&= (\sum_{i=1}^{n} y_i x_i)(z_1 - c) + y_{n+1} x_{n+1}(z_2 - d) \\
&= (\sum_{i=1}^{n+1} y_i x_i)a,
\end{aligned}
$$

as required. ■

We know, from Corollary 1.6, that A is a Frobenius algebra if and only if $A \cong A^*$. A weaker requirement is that if $A_1, \ldots, A_n$ and $B_1, \ldots, B_k$ are all nonisomorphic indecomposable direct summands of A and A^*, respectively, then $n = k$ and, after possibly reordering the B_i, $A_i \cong B_i$ for all $i \in \{1, \ldots, n\}$. If this requirement holds, then we say that A is a *quasi-Frobenius algebra*. Our first aim is to provide a number of characterizations of quasi-Frobenius algebras.

8.8. Theorem. *The following conditions are equivalent:*
(i) A is a quasi-Frobenius algebra.
(ii) $_AA$ is a cogenerator.
(iii) $_AA$ is injective.
(iv) A_A is a cogenerator.
(v) A_A is injective.
(vi) For any left ideal X of A and any right ideal Y of A,

$$l(r(X)) = X \quad and \quad r(l(Y)) = Y$$

Proof. (i) $\Rightarrow$ (ii): By Lemma 8.5(ii), A^* is an injective cogenerator. Hence, in the notation of Lemma 8.3,

$$A^* \cong \oplus_{i=1}^r n_i I(V_i) \tag{1}$$

for some positive integers n_i, $1 \leq i \leq r$. By hypothesis, A is a quasi-Frobenius algebra, hence each $I(V_i)$ is isomorphic to a submodule of $_AA$. Thus, by Lemma 8.1, $_AA$ is a cogenerator.

(ii) $\Rightarrow$ (iii): By Lemma 8.3(i), each $I(V_i)$ is isomorphic to an indecomposable direct summand of V. Since the number of nonisomorphic indecomposable summands of $_AA$ is precisely r (Theorem 5.4(ii)), we conclude that all indecomposable direct summands of $_AA$ are injective. Hence $_AA$ is also injective.

(iii) $\Rightarrow$ (i): Since $_AA$ is injective, so is any indecomposable direct summand of A. But the number of such nonisomorphic summands is precisely r, hence by Lemma 8.2, $I(V_1), \ldots, I(V_r)$ are all nonisomorphic indecomposable direct summands of A. Now apply (1).

(ii) $\Rightarrow$ (vi): Let X be a left ideal of A. Then X is a submodule of

$_AA$ and we obviously have $X \subseteq l(r(X))$. In particular, we may assume that $A \neq X$, in which case we can choose $a \in A, a \notin X$. Since $_AA$ is a cogenerator, there exists a homomorphism $f : A/X \to A$ for which $f(a + X) \neq 0$. If $\pi : A \to A/X$ is the natural homomorphism, then $0 = f \circ \pi(X) = Xf \circ \pi(1)$. Hence $(f \circ \pi)(1) \in r(X)$. Furthermore,

$$a(f \circ \pi)(1) = (f \circ \pi)(a) = f(a + X) \neq 0$$

and so $a \notin l(r(X))$. This proves that $l(r(X)) = X$.

Let Y be any right ideal of A and write $Y = \sum_{i=1}^{n} y_i A$ for some $y_i \in Y$. Then $l(Y) = \cap_{i=1}^{n} l(y_i A)$ and, applying Lemma 8.6, we have

$$r(l(Y)) = r(l(\sum_{i=1}^{n} y_i A)) = r(\cap_{i=1}^{n} l(y_i A)) = \sum_{i=1}^{n} r(l(y_i A))$$

It therefore suffices to show that

$$r(l(yA)) = yA \quad \text{for all} \quad y \in A$$

It is clear that $yA \subseteq r(l(yA))$. Assume that $x \in r(l(yA))$. Then $l(y) \subseteq l(x)$ and so the map $Ay \to A$, $ay \mapsto ax, a \in A$, is an A-homomorphism. Since A is injective, there exists $z \in A$ such that $yz = x$, i.e. $x \in yA$. Thus $r(l(yA)) \subseteq yA$, as required.

(vi) $\Rightarrow$ (iii): By Lemma 8.7, it suffices to verify that for any left ideals X, Y of A,

$$r(X \cap Y) = r(X) + r(Y)$$

By hypothesis,

$$l(r(X \cap Y)) = X \cap Y = l(r(X)) \cap l(r(Y))$$
$$= l(r(X) + r(Y))$$

Applying right annihilators to the equality

$$l(r(X \cap Y)) = l(r(X) + r(Y)),$$

the required assertion follows.

(iii) $\Leftrightarrow$ (v): Let $e_1, \ldots, e_n$ be primitive idempotents of A such that $Ae_1, \ldots, Ae_n$ are all nonisomorphic principal indecomposable A-modules. Then, by Lemma 1.6.9, $e_1 A, \ldots, e_n A$ are all nonisomorphic principal indecomposable right A-modules. By Lemma 6.6,

$$(Ae_1)^*, \ldots, (Ae_n)^*$$

are nonisomorphic indecomposable right A-modules and any given Ae_i is injective if and only if $(Ae_i)^*$ is projective. This shows that (iii) implies (v) and the reverse implication follows by a similar argument.

(iv) $\Leftrightarrow$ (v): Apply the right analogue of arguments employed in the proof of (ii)$\Longleftrightarrow$(iii). ∎

8.9. Corollary. *The following conditions are equivalent:*
(i) A is a quasi-Frobenius algebra.
(ii) Every projective left A-module is injective.
(iii) Every injective left A-module is projective.
(iv) Every projective right A-module is injective.
(v) Every injective right A-module is projective.
(vi) For any left ideals X, Y of A and for any right ideal Z of A,

$$r(X \cap Y) = r(X) + r(Y) \quad and \quad r(l(Z)) = Z$$

Proof. Note that $_AA$ (respectively $_AA$) is injective if and only if every projective left(respectively, right) A-module is injective. Hence, by Theorem 8.8, (i), (ii) and (iv) are equivalent. By Lemma 8.2, the number of nonisomorphic (left or right) injective indecomposable A-modules is equal to the number of nonisomorphic projective indecomposable A-modules. This shows that (ii) $\Longleftrightarrow$ (iii) and (iv) $\Longleftrightarrow$ (v). The implication (ii) $\Rightarrow$ (vi) is a consequence of Lemma 8.6 and Theorem 8.8, while the implication (vi)$\Rightarrow$ (ii) follows from Lemma 8.7. ∎

Let V be a (left) A-module. Then $Hom_A(V, A)$ is a right A-module via

$$(fa)(v) = f(v)a \qquad (f \in Hom_A(V, A), v \in V, a \in A)$$

Similarly, if V is a right A-module, then $Hom_A(V, A)$ is a left A-module via

$$(af)(v) = af(v) \qquad (f \in Hom_A(V, A), v \in V, a \in A)$$

8.10. Lemma. *Let A be a Frobenius algebra, let V be a (left or right) A-module and let U be a submodule of V. If*

$$U^\perp = \{f \in Hom_A(V, A) | f(U) = 0\},$$

then the map

$$\begin{cases} Hom_A(V, A)/U^\perp & \xrightarrow{\lambda} & Hom_A(U, A) \\ f + U^\perp & \mapsto & f|U \end{cases}$$

is an A-isomorphism.

Proof. Assume for definiteness that V is a left A-module (the same argument will establish the case, where V is a right A-module). Then, by definition, λ is an injective A-homomorphism. By Theorem 8.8, $_AA$ is injective and so, by Lemma 5.5, any A-homomorphism $\psi : U \to A$ can be extended to an A-homomorphism $f : V \to A$. Since $\lambda(f + U^\perp) = f|U = \psi$, the result follows. ■

8.11. Theorem. *The following conditions are equivalent:*
(i) A is a quasi-Frobenius algebra.
(ii) For any left and right simple A-modules V_1 and V_2, $Hom_A(V_i, A)$ is simple, $i = 1, 2$.
(iii) For any left and right A-modules V_1 and V_2, respectively, the composition length of V_i is equal to that of $Hom_A(V_i, A)$, $i = 1, 2$.

Proof. (i) $\Rightarrow$ (ii): Let V be a left simple A-module. By Theorem 8.8, $_AA$ is a cogenerator. Hence there exists an injective A-homomorphism $f : V \to A$ and so $Hom_A(V, A) \neq 0$. Fix $0 \neq \alpha \in Hom_A(V, A)$. We will show the right A-module $Hom_A(V, A)$ is generated by α, from which the required assertion will follow.

Since $_AA$ is injective (Theorem 8.8), for any $\beta \in Hom_A(V, A)$ there exists an A-homomorphism $\beta' : A \to A$ such that $\beta = \beta' \circ \alpha$. Since β' is a right multiplication by some element $a \in A$, we have $\beta = \alpha a$. A similar argument establishes the case where V is a right simple A-module.

(ii) $\Rightarrow$ (i): Let X and Y be left and right ideals of A, respectively. By Theorem 8.8, it suffices to show that

$$l(r(X)) = X \quad \text{and} \quad r(l(Y)) = Y$$

Let $X_1 \subseteq X_2$ be left ideals of A such that X_2/X_1 is a simple A-module. We first claim that either $r(X_1)/r(X_2)$ is simple or equal to zero. The same argument will prove an analogous assertion for right ideals of A.

Consider the map $f : r(X_1)/r(X_2) \to Hom_A(X_2/X_1, A)$ given by

$$f(a + r(X_2))(b + X_1) = ba \qquad (a \in r(X_1), b \in X_2)$$

This map is clearly well defined and is an injective homomorphism of right A-modules. Since, by hypothesis, $Hom_A(X_2/X_1, A)$ is simple, the claim is verified.

There exists a composition series of the left A-module A containing X, say

$$0 = X_0 \subset X_1 \subset \cdots \subset X_t = A \qquad (2)$$

Now consider the chain

$$A = r(X_0) \supseteq r(X_1) \supseteq \cdots \supseteq r(X_t) = 0 \qquad (3)$$

By the claim proved above, $c(A_A) \leq c(_A A)$, where $c(V)$ is the composition length of an A-module V. By symmetry, $c(_A A) \leq c(A_A)$ and so $c(_A A) = c(A_A)$. Hence (3) is a composition series for A_A. But then

$$0 = l(r(X_0)) \subset \cdots \subset l(r(X_t)) = A$$

is also a composition series of $_A A$. Since (2) is a composition series for $_A A$ and $X_i \subseteq l(r(X_i))$, we have $X_i = l(r(X_i))$, $1 \leq i \leq t$. In particular, $l(r(X)) = X$. A similar argument proves that $r(l(Y)) = Y$ for any right ideal Y of A.

(iii) $\Rightarrow$ (ii): Obvious.

(ii) $\Rightarrow$ (iii): Let V be a left A-module. To prove that V and $Hom_A(V, A)$ have the same composition length, we argue by induction on $n = c(V)$. By hypothesis, the required assertion is true for $n = 1$. Assume that the result is true for all V with $c(V) \leq n$ and let W be a left A-module with $c(W) = n + 1$. Denote by U a simple submodule of W. Then, by induction hypothesis, $c(Hom_A(W/U, A)) = n$. Put

$$U^\perp = \{f \in Hom_A(W, A) | f(U) = 0\}$$

Then obviously

$$Hom_A(W/U, A) \cong U^\perp$$

and hence $c(U^\perp) = n$.

Now A is a quasi-Frobenius algebra (by (ii) $\Rightarrow$ (i)) and hence, by Lemma 8.10, the map

$$\begin{cases} Hom_A(W, A)/U^\perp & \to & Hom_A(U, A) \\ f + U^\perp & \mapsto & f|U \end{cases}$$

is an isomorphism. Since, by hypothesis, $Hom_A(U, A)$ is simple we conclude that $c(Hom_A(W, A)) = n + 1$. A similar argument establishes the case where V is a right A-module. $\blacksquare$

We are now ready to provide further characterizations of quasi-Frobenius algebras.

8.12. Theorem. *The following conditions are equivalent:*
(i) A is a quasi-Frobenius algebra.
(ii) For any primitive idempotent e of A, $Soc(Ae)$ and $Soc(eA)$ are simple and $Soc(_AA)$ (respectively, $Soc(A_A))$) contains an isomorphic copy of all simple left (respectively, right) A-modules.
(iii) For any primitive idempotent e of A, $Soc(Ae)$ and $Soc(eA)$ are simple and $Soc(_AA) = Soc(A_A)$.
(iv) If $e_1, \ldots, e_k$ are primitive idempotents of A such that $Ae_1, \ldots, Ae_k$ are all nonisomorphic principal indecomposable A-modules, then there exists a permutation π of $\{1, \ldots, k\}$ such that

$$Soc(Ae_i) \cong \bar{A}\bar{e}_{\pi(i)} \quad and \quad Soc(e_{\pi(i)}A) \cong \bar{e}_i\bar{A} \qquad (1 \le i \le k)$$

where $\bar{e}_i$ is the image of e_i in $\bar{A} = A/J(A)$.

Proof. (i)$\Rightarrow$ (ii): Let e be a primitive idempotent of A and let V be a simple submodule of Ae. Since Ae is an indecomposable and, by Corollary 8.9, is also injective, it follows from Theorem 5.8(ii) that Ae is an injective hull of V. Hence V is an essential submodule of Ae, which implies that $V = Soc(Ae)$. A similar argument shows that $Soc(eA)$ is simple. Since, by Theorem 8.8, both $_AA$ and A_A are cogenerators, the remaining assertion follows by virtue of Lemma 8.1.

(ii)$\Rightarrow$ (iii): Let e be a primitive idempotent of A. We will show that $Soc(_AA)$ contains $Soc(eA)$, which will imply that $Soc(A_A) \subseteq Soc(_AA)$.

A similar argument will show the opposite inclusion.

By hypothesis, $Soc(_AA)$ contains an isomorphic copy of $\bar{A}\bar{e}$ and hence $eSoc(_AA) \neq 0$. Since $Soc(_AA)$ is a two-sided ideal, it follows that $Soc(_AA) \supseteq eSoc(_AA) \neq 0$. But $Soc(eA)$ is simple and is an essential submodule of eA. Hence

$$Soc(_AA) \supseteq eSoc(_AA) \supseteq Soc(eA)$$

as required.

(iii) $\Rightarrow$ (ii): Let e be a primitive idempotent of A. Since

$$0 \neq Soc(eA) = eSoc(eA)$$

it follows that

$$0 \neq eSoc(A_A) = eSoc(_AA)$$

Hence there exists $x \in Soc(_AA)$ such that Aex is simple. Therefore $\bar{A}\bar{e} \cong Aex$ which shows that $Soc(_AA)$ contains an isomorphic copy of all simple left A-modules. A similar argument shows that $Soc(A_A)$ contains an isomorphic copy of all simple right A-modules.

(iii) $\Rightarrow$ (iv): Since $Soc(Ae_i)$ is simple for each $i \in \{1, \ldots, k\}$, there exists $\pi(i) \in \{1, \ldots, k\}$ such that

$$Soc(Ae_i) \cong \bar{A}\bar{e}_{\pi(i)} \qquad (1 \leq i \leq k) \qquad (4)$$

By (ii), $Soc(_AA)$ contains an isomorphic copy of all simple left A-modules which readily implies that π is a permutation of $\{1, \ldots, k\}$.

Choose $a_i \in A$ such that $Soc(Ae_i) = Aa_ie_i$, $1 \leq i \leq k$. It follows from (4) that $e_{\pi(i)}a_ie_i \neq 0$ and hence

$$Soc(Ae_i) = Ae_{\pi(i)}a_ie_i \qquad (5)$$

Because

$$0 \neq e_{\pi(i)}a_ie_i \in Soc(_AA) = Soc(A_A),$$

we have

$$e_{\pi(i)}a_ie_iA \subseteq Soc(A_A) \cap e_{\pi(i)}A = Soc(e_{\pi(i)}A)$$

and, since $Soc(e_{\pi(i)}A)$ is simple, we deduce that

$$Soc(e_{\pi(i)}A) = e_{\pi(i)}a_ie_iA \qquad (6)$$

But then the surjective homomorphism

$$\begin{cases} e_i A & \to & Soc(e_{\pi(i)} A) \\ e_i a & \mapsto & e_{\pi(i)} a_i e_i a \end{cases}$$

induces an isomorphism

$$\bar{e}_i \bar{A} \cong Soc(e_{\pi(i)} A),$$

as required.

(iv) $\Rightarrow$ (ii): By the Krull-Schmidt theorem, we may assume that the idempotent e in (ii) coincides with e_i in (iv) for some $i \in \{1, \ldots, k\}$. Thus (ii) is a consequence of (iv).

(ii) $\Rightarrow$ (i): Owing to Theorem 8.11, it suffices to show that for any left and right simple A-modules V and W, $Hom_A(V, A)$ are simple. By hypothesis, the latter will follow provided (in the notation of (iv)) we show that the modules $Hom_A(Soc(Ae_i), A)$ and $Hom_A(Soc(e_i A), A)$ are simple, $1 \leq i \leq k$. By symmetry, it suffices to show that

$$Hom_A(Soc(Ae_i), A)$$

is simple.

Consider a nonzero A-homomorphism

$$f : Soc(Ae_i) \to A$$

Since (ii) $\Rightarrow$ (iii), it follows from (5) that for any $a \in A$,

$$f(ae_{\pi(i)} a_i e_i) = ae_{\pi(i)} f(e_{\pi(i)} a_i e_i)$$

and that for $x = f(e_{\pi(i)} a_i e_i)$, $Ae_{\pi(i)} x = f(Soc(Ae_i))$ is simple. Now

$$0 \neq e_{\pi(i)} x A \subseteq Soc(A_A) \cap e_{\pi(i)} A = Soc(e_{\pi(i)} A)$$

and, by hypothesis, $Soc(e_{\pi(i)} A)$ is simple. Hence, by (6),

$$e_{\pi(i)} x A = Soc(e_{\pi(i)} A) = e_{\pi(i)} a_i e_i A$$

and therefore there exists $e_i a_0 \in e_i A$ such that

$$e_{\pi(i)} x = e_{\pi(i)} a_i e_i a_0$$

It follows that, for all $a \in A$,

$$f(ae_{\pi(i)}a_ie_i) = ae_{\pi(i)}x = ae_{\pi(i)}a_ie_ia_0$$

which means that $f(y) = ye_ia_0$ for all $y \in Soc(Ae_i) = Ae_{\pi(i)}a_ie_i$. Hence the map

$$\begin{cases} e_iA \xrightarrow{\psi} Hom_A(Soc(Ae_i), A) \\ e_ia \mapsto r_a \end{cases}$$

where $r_a(y) = ye_ia$, $y \in Soc(Ae)$ is a surjective A-homomorphism. Since

$$0 = Soc(A_A)J(A) = Soc(_AA)J(A)$$

we have $e_iJ(A) \subseteq Ker\psi$. But $e_iJ(A)$ is a maximal ideal of e_iA, hence $Hom_A(Soc(Ae_i), A)$ is simple, as required. ∎

We have now come to the demonstration for which this section has been developed.

8.13. Theorem. *The following conditions are equivalent:*
(i) A is a Frobenius algebra.
(ii) A is a quasi-Frobenius algebra such that

$$Soc(_AA) \cong A/J(A) \quad and \quad Soc(A_A) \cong A/J(A)$$

(iii) $Soc(_AA) \cong A/J(A)$ and $Soc(A_A) \cong A/J(A)$.
(iv) A is a quasi-Frobenius algebra such that

$$Soc(_AA) \cong A/J(A) \quad or \quad Soc(A_A) \cong A/J(A)$$

(v) For all left ideals X and right ideals Y in A, we have

$$l(r(X)) = X, r(l(Y)) = Y$$

and

$$dim_FA = dim_Fr(X) + dim_FX = dim_Fl(Y) + dim_FY$$

Proof.   (ii) ⇒ (iii): Obvious.
(ii) ⇒ (iv): Obvious.
(iii) ⇒ (ii): Our hypotheses ensure that $Soc(_AA)$ contains a copy of

any simple left A-module and that $Soc(A_A)$ satisfies the same property for simple right A-modules. By Theorem 8.12(ii), we are therefore left to verify that (in the notation of Theorem 8.12(iv)), $Soc(Ae_i)$ and $Soc(e_iA)$ are simple for all $i \in \{1, \ldots, k\}$. We have

$$_AA \cong \oplus_{i=1}^{k} n_i Ae_i \quad \text{and} \quad \bar{A} \cong \oplus_{i=1}^{k} n_i \bar{A}\bar{e}_i$$

for some positive integers n_i, where $\bar{e}_i$ is the image of e_i in $\bar{A} = A/J(A)$. Since $\bar{A} \cong Soc(_AA)$, we then have

$$Soc(_AA) \cong \oplus_{i=1}^{k} n_i Soc(Ae_i) \cong \oplus_{i=1}^{k} n_i \bar{A}\bar{e}_i$$

and so each $Soc(Ae_i)$ is simple. A similar argument shows that each $Soc(e_iA)$ is also simple, as required.

(iv) $\Rightarrow$ (ii): In the notation above, we have

$$\bar{A} \cong \oplus_{i=1}^{k} n_i \bar{A}\bar{e}_i \quad \text{and} \quad Soc(_AA) \cong \oplus_{i=1}^{k} n_i Soc(Ae_i)$$

and

$$\bar{A} \cong \oplus_{i=1}^{k} n_i \bar{e}_i \bar{A} \quad \text{and} \quad Soc(A_A) \cong \oplus_{i=1}^{k} n_i Soc(e_iA)$$

where, by Theorem 8.12, all $Soc(Ae_i)$, $Soc(e_iA)$ are simple. Hence, by Theorem 8.12(iv),

$$\bar{A} \cong Soc(_AA) \quad \text{if and only if} \quad n_{\pi(i)} = n_i, 1 \leq i \leq k \qquad (7)$$

and

$$\bar{A} \cong Soc(A_A) \quad \text{if and only if} \quad n_{\pi(i)} = n_i, 1 \leq i \leq k$$

and the required assertion follows.

(i) $\Rightarrow$ (iv): Since A is a Frobenius algebra, it is also a quasi-Frobenius algebra. Furthermore, since the left A-modules A and A^* are isomorphic, it follows from Lemma 8.5(i) that $A/J(A) \cong Soc(_AA)$, as required.

(ii) $\Rightarrow$ (i): By Theorem 8.8, $_AA$ is injective. Since $Soc(_AA)$ is an essential submodule of A, it follows from Theorem 5.8(i) that A is an injective hull of $Soc(_AA)$. On the other hand, by Lemma 8.5, A^* is an injective hull of a copy of $A/J(A)$. Since, by hypothesis, $Soc(_AA) \cong A/J(A)$ we deduce that the left A-modules A and A^* are isomorphic. Thus A is a Frobenius algebra.

(i) $\Rightarrow$ (v): Direct consequence of Theorem 6.1.

(v) $\Rightarrow$ (i): By Theorem 8.8, A is a quasi-Frobenius algebra. Hence, by the equivalence of (i) and (iv), it suffices to show that $Soc(_AA) \cong A/J(A)$ or, by (7), that $n_{\pi(i)} = n_i$ for all $i \in \{1, \dots, k\}$.

To this end, we fix $i \in \{1, \dots, k\}$ and put

$$S = Soc(A), D = \bar{e}_i \bar{A} \bar{e}_i \quad \text{and} \quad D' = \bar{e}_{\pi(i)} \bar{A} \bar{e}_{\pi(i)}$$

Then D and D' are division algebras such that n_i is equal to the right dimension of $\bar{A}\bar{e}_i$ over D and also to the left dimension of $\bar{e}_i\bar{A}$ over D. Note also that, by Theorem 8.12,

$$Se_i \cong \bar{A}\bar{e}_{\pi(i)} \quad \text{and} \quad e_{\pi(i)}S \cong \bar{e}_i\bar{A}$$

and therefore

$$e_{\pi(i)}Se_i \cong D' \quad \text{and} \quad e_{\pi(i)}Se_i \cong D$$

as F-spaces. Thus

$$dim_F D = dim_F D' \tag{8}$$

Using the hypothesis concerning the dimensions of the annihilators, we also have

$$\begin{aligned}
dim_F \bar{e}_i \bar{A} &= dim_F e_{\pi(i)} S \\
&= dim_F A - dim_F(l(e_{\pi(i)}S)) \\
&= dim_F A - dim_F(J(A)e_{\pi(i)} + A(1 - e_{\pi(i)})) \\
&= dim_F(\bar{A}\bar{e}_{\pi(i)}) \tag{9}
\end{aligned}$$

On the other hand,

$$dim_F(\bar{A}\bar{e}_{\pi(i)}) = dim_{D'}(\bar{A}\bar{e}_{\pi(i)})dim_F D' = n_{\pi(i)}dim_F D'$$

and

$$dim_F(\bar{e}_i\bar{A}) = dim_D(\bar{e}_i\bar{A})dim_F D = n_i dim_F D$$

Thus, by (8) and (9), $n_i = n_{\pi(i)}$ and the result follows. ■

9. Characters of symmetric algebras

Throughout this section, A denotes a finite-dimensional algebra over a a field F. As usual, all A-modules are assumed to be left and finitely generated.

Let V be a vector space over F having a basis $v_1, \ldots, v_n$ of n elements, and let $f \in End_F(V)$. With respect to this basis, f can be represented as an $n \times n$-matrix (a_{ij}) over F. By the *trace* of f, written trf, we mean

$$trf = tr(a_{ij}) = \sum_{i=1}^{n} a_{ii}$$

If a different basis for V is chosen, then the two different matrices for f are similar, and hence have the same trace. Thus $f \mapsto trf$ is a well-defined F-valued function on $End_F(V)$.

Now assume that V is an A-module. Then V determines a homomorphism

$$\rho : A \to End_F(V)$$

By the *character* of V, we understand the map

$$\chi : A \to F$$

given by

$$\chi(a) = tr(\rho(a)) \qquad \text{for all} \quad a \in A$$

In general, an element $\chi \in A^* = Hom_F(A, F)$ is called a *character* of A (or an A-character) if χ is the character of a suitable A-module V. We say that an A-character χ is *irreducible* if χ is the character of a simple A-module.

The following terminology is extracted from Fossum (1970). Given $f \in A^*$, we say that f is a *class function* if

$$f(ab) = f(ba) \qquad \text{for all} \quad a, b \in A$$

We refer to f as a *quasicharacter* if f is a class function such that

$$J(A) \subseteq Kerf$$

In what follows, we write

$$cf(A), qch(A) \quad \text{and} \quad ch(A)$$

to denote the F-subspaces of A^* generated, respectively, by all class functions, quasicharacters and characters of A.

9.1. Lemma. *We have*

$$ch(A) \subseteq qch(A) \subseteq cf(A)$$

Proof. The inclusion $qch(A) \subseteq cf(A)$ is a consequence of the definition. Let χ be an A-character afforded by the A-module V. Then χ is obviously a class function. Furthermore, χ is the sum of the characters of the composition factors of V (counting multiplicities). The character of each composition factor is irreducible. Since $J(A)$ annihilates all simple A-modules, we conclude that $J(A) \subseteq Ker\chi$, i.e. χ is a quasicharacter. ■

9.2. Lemma. *Ler $\bar{A} = A/J(A)$ and, for any $f \in qch(A)$, let $\bar{f} \in \bar{A}^*$ be defined by*

$$\bar{f}(a + J(A)) = f(a) \qquad (a \in A)$$

Then the map $f \mapsto \bar{f}$ is an F-isomorphism from $qch(A)$ onto $qch(\bar{A})$ which maps $ch(A)$ onto $ch(\bar{A})$.

Proof. Since $J(A) \subseteq Kerf$, $\bar{f}$ is well defined. Furthermore, it is clear that $\bar{f}$ is a quasicharacter of $\bar{A}$. By definition, the map $f \mapsto \bar{f}$ is obviously F-linear. Now assume that $\lambda \in qch(\bar{A})$. Then the composite map $\lambda \circ \pi$, where $\pi : A \to \bar{A}$ is the natural map, is in $cf(A)$ and $\overline{\lambda \circ \pi} = \lambda$. Moreover, since

$$J(A) = Ker\pi \subseteq Kerf$$

$\lambda \circ \pi \in qch(A)$, proving that $f \mapsto \bar{f}$ is a surjection. Since $\bar{f} = 0$ implies $f = 0$, we finally conclude that the given map is an F-isomorphism.

To prove the second assertion, we first note that $ch(A)$ is the F-linear span of all irreducible A-characters. But if V is a simple A-module with character χ, then $J(A)V = 0$ and so V is a simple $\bar{A}$-module with character $\bar{\chi}$. Conversely, if V is a simple $\bar{A}$-module with character χ, then V is also a simple A-module with character $\chi \circ \pi$

and $\overline{\chi \circ \pi} = \chi$. This shows that $f \mapsto \bar{f}$ maps $ch(A)$ onto $ch(\bar{A})$, thus completing the proof. ■

Turning to symmetric algebras, we now prove

9.3. Theorem. *(Fossum (1970)). Let A be a symmetric algebra and let $f : A \to A^*$ be an isomorphism of (A, A)-bimodules (see Lemma 1.5).*

(i) $Z(A) = \{a \in A | f(a) \in cf(A)\}$.
(ii) $Soc(A) = \{a \in A | f(a)(J(A)) = 0\}$.
(iii) $Soc(A) \cap Z(A) = \{a \in A | f(a) \in qch(A)\}$.
(iv) If F is a splitting field for A, then $qch(A) = ch(A)$.
(v) If e is a central idempotent of A, then $f(e) \in qch(A)$ if and only if Ae is semisimple.

Proof. (i) Assume that $a \in Z(A)$. Then, for all $x, y \in A$,

$$f(a)(xy) = f(ya)(x) = f(ay)(x) = f(a)(yx),$$

proving that $f(a) \in cf(A)$. Conversely, assume that $\chi \in cf(A)$ and put $b = f^{-1}(\chi)$. Then, for any given $a, x \in A$,

$$(a\chi)(x) = \chi(xa) = \chi(ax) = (\chi a)(x)$$

and so

$$ab = af^{-1}(\chi) = f^{-1}(a\chi) = f^{-1}(\chi a) = f^{-1}(\chi)a = ba,$$

as required.

(ii) Assume that $a \in A$ is such that $f(a)(J(A)) = 0$. Then, given $b \in J(A), x \in A$,

$$f(ba)(x) = (bf(a))(x) = f(a)(xb) = 0$$

proving that $f(ba) = 0$ and so $ba = 0$. Hence $a \in Soc(A)$. Similarly, if $a \in Soc(A)$, then $f(a)(J(A)) = 0$, as required.

(iii) This is a direct consequence of (i) and (ii).

(iv) Owing to Lemmas 9.1 and 9.2, it suffices to show that

$$qch(\bar{A}) = ch(\bar{A})$$

where $\bar{A} = A/J(A)$. Since $\bar{A}$ is semisimple, it is symmetric by virtue of Proposition 1.9. By (i), we have

$$dim_F Z(\bar{A}) = dim_F cf(\bar{A}) = dim_F qch(\bar{A})$$

Since F is a splitting field for A, the number of nonisomorphic simple A-modules is equal to $dim_F Z(\bar{A})$. As is well known (see Curtis and Reiner (1962, p.183)), the characters of these modules are F-linearly independent and thus form a basis for $ch(\bar{A})$. Hence

$$dim_F ch(\bar{A}) = dim_F Z(\bar{A}) = dim_F qch(\bar{A})$$

and the required assertion follows by virtue of Lemma 9.1.

(v) We first observe that $J(Ae) = J(A)e$. Now assume that Ae is semisimple. Then $J(A)e = 0$ and so $e \in Soc(A) \cap Z(A)$. Hence, by (iii), $f(e) \in qch(A)$. Conversely, assume that $f(e) \in qch(A)$. Then, by (iii), $e \in Soc(A)$ and so $J(A)e = 0$. Thus Ae is semisimple, as we wished to show. ∎

9.4. Corollary. *In the notation of Theorem 9.3, the following conditions are equivalent:*

(i) A is semisimple.
(ii) $cf(A) = qch(A)$.
(iii) $f(1) \in qch(A)$.

Proof. (i) $\Rightarrow$ (ii): Since A is semisimple, $J(A) = 0$ and so $cf(A) = qch(A)$.

(ii) $\Rightarrow$ (iii): By Theorem 9.3(i), $f(1) \in cf(A) = qch(A)$.

(iii) $\Rightarrow$ (i): If $f(1) \in qch(A)$, then by Theorem 9.3(v), A is semisimple. ∎

Let $B = B(e)$ be a block of an F-algebra A. If $\chi \in A^*$ (and in particular if χ is an A-character) we say that χ *belongs* to B (or to e) if $e\chi = \chi$. Note that for $\chi \in A^*$, $e\chi = \chi$ is equivalent to

$$\chi(A(1 - e)) = 0$$

9.5. Lemma. *Let $B = B(e)$ be a block of A and let V be an A-module with character χ. If V belongs to B, then χ belongs to B.*

Proof.   By definition, $eV = V$ and therefore $A(1 - e)V = 0$. Thus $\chi(A(1 - e)) = 0$ as required. ∎

Note that the converse to the above is also true provided either $\chi \neq 0$ and V is simple or the characteristic of F is zero. As a preliminary to our next result, let us recall the following piece of information.

Assume that (A, ψ) is a symmetric algebra and let $f : A \to A^*$ be the corresponding (A, A)-bimodule isomorphism given by

$$f(a)(x) = \psi(xa) \qquad \text{for all} \quad x, a \in A \tag{1}$$

Recall that if $a_1, \ldots, a_n$ is an F-basis of A, then the ψ-dual basis $b_1, \ldots, b_n$ of A satisfies

$$\psi(a_i b_j) = \delta_{ij} \quad \text{for all} \quad i, j \in \{1, \ldots, n\} \tag{2}$$

In view of (1) and the assumption that A is symmetric, (2) can be written as

$$f(b_j)(a_i) = \delta_{ij} = f(a_i)(b_j) \tag{3}$$

9.6. Lemma.   *With the assumptions and notation above, the following properties hold:*
   (i) For any $\lambda \in A^*$, $f^{-1}(\lambda) = \sum_{i=1}^{n} \lambda(a_i) b_i = \sum_{i=1}^{n} \lambda(b_i) a_i$.
   (ii) if χ is the character of $_A A$, then

$$f^{-1}(\chi) = \sum_{i=1}^{n} a_i b_i$$

Proof.   (i) It follows from (3) that

$$\lambda = \sum_{i=1}^{n} \lambda(a_i) f(b_i) = \sum_{i=1}^{n} \lambda(b_i) f(a_i)$$

as required.
   (ii) For each $i, j \in \{1, \ldots, n\}$ write $a_i a_j = \sum_{k=1}^{n} \alpha_{ikj} a_k$. Then $\chi(a_i) = \sum_{k=1}^{n} \alpha_{ikk}$. On the other hand, for any $j \in \{1, \ldots, n\}$,

$$f\left(\sum_{i=1}^{n} a_i b_i\right)(a_j) \;=\; \sum_{i=1}^{n} f(a_i b_i)(a_j) = \sum_{i=1}^{n} f(b_i)(a_j a_i)$$

$$= \sum_{i=1}^{n} f(b_i)(\sum_{k=1}^{n} \alpha_{jki}a_k)$$

$$= \sum_{i,k=1}^{n} \alpha_{jki} f(b_i)(a_k) = \sum_{i,k=1}^{n} \alpha_{jki}\delta_{ik}$$

$$= \sum_{k=1}^{n} \alpha_{jkk} = \chi(a_j),$$

as required. ∎

9.7. Theorem. *(Fossum (1970)). Let (A, ψ) be a symmetric F-algebra, let F be a splitting field for A and let $b_1, \ldots, b_n$ be an F-basis of A which is ψ-dual to some basis $a_1, \ldots, a_n$. Let $f : A \to A^*$ be the (A, A)-bimodule isomorphism given by (1), let χ be the character of a simple A-module V and let ρ be any matrix representation of A afforded by V.*

$$(i) \quad f^{-1}(\chi) = \sum_{i=1}^{n} \chi(a_i)b_i \in Z(A) \quad and \quad \rho(f^{-1}(\chi)) = \alpha(\chi)I_d$$

for some $\alpha(\chi) \in F$ which does not depend on the choice of ρ (here I_d is the $d \times d$ identity matrix).

(ii) $\sum_{i=1}^{n} \chi(a_i)\chi(b_i) = \alpha(\chi)\chi(1)$
and, in particular, if $\chi(1) \neq 0$, then

$$\alpha(\chi) = \chi(1)^{-1} \sum_{i=1}^{n} \chi(a_i)\chi(b_i)$$

(iii) If A is semisimple, then $\alpha(\chi) \neq 0$ and the block idempotent $e = e(\chi)$ to which χ belongs is given by

$$e(\chi) = \alpha(\chi)^{-1} \sum_{i=1}^{n} \chi(a_i)b_i$$

Proof. (i) By Lemma 9.6(i),

$$f^{-1}(\chi) = \sum_{i=1}^{n} \chi(a_i)b_i \tag{4}$$

and, by Theorem 9.3(i), $f^{-1}(\chi) \in Z(A)$. Since F is a splitting field for A, it follows that

$$\rho(f^{-1}(\chi)) = \alpha(\chi)I_d \qquad (5)$$

for some $\alpha(\chi) \in F$ and clearly $\alpha(\chi)$ does not depend upon the choice of ρ afforded by V.

(ii) This follows by taking traces on both sides of (5) and using (4).

(iii) The irreducible A-character χ belongs to a unique block idempotent, say e. Then $0 \neq \chi \in e\,ch(A)$, and applying f^{-1} this gives

$$0 \neq f^{-1}(\chi) \in eZ(A)$$

Because $eZ(A) = Fe$, we have $f^{-1}(\chi) = \beta(\chi)e$ for some $\beta(\chi) \in F$, $\beta(\chi) \neq 0$. But then

$$\rho(f^{-1}(\chi)) = \beta(\chi)\rho(e) = \beta(\chi)I_d = \alpha(\chi)I_d$$

and hence $\beta(\chi) = \alpha(\chi) \neq 0$. Thus, by (4),

$$e = \alpha(\chi)^{-1}f^{-1}(\chi) = \alpha(\chi)^{-1}\sum_{i=1}^{n}\chi(a_i)b_i$$

and the result follows. ∎

9.8. Theorem. *(Fossum (1970)). Further to notation and assumptions of Theorem 9.7, suppose that A is semisimple.*

(i) If $\chi(x) \neq 0$ for some $x \in A$, then

$$\alpha(\chi) = \chi(x)^{-1}\sum_{i=1}^{n}\chi(a_i)\chi(b_ix)$$

(ii) If $\chi(x) \neq 0$ for some $x \in A$, then

$$\sum_{i=1}^{n}\chi(a_i)\chi(b_ix) \neq 0$$

and

$$e(\chi) = \chi(x)(\sum_{i=1}^{n}\chi(a_i)\chi(b_ix))^{-1}\sum_{i=1}^{n}\chi(a_i)b_i$$

(iii) If χ_1, χ_2 are distinct irreducible A-characters, then

$$\sum_{i=1}^{n} \chi_1(a_i)\chi_2(b_i) = 0$$

(iv) For any irreducible A-character χ,

$$\chi(\sum_{i=1}^{n} a_i b_i) = \chi(1)\sum_{i=1}^{n} \chi(a_i)\chi(b_i)$$

Proof.   (i) Owing to Theorem 9.7(iii),

$$\alpha(\chi)e(\chi) = \sum_{i=1}^{n} \chi(a_i)b_i \qquad (6)$$

If $\chi(x) \neq 0$, then

$$\chi(e(\chi)x) = \chi(x) \neq 0$$

and therefore, by applying χ to both sides of (6), we obtain

$$\alpha(\chi)\chi(x) = \sum_{i=1}^{n} \chi(a_i)\chi(b_i x) \qquad (7)$$

Since $\chi(x) \neq 0$, the desired conclusion follows.

(ii) The first assertion is a consequence of (7) and the fact that both $\chi(x)$ and $\alpha(\chi)$ are nonzero. The second follows from the first, by applying Theorem 9.7(iii).

(iii) Owing to Theorem 9.7(i), $f^{-1}(\chi_1) = \sum_{i=1}^{n} \chi_1(a_i)b_i$. Let χ_1 and χ_2 belong to the block idempotents e_1 and e_2, respectively. By hypothesis, $\chi_1 \neq \chi_2$ and therefore $e_2 e_1 = 0$. But $\chi_1 = e_1\chi_1$ and $\chi_2 = e_2\chi_2 = \chi_2 e_2$ and so

$$\begin{aligned} \chi_2(f^{-1}(\chi_1)) &= (\chi_2 e_2)(f^{-1}(e_1\chi_1)) \\ &= \chi_2(e_2 e_1 f^{-1}(\chi_1)) \\ &= 0, \end{aligned}$$

proving (iii).

(iv) Let $\chi_1, \ldots, \chi_r$ be the distinct irreducible A-characters. Since F is a splitting field for A and A is semisimple, each χ_i appears in the character λ of the regular module with multiplicity equal to its degree. Thus

$$\lambda = \sum_{j=1}^{r} \chi_j(1)\chi_j$$

Applying Lemma 9.6(ii), we deduce that

$$
\begin{aligned}
\sum_{i=1}^{n} a_i b_i &= f^{-1}(\lambda) \\
&= \sum_{j=1}^{r} \chi_j(1) f^{-1}(\chi_j) \\
&= \sum_{j=1}^{r} \chi_j(1)[\sum_{i=1}^{n} \chi_j(a_i)b_i]
\end{aligned}
$$

where the last equality is a consequence of Theorem 9.7(i). Thus

$$\chi(\sum_{i=1}^{n} a_i b_i) = \sum_{j=1}^{r} \chi_j(1)[\sum_{i=1}^{n} \chi_j(a_i)\chi(b_i)]$$

and therefore, by (iii),

$$\chi(\sum_{i=1}^{n} a_i b_i) = \sum_{i=1}^{n} \chi(a_i)\chi(b_i),$$

as asserted. ∎

10. Applications to projective modular representations

All conventions and notation employed in the previous section remain in force. In particular, A denotes a finite-dimensional algebra over a field F. Our aim is twofold: first to prove a general result concerning the degrees of irreducible characters of symmetric algebras (Corollary 10.2) and second to apply the information obtained to investigate the degrees of irreducible projective modular representations of finite groups.

10.1. Theorem. *(Fossum (1970)). Let (A, ψ) be a symmetric F-algebra, where F is an algebraic number field which is a splitting field* *for A. Let R denote the ring of algebraic integers in F and assume that there exist ψ-dual bases $a_1, \ldots, a_n$ and $b_1, \ldots, b_n$ of A such that $\sum_{i=1}^{n} Ra_i = \sum_{i=1}^{n} Rb_i$ is an R-subalgebra of A. If χ is an irreducible A-character such that*

$$\sum_{i=1}^{n} \chi(a_i)\chi(b_i) \in \mathbb{Q}$$

then

$$\chi(1)^{-1} \sum_{i=1}^{n} \chi(a_i)\chi(b_i) \in \mathbb{Z}$$

Proof. Let V be a simple A-module with character χ and let $\alpha(\chi)$ and f be as in Theorem 9.7(i). Let p be a rational prime and let ν_p be a valuation on F extending the p-adic valuation of $\mathbb{Q}$. Let R_p be the valuation ring in F corresponding to ν_p.

Now $\chi(1) \neq 0$ since F has characteristic zero, so by Theorem 9.7(ii)

$$\alpha(\chi) = \chi(1)^{-1} \sum_{i=1}^{n} \chi(a_i)\chi(b_i)$$

is rational since $\sum_{i=1}^{n} \chi(a_i)\chi(b_i)$ is rational by hypothesis. We claim that $\alpha(\chi) \in R_p$; if sustained, it will follow that $\alpha(\chi) \in \mathbb{Q} \cap R_p$ for all rational primes p and hence $\alpha(\chi) \in \mathbb{Z}$, as required.

To substantiate our claim, put $A_0 = \sum_{i=1}^{n} Ra_i$ and $R_p A_0 = \sum_{i=1}^{n} R_p a_i$ Since V is a simple A-module, $V = Ax$ for some $0 \neq x \in V$ and so setting $W = R_p A_0 x$, it follows that W is a finitely generated $R_p A_0$-module. Since $R_p A_0$ is R_p-finitely generated, so is W. But R_p is a principal ideal domain and W is R_p-torsion-free (since $W \subseteq V$ and $R_p \subseteq F$), so W is R_p-free, say with R_p-basis B. It is clear that B is also an F-basis for V.

Finally, let ρ be a matrix representation of A afforded by this F-basis B of V. Since B is an R_p-basis for W and since $R_p A_0 W \subseteq W$, the matrix $\rho(y)$ has all its entries in R_p for any $y \in R_p A_0$. In particular,

$$\chi(a_i) = trace \rho(a_i) \in R_p \quad \text{for all} \quad i$$

Hence, by Theorem 9.7(i),

$$f^{-1}(\chi) = \sum_{i=1}^{n} \chi(a_i)b_i \in R_p A_0$$

and therefore $\alpha(\chi) \in R_p$, as required. ∎

10.2.   Corollary. *Further to the assumptions and notation of Theorem 10.1, assume that*

$$\sum_{i=1}^{n} a_i b_i = (dim_F A) \cdot 1$$

Then A is semisimple and $\chi(1)$ divides $dim_F A$ for any irreducible A-character χ.

Proof.   By assumption and Lemma 9.6(ii),

$$(dim_F A) \cdot 1 = \sum_{i=1}^{n} a_i b_i = f^{-1}(\lambda)$$

where λ is the character of $_A A$. Therefore $1 = (dim_F A)^{-1} f^{-1}(\lambda)$ and

$$f(1) = (dim_F A)^{-1} \lambda \in qch(A)$$

Thus, by Corollary 9.4, A is semisimple.

By Theorem 9.8(iv), we have

$$(dim_F A)\chi(1) = \chi(\sum_{i=1}^{n} a_i b_i) = \chi(1) \sum_{i=1}^{n} \chi(a_i)\chi(b_i)$$

for any irreducible A-character χ. Hence

$$dim_F A = \sum_{i=1}^{n} \chi(a_i)\chi(b_i)$$

and therefore, by Theorem 10.1, $\chi(1)$ divides $dim_F A$. ∎

The rest of the section will be devoted to providing an application of the result above to the degrees of irreducible projective modular representations of finite groups. We begin by sketching some background information.

Let G be a finite group and let F^* be the multiplicative group of a field F. Denote by $Z^2(G, F^*)$ the set of all functions

$$\alpha : G \times G \to F^*$$

which satisfy the following identities

$$\alpha(x, 1) = \alpha(1, x) = 1 \qquad (x \in G)$$
$$\alpha(x, y)\alpha(xy, z) = \alpha(y, z)\alpha(x, yz) \quad (x, y, z \in G)$$

We shall refer to the elements of $Z^2(G, F^*)$ as *cocycles*. Given $\alpha, \beta \in Z^2(G, F^*)$, define $\alpha\beta$ by the rule

$$(\alpha\beta)(x, y) = \alpha(x, y)\beta(x, y) \qquad (x, y \in G)$$

It is then obvious that $\alpha\beta$ is also a cocycle and that $Z^2(G, F^*)$ becomes an abelian group.

Let $t : G \to F^*$ be such that $t(1) = 1$ and let $\delta t : G \times G \to F^*$ be defined by

$$(\delta t)(x, y) = t(x)t(y)t(xy)^{-1} \qquad (x, y \in G)$$

We shall refer to δt as a *coboundary*. It is routine to verify that the set $B^2(G, F^*)$ of all coboundaries constitutes a subgroup of $Z^2(G, F^*)$. The factor group

$$H^2(G, F^*) = Z^2(G, F^*)/B^2(G, F^*)$$

is called the *second cohomology group* of G over F^*. The elements of $H^2(G, F^*)$ are called *cohomology classes*; any two cocycles contained in the same cohomology class are said to be *cohomologous*.

Given $\alpha \in Z^2(G, F^*)$, denote by $F^\alpha G$ the vector space over F with basis $\{\bar{g} | g \in G\}$ which is in one-to-one correspondence with G. Define multiplication in $F^\alpha G$ distributively, using for all $x, y \in G$

$$\bar{x}\bar{y} = \alpha(x, y)\overline{xy}$$

Then $F^\alpha G$ becomes an F-algebra which is a twisted group algebra of G over F. Conversely, any twisted group algebra of G over F arises in this manner. Note that if $\alpha(x, y) = 1$ for all $x, y \in G$, then $F^\alpha G \cong FG$. More generally, we have

10.3. Lemma. $F^\alpha G \cong FG$ *if and only if α is a coboundary.*

Proof. Assume that $F^\alpha G \cong FG$. Because FG admits an F-algebra homomorphism into F, so does $F^\alpha G$, say $f : F^\alpha G \to F$. Then, for all $x, y \in G$

$$f(\bar{x})f(\bar{y}) = f(\bar{x}\bar{y}) = f(\alpha(x,y)\overline{xy}) = \alpha(x,y)f(\overline{xy})$$

and so $\alpha = \delta t$, where $t : G \to F^*$ is defined by $t(g) = f(\bar{g})$.

Conversely, suppose that $\alpha = \delta t$ for some $t : G \to F^*$ with $t(1) = 1$. Then the map $\psi : F^\alpha G \to FG$ which is the extension of $\bar{g} \mapsto t(g)g$ by F-linearity gives the desired isomorphism. ∎

While on the subject, we also record the following elementary results.

10.4. Lemma. *Let F be an algebraically closed field and let G be a group of order n.*

(i) $H^2(G, F^*)$ is a finite group, all elements of which have order dividing n.

(ii) If $\alpha \in Z^2(G, F^*)$ and m is the order of the cohomology class of α, then α is cohomologous to a cocycle of order m.

(iii) If $char F = p > 0$, then p does not divide the order of $H^2(G, F^*)$.

Proof. (i) Assume that $\alpha \in Z^2(G, F^*)$. Since $\alpha(x,y)\alpha(xy,z) = \alpha(y,z)\alpha(x,yz)$, we have

$$\prod_{x \in G} \alpha(y,z)\alpha(x,yz) = \prod_{x \in G} \alpha(x,y)\alpha(xy,z)$$

Setting $t(y) = \prod_{x \in G} \alpha(x,y)$ it now follows that

$$\alpha(y,z)^n t(yz) = t(y)t(z) \qquad (x,y,z \in G)$$

Thus $\alpha^n = \delta t$, proving that each element of $H^2(G, F^*)$ has order dividing n. Thus (i) will follow from (ii) to be proved below. Indeed, if (ii) is true, then $H^2(G, F^*)$ can be regarded as a subset of all mappings from $G \times G$ to the group of n-th roots of 1 in F. But the latter is finite, hence so is $H^2(G, F^*)$.

(ii) Write $\alpha(x,y)^m = t(x)t(y)t(xy)^{-1}$, $x,y \in G$ for some $t : G \to F^*$ with $t(1) = 1$. Since F is algebraically closed, for any $x \in G$, there exists $\mu(x) \in F^*$ such that $\mu(x)^m = t(x)^{-1}$. Setting $\beta = \alpha(\delta\mu)$, we deduce that $\beta^m = 1$, as required.

(iii) Suppose that $char F = p > 0$. Then, for all $a, b \in F$, $(a - b)^p = a^p - b^p$, so $a^p = b^p$ implies $a = b$. Assume that $\alpha \in Z^2(G, F^*)$ is such that

$$\alpha(x, y)^p = t(x)t(y)t(xy)^{-1}, \qquad x, y \in G$$

for some $t : G \to F^*$ with $t(1) = 1$. Then $\alpha(x, y)^p = \mu(x)^p \mu(y)^p \mu(xy)^{-p}$ for some $\mu : G \to F^*$ and hence $\alpha = \delta\mu$, as required. ∎

10.5. Corollary. *Let F be an algebraically closed field of characteristic $p \geq 0$ and let $\alpha \in Z^2(G, F^*)$. Then $F^\alpha G \cong FG$ under either of the following hypothesis:*
(i) G is a cyclic group.
(ii) $p > 0$ and G is a cyclic extension of a p-group.

Proof. (i) Let g be a generator of G, say of order m. Then $\bar{g}^m = \lambda \cdot \bar{1}$ for some $\lambda \in F^*$. Chose $\mu \in F$ with $\mu^m = \lambda^{-1}$ so that $(\mu\bar{g})^m = \bar{1}$. Then the elements $\bar{1}, \mu\bar{g}, \ldots, (\mu\bar{g})^{m-1}$ form an F-basis of $F^\alpha G$. Thus $F^\alpha G \cong FG$.

(ii) Let q be a prime and let S be a Sylow q-subgroup of G. It is a standard fact of cohomology theory (e.g. see M. Hall(1959)) that the q-component of $H^2(G, F^*)$ is isomorphic to a subgroup of $H^2(S, F^*)$. Hence, by Lemma 10.4(iii), we may assume that G is cyclic, in which case the result follows by virtue of (i). ∎

We now end this digression and return to projective representations defined below.

Let V be a finite-dimensional vector space over a field F. A map

$$\rho : G \to GL(V)$$

is called a *projective representation* of G over F if there exists a mapping $\alpha : G \times G \to F^*$ such that

$$\rho(x)\rho(y) = \alpha(x, y)\rho(xy) \qquad \text{for all} \quad x, y \in G \tag{1}$$

$$\rho(1) = 1_V \tag{2}$$

Thus an ordinary representation is a projective representation with $\alpha(x,y) = 1$ for all $x, y \in G$. If we identify $GL(V)$ with $GL(n, F)$, $n = dim_F V$, the resulting map is called a *projective matrix representation* of G over F. As in the case of ordinary representations, we shall treat the terms "projective representation" and "projective matrix representation" as interchangeable. In view of the associativity of the multiplication in G, conditions (1) and (2) imply that $\alpha \in Z^2(G, F^*)$. To stress the dependence of ρ on V and α, we shall often refer to ρ as an α-*representation* on the space V.

Two projective representations $\rho_i : G \to GL(V_i)$, $i = 1, 2$, are called *linearly equivalent* if there exists a vector space isomorphism $f : V_1 \to V_2$ such that

$$\rho_2(g) = f\rho_1(g)f^{-1} \tag{3}$$

for all $g \in G$.

The projective representation ρ on the space V is said to be *irreducible* if 0 and V are the only subspaces of V which are sent into themselves by all the transformations $\rho(g), g \in G$.

An α-representation ρ on the space V is *completely reducible* if for any subspace W invariant under all transformations $\rho(g), g \in G$, there exists another such subspace W' with $V = W \oplus W'$.

The following result shows that the study of α-representations is equivalent to the study of $F^\alpha G$-modules.

10.6. Lemma. *There is a bijective correspondence between α-representations of G and $F^\alpha G$-modules. This correspondence maps bijectively linearly equivalent (irreducible, completely reducible) representations onto isomorphic (simple, semisimple) modules.*

Proof. Let ρ be an α-representation of G on the space V. Then we can define a homomorphism $f : F^\alpha G \to End_F(V)$ by setting $f(\bar{g}) = \rho(g)$ and extending by linearity. Thus V becomes an $F^\alpha G$-module by setting

$$\left(\sum x_g \bar{g}\right) v = \sum x_g \rho(g) v \qquad (x_g \in F, g \in G, v \in V)$$

Conversely, given an $F^\alpha G$-module V, and hence a homomorphism $f : F^\alpha G \to End_F(V)$, define $\rho(g) = f(\bar{g})$. Then $\rho(g) \in GL(V)$ since $\bar{g}$ is a unit of $F^\alpha G$. Furthermore,

$$\rho(x)\rho(y) = f(\bar{x})f(\bar{y}) = f(\bar{x}\bar{y}) = f(\alpha(x,y)\overline{xy}) = \alpha(x,y)\rho(xy)$$

so that ρ is an α-representation on V. This sets up a bijective correspondence between α-representations and $F^\alpha G$-modules.

A subspace W of V is invariant under all $\rho(g)$ if and only if W is an $F^\alpha G$-submodule of V. Therefore the correspondence maps bijectively irreducible (completely reducible) representations onto simple (semisimple) modules.

Note that an F-isomorphism $f : V_1 \to V_2$ of $F^\alpha G$-modules is an $F^\alpha G$-isomorphism if and only if $\bar{g}f(v) = f(\bar{g}v)$ for all $g \in G, v \in V_1$. Assume that $\rho_i : G \to GL(V_i), i = 1, 2$, are two α-representations. Then ρ is linearly equivalent to ρ_2 if and only if there is an F-isomorphism $f : V_1 \to V_2$ such that $\rho_2(g)f = f\rho_1(g)$, for all $g \in G$. The latter is equivalent to

$$\rho_2(g)f(v) = f\rho_1(g)v$$

or to $\bar{g}f(v) = f(\bar{g}v)$, for all $g \in G, v \in V_1$. Thus two α-representations are linearly equivalent if and only if the corresponding modules are isomorphic. ∎

Let H be a subgroup of G. Given a cocycle $\alpha : G \times G \to F^*$ its restriction to $H \times H$ is also a cocycle. To prevent our expressions from becoming too cumbersome, from now on we shall use the same symbol for an element of $Z^2(G, F^*)$ and its restriction to $Z^2(H, F^*)$. With this convention, we may identify $F^\alpha H$ with the subalgebra of $F^\alpha G$ consisting of all F-linear combinations of the elements $\{\bar{h}|h \in H\}$.

If V is an $F^\alpha G$-module, then we shall denote by V_H the $F^\alpha H$-module obtained by the restriction of algebra. This process is called *restriction* and it permits us to go from any $F^\alpha G$-module V to a uniquely determined $F^\alpha H$-module V_H.

As in the case of modules over group algebras, there is a dual process of induction.

Let V be any $F^\alpha H$-module. Since we may consider $F^\alpha H$ as a subalgebra of $F^\alpha G$, we can define an $F^\alpha G$-module structure on the tensor product

$$F^\alpha G \otimes_{F^\alpha H} V$$

This is the *induced module* and we denote it by V^G.

Now assume that N is a normal subgroup of G and let U be an $F^\alpha N$-module. Then, for any $g \in G$, $\bar{g} \otimes U$ is an $F^\alpha N$-submodule of U^G. It is clear that

$$H = \{g \in G | \bar{g} \otimes U \cong U\}$$

is a subgroup of G containing N. We shall refer to H as the *inertia group* of U; in case $H = G$ we shall say that U is *G-invariant*.

10.7. Theorem. *(Clifford (1937)). Let V be a simple $F^\alpha G$-module and let U be a simple $F^\alpha N$-submodule of V_N. Let H and W denote, respectively, the inertia group of U and the sum of all submodules of V_N isomorphic to U.*

(i) $V_N \cong \oplus_{t \in T} \bar{t} \otimes U$
where T is a left transversal for H in G and the $\bar{t} \otimes U$, $t \in T$, are pairwise nonisomorphic simple $F^\alpha N$-modules. In particular, V_N is semisimple.

(ii) W is a simple $F^\alpha H$-module such that

$$W_N \cong eU \quad and \quad V \cong W^G$$

for some positive integer e.

Proof. First of all, it is obvious that $V = \sum_{g \in G} \bar{g}U$ and that $\bar{g}U$ is an $F^\alpha N$-module isomorphic to $\bar{g} \otimes U$. Therefore each $\bar{g}U$ is a simple $F^\alpha N$-module. Thus V_N is semisimple. It follows from the definition of $\bar{g} \otimes U$ that $\bar{x} \otimes (\bar{y} \otimes U) \cong \overline{xy} \otimes U$ for all $x, y \in G$ and so $\bar{x} \otimes U \cong \bar{y} \otimes U$ if and only if $xH = yH$. Setting $W_i = \sum_{g \in g_i H} \bar{g}U$, $1 \leq i \leq n$, we deduce from Proposition 1.3.6 that

$$V_N = W_1 \oplus \cdots \oplus W_n \qquad (W_1 = W)$$

and that

$$W_i \cong e_i(\bar{g}_i \otimes U) \qquad (1 \leq i \leq n)$$

for some positive integer e_i.

It is clear that W is an $F^\alpha H$-module. Now fix $g \in G, i \in \{1, \ldots, n\}$, and write $gg_i = g_j s$ for some $j \in \{1, \ldots, n\}$ and $s \in H$. Then

$$\bar{g}W_i = \bar{g}\left(\sum_{h \in H} \overline{g_i h}U\right) = \sum_{h \in H} \overline{g_i sh}U = W_j$$

and obviously $\bar{x}(\bar{y}W_i) = \overline{xy}W_i$ for all $x, y \in G$. Thus the group G acts on the set $\{W_1 = W, W_2, \ldots, W_n\}$ transitively and H is the stabilizer of W under this action. It follows that $dim_F W_i = dim_F W$ and hence that all the e_i are equal. The transitivity of action also implies that $V \cong W^G$. Finally, because V is simple, so is W and the result is established. ∎

Given $\alpha \in Z^2(G/N, F^*)$, we denote by $inf\,\alpha$ the element of $Z^2(G, F^*)$ defined by

$$(inf\,\alpha)(x, y) = \alpha(xN, yN) \qquad \text{for all} \quad x, y \in G$$

10.8. Theorem. *(Mackey (1958)). Let F be an algebraically closed field, let $\alpha \in Z^2(G, F^*)$ and let V be a simple G-invariant $F^\alpha N$-module. Then there exists a cocycle $\omega = \omega_G(V) \in Z^2(G/N, F^*)$ such that for $\beta = \alpha inf(\omega)$, V can be extended to an $F^\beta G$-module.*

Proof.   Given $g \in G$, define ^gV to be the $F^\alpha N$-module whose underlying space is V and on which the elements $\bar{n}, n \in N$, act according to the rule

$$\bar{n} * v = (\bar{g}^{-1}\bar{n}\bar{g})v \qquad \text{for all} \quad v \in V$$

Then $^gV \cong \bar{g} \otimes V$ and so, for any $g \in G$, there exists an $F^\alpha N$-isomorphism $f_g : {}^gV \to V$. Hence, for all $g \in G, n \in N$ and $v \in V$,

$$f_g(\bar{g}^{-1}\bar{n}\bar{g}v) = \bar{n}f_g(v) \tag{4}$$

or, equivalently,

$$\alpha(n, g)\alpha^{-1}(g, g^{-1}ng)f_g(\overline{g^{-1}ng}v) = \bar{n}f_g(v) \tag{5}$$

Now, for any $n \in N$, the map $^nV \to V, v \mapsto \bar{n}v$ is an $F^\alpha N$-isomorphism. Hence we may assume that

$$f_n(v) = \bar{n}v \qquad \text{for all} \quad n \in N, v \in V \tag{6}$$

It is an easy consequence of (4) that $f_x f_y f_{xy}^{-1} \in End_{F^\alpha N}(V)$, for all $x, y \in G$. Since F is algebraically closed, there exists $\beta : G \times G \to F^*$ such that $f_x f_y f_{xy}^{-1} = \beta(x,y) 1_V$ for all $x, y \in G$. It follows that the mapping $\rho : G \to GL(V)$ defined by $\rho(g) = f_g$ is a β-representation of G. Applying (6) and Lemma 10.6, it follows that V extends to an $F^\beta G$-module where the elements $\bar{g}, g \in G$ act accordingly to the rule $\bar{g}v = f_g(v), v \in V$.

By the foregoing, we are left to verify that β can always be chosen so as to be of the form $\alpha inf(\omega)$, for some $\omega \in Z^2(G/N, F^*)$. Invoking (5), we have

$$\alpha(n,g)\alpha^{-1}(g, g^{-1}ng)\bar{g}\overline{g^{-1}ng}v = \bar{n}\bar{g}v \quad (in\ F^\beta G)$$

whence, since $\bar{g}\overline{g^{-1}ng} = \beta(g, g^{-1}ng)\beta^{-1}(n,g)\bar{n}\bar{g}$

$$\alpha^{-1}(n,g)\beta(n,g) = \alpha^{-1}(g, g^{-1}ng)\beta(g, g^{-1}ng) \quad (n \in N, g \in G) \quad (7)$$

We now put $\lambda = \alpha^{-1}\beta$ so that $\lambda(x,y) = 1$ for all $x, y \in N$. Then, by (7),

$$\lambda(n,g) = \lambda(g, g^{-1}ng) \quad \text{for all} \quad n \in N, g \in G \quad (8)$$

Let Γ be the λ-representation of G afforded by W^G, where W is the trivial $F^\lambda N$-module. Then, by (8), $\Gamma(n) = 1$ for all $n \in N$ and therefore

$$\Gamma(g) = \lambda(n,g)\Gamma(ng) \quad \text{for all} \quad n \in N, g \in G \quad (9)$$

Let T denote a transversal for N in G with $1 \in T$ and, for any $g \in G$, let $c(g) \in T$ be such that $g \in c(g)N$. Let $L(g) = \Gamma(c(g))$ for all $g \in G$. Then, setting $\mu(g) = \lambda^{-1}(c(g)g^{-1}, g)$ we have, by (9),

$$L(g) = \Gamma(c(g)g^{-1}g) = \mu(g)\Gamma(g)$$

so L is a γ-representation of G where

$$\gamma = (\delta\mu)\lambda = \alpha^{-1}(\beta(\delta\mu)) \quad (10)$$

Since L is constant on the cosets of N, it follows that $\gamma = inf(\omega)$ for some $\omega \in Z^2(G/N, F^*)$. Observe also that $\mu(n) = 1$ for all $n \in N$. Hence, replacing f_g by $\mu(g)f_g$, we conclude from (10) that β can always be chosen to be of the desired form. ∎

We shall refer to $\omega = \omega_G(V)$ in Theorem 10.8 as the *obstruction* to the extension of V to $F^\alpha G$. As a preliminary to the next result, we need the following lemma.

10.9. Lemma. *Let A be a finite-dimensional algebra over an algebraically closed field F, let V be a simple A-module, and let $W = U \otimes_F V$ where U is some finite-dimensional vector space over F. For each $a \in A$, let $\varphi_a \in End_F(V)$ be defined by $\varphi_a(v) = av, v \in V$.*
(i) If $\theta \in End_F(W)$ is such that for all $a \in A$,

$$\theta(1 \otimes \varphi_a) = (1 \otimes \varphi_a)\theta,$$

then $\theta = \psi \otimes 1$ for some $\psi \in End_F(U)$.
(ii) If $\theta \in GL(W)$ and $\tau \in GL(V)$ are such that for all $a \in A$,

$$\theta^{-1}(1 \otimes \varphi_a)\theta = 1 \otimes \tau^{-1}\varphi_a\tau,$$

then $\theta = \psi \otimes \tau$ for some $\psi \in GL(U)$.

Proof. (i) Let $\{u_1, \ldots, u_m\}$ be an F-basis for U. Then, for all $v \in V$. we have

$$\theta(u_i \otimes v) = \sum_{j=1}^{m} u_j \otimes \theta_{ji}(v) \tag{11}$$

for some $\theta_{ji} \in End_F(V)$. For all $a \in A$, we have

$$
\begin{aligned}
\theta(1 \otimes \varphi_a)(u_i \otimes v) &= \theta(u_i \otimes \varphi_a(v)) \\
&= \sum_{j=1}^{m} u_j \otimes \theta_{ji}\varphi_a(v) \\
&= (1 \otimes \varphi_a)\theta(u_i \otimes v) \\
&= \sum_{j=1}^{m} u_j \otimes \varphi_a\theta_{ji}(v)
\end{aligned}
$$

Because the $\{u_j\}$ are linearly independent, we conclude that

$$\varphi_a\theta_{ji}(v) = \theta_{ji}\varphi_a(v) \quad (a \in A, v \in V, 1 \le i, j \le m)$$

or, equivalently, that $\theta_{ji} \in End_A(V)$. Since F is algebraically closed, we have $\theta_{ji} = \lambda_{ji} \cdot 1_V$ for some $\lambda_{ji} \in F$. Define $\psi \in End_F(U)$ by

$$\psi(u_i) = \sum_{j=1}^{m} \lambda_{ji}u_j \quad (1 \le i \le m)$$

Then $\psi \in GL(U)$ and, by (11), $\theta = \psi \otimes 1$, proving (i).

(ii) Put $\gamma = 1 \otimes \tau$. Then

$$\theta\gamma^{-1}(1 \otimes \varphi_a)\gamma\theta^{-1} = \theta(1 \otimes \tau^{-1}\varphi_a\tau)\theta^{-1} = 1 \otimes \varphi_a$$

and thus, by (i), $\theta\gamma^{-1} = \psi \otimes 1$ for some $\psi \in GL(U)$. Therefore

$$\theta = (\psi \otimes 1)\gamma = \varphi \otimes \tau$$

as asserted. ∎

Assume that $\rho_i : G \to GL(V_i), i = 1,2$, is an α_i-representation of G. Consider the map

$$\rho_1 \otimes \rho_2 : G \to GL(V_1 \otimes V_2)$$

given by

$$(\rho_1 \otimes \rho_2)(g) = \rho_1(g) \otimes \rho_2(g) \quad \text{for all} \quad g \in G$$

Then $\rho_1 \otimes \rho_2$ is obviously an $\alpha_1\alpha_2$-representation. We shall refer to $\rho_1 \otimes \rho_2$ as the *tensor product* of the projective representations ρ_1 and ρ_2. In module-theoretic language, the tensor product can be defined as follows. Let V and W be $F^\alpha G$ and $F^\beta G$-modules, respectively. Then the vector space $V \otimes_F W$ is an $F^{\alpha\beta}G$-module where the action of the elements $\bar{g}, g \in G$, is given by

$$\bar{g}(v \otimes w) = \bar{g}v \otimes \bar{g}w \quad (v \in V, w \in W)$$

and then extended to $V \otimes_F W$ and $F^{\alpha\beta}G$ by F-linearity.

Let N be a normal subgroup of G, let $\alpha \in Z^2(G/N, F^*)$ and let V be an $F^\alpha(G/N)$-module. Then one can form an $F^{inf(\alpha)}G$-module $inf(V)$ whose underlying space is V on which the elements $\bar{g}, g \in G$, act according to the rule:

$$\bar{g}v = \overline{gN}v \quad (v \in V)$$

We shall refer to $inf(V)$ as being *inflated* from V.

10.10. Theorem. *Let F be an algebraically closed field, let $\alpha \in Z^2(G, F^*)$ and let M be a simple FG-module. Let V be a simple submodule of M_N, let H be the inertia group of V, let $\omega = \omega_H(V) \in$*

$Z^2(H/N, F^*)$ be an obstruction to the extension of V to $F^\alpha H$ and let $ext(V)$ be any extension of V to $F^{\alpha inf(\omega)} H$ (by Theorem 10.8, such an extension always exists). Then there exists a simple $F^{\omega^{-1}}(H/N)$-module U such that

$$M \cong (inf(U) \otimes ext(V))^G$$

Proof. By Theorem 10.7(ii), we may assume that $H = G$, in which case V is G-invariant. By Theorem 10.7(i), M_N is a direct sum of copies of V. We may therefore assume that $M = W \otimes V$, where W is an F-space such that for all $n \in N, w \in W$ and $v \in V$

$$\bar{n}(w \otimes v) = w \otimes \bar{n}v \qquad (12)$$

Let $\{\tilde{g}|g \in G\}$ and $\{\hat{g}|g \in G\}$ denote F-bases of $F^\beta G$ and $F^\gamma G$, where $\gamma = inf(\omega^{-1}), \beta = \alpha inf(\omega)$ and

$$\tilde{x}\tilde{y} = \beta(x,y)\widetilde{xy}, \bar{n} = \tilde{n}$$

and

$$\hat{x}\hat{y} = \gamma(x,y)\widehat{xy}$$

for all $x, y \in G, n \in N$. Let

$$\eta : F^\alpha N \to End_F(V), \rho : F^\alpha G \to End_F(W), \tau : F^\beta G \to End_F(L)$$

be the homomorphisms determined by the modules V, W, and $L = ext(V)$. Then, by (12) and the equality $L_N = V$, we derive, respectively

$$\rho(\bar{n}) = 1 \otimes \eta(\bar{n}) \qquad (n \in N) \qquad (13)$$

$$\tau(\bar{n}) = \eta(\bar{n}) \qquad (n \in N) \qquad (14)$$

Since $(inf\omega)(g, n) = (inf\omega)(n, g) = 1$ for all $n \in N, g \in G$, and since $\beta = \alpha inf\omega$, we have

$$\alpha(n, g)\alpha^{-1}(g, g^{-1}ng) = \beta(n, g)\beta^{-1}(g, g^{-1}ng)$$

Taking into account that

$$\bar{n}\bar{g} = \alpha(n,g)\overline{ng} = \alpha(n,g)\overline{g(g^{-1}ng)}$$
$$= \alpha(n,g)\alpha^{-1}(g, g^{-1}ng)\bar{g}\overline{g^{-1}ng}$$

we therefore deduce that $\bar{g}^{-1}\bar{n}\bar{g} = \tilde{g}^{-1}\bar{n}\tilde{g}$. Applying (14) and (13), we infer that for all $g \in G, n \in N$,

$$
\begin{aligned}
\rho(\bar{g})^{-1}(1 \otimes \eta(\bar{n}))\rho(\bar{g}) &= \rho(\bar{g})^{-1}\rho(\bar{n})\rho(\bar{g}) = \rho(\bar{g}^{-1}\bar{n}\bar{g}) \\
&= 1 \otimes \eta(\bar{g}^{-1}\bar{n}\bar{g}) = 1 \otimes \tau(\tilde{g}^{-1}\bar{n}\tilde{g}) \\
&= 1 \otimes \tau(\tilde{g})^{-1}\eta(\bar{n})\tau(\tilde{g})
\end{aligned}
$$

Hence, by Lemma 10.9(ii), for each $g \in G$ there exists $\psi(\hat{g}) \in GL(W)$ such that

$$
\rho(\bar{g}) = \psi(\hat{g}) \otimes \tau(\tilde{g})
$$

Setting $\hat{g}w = \psi(\hat{g})(w), w \in W$, it is now straightforward to verify that W becomes an $F^\gamma G$-module and consequently

$$
M \cong W \otimes ext(V)
$$

Furthermore, because M is simple, W must also be simple. Finally, bearing in mind that by (13), $\psi(\hat{n}) = 1$ for all $n \in N$, we deduce that $W \cong inf(U)$ for some simple $F^{\omega^{-1}}(G/N)$-module U. This completes the proof of the theorem. ∎

The following lemma will enable us to take full advantage of the results so far obtained.

10.11.   Lemma.   *Let G be a finite group, let F be a field and let $\alpha \in Z^2(G, F^*)$. Define $\psi : F^\alpha G \to F$ by $\psi(\sum x_g\bar{g}) = x_1$. Then $(F^\alpha G, \psi)$ is a symmetric F-algebra such that $\{\bar{g}|g \in G\}$ and $\{\bar{g}^{-1}|g \in G\}$ are ψ-dual F-bases of $F^\alpha G$.*

Proof.   It is clear that ψ is an F-linear map. Fix $x = \sum x_g\bar{g}$ and $y = \sum y_g\bar{g}$ in FG, $x_g, y_g \in F, g \in G$. Then we have

$$
\begin{aligned}
\psi(xy) &= \sum_{g \in G} x_g y_{g^{-1}}\alpha(g, g^{-1}) \\
&= \sum_{g \in G} y_g x_{g^{-1}}\alpha(g, g^{-1}) \\
&= \psi(yx)
\end{aligned}
$$

and

$$
\psi(\bar{g}^{-1}x) = x_g \qquad \text{for all} \quad g \in G
$$

In particular, if $\psi(F^\alpha Gx) = 0$, then $x = 0$ which shows that $(F^\alpha G, \psi)$ is a symmetric F-algebra. That the given F-bases are ψ-dual is a consequence of the definition of ψ. ∎

10.12. Proposition. *Let G be a finite group, let F be an algebraically closed field such that $char F \nmid |G|$ and let $\alpha \in Z^2(G, F^*)$. If V is a simple $F^\alpha G$-module, then $\dim_F V$ divides the order of G.*

Proof. By making a diagonal change of basis $\{\bar{g}|g \in G\}$ of $F^\alpha G$, we may repalce α by any cohomologous cocycle. Thus, by Lemma 10.4(ii), we may assume that α is of finite order m dividing $|G|$. Let ε be a primitive m-th root of unity and let $K = L(\varepsilon)$ where L is the prime subfield of F. Then $F^\alpha G \cong F \otimes_K A$, where A is a twisted group algebra of G over K. Thus we may harmlessly assume that F is a finite extension of L and F is a splitting field for $F^\alpha G$.

If $char F = 0$, then F is an algebraic number field and so the required assertion follows by virtue of Lemma 10.11 and Corollary 10.2. Note also that if Z is a central subgroup of G and ρ is an ordinary irreducible F-representation of G, then ρ determines an irreducible projective representation of G/Z of the same degree, hence $deg\rho$ divides $(G : Z)$. Since $char F \nmid |G|$, the same is also true for $char F \neq 0$ (see Curtis and Reiner (1962)).

Returning to the general case, let $G^* =< \varepsilon^i\bar{g}|g \in G, 1 \le i \le m >$. Then the map $f : G^* \to G$ defined by $f(\varepsilon^i\bar{g}) = g$ is a surjective homomorphism whose kernel is a central subgroup $< \varepsilon >$. Furthermore, if $\rho : G \to GL(V)$ is an α-representation of G afforded by V, then the map $\rho^* : G \to GL(V)$ given by

$$\rho^*(\varepsilon^i\bar{g}) = \varepsilon^i\rho(g) \quad \text{for all} \quad g \in G$$

is an irreducible representation of G^*. Since $deg\rho = deg\rho^*$, the result follows by the previous paragraph. ∎

The result above can be significantly improved. To this end, we first introduce the following terminology.

Let p be a prime. The group G is said to be *p-solvable* if the composition factors of G are either p-groups or p'-groups. It is an easy consequence of the definition that

(i) Any solvable group is p-solvable.

(ii) Any extension of a p-solvable group by a p-solvable group is p-solvable.

(iii) Subgroups and homomorphic images of p-solvable groups are p-solvable.

We are now ready to prove our main result.

10.13. Theorem. *(Karpilovsky (1985)). Let N be a normal subgroup of a finite group G, let F be an algebraically closed field of an arbitrary characteristic and let $\alpha \in Z^2(G, F^*)$. If $char F = p > 0$ divides $(G : N)$, assume that G/N is p-solvable. Then, for any simple $F^\alpha G$-module V, $dim_F V$ divides $(G : N)d$, where d is the dimension of a simple submodule of V_N.*

Proof. We first demonstrate that the result is true under either of the following hypotheses:

(i) $char F \nmid (G : N)$.

(ii) $char F = p > 0$ and G/N is a p-group.

Let W be a simple submodule of V_N, let H be the inertia group of W, and let $\omega \in Z^2(H/N, F^*)$ be the obstruction to the extension of W to $F^\alpha H$. Owing to Theorem 10.10, there exists a simple $F^{\omega^{-1}}(H/N)$-module S such that

$$V \cong (inf(S) \otimes ext(W))^G$$

where $inf(S)$ is the $F^{inf(\omega^{-1})}H$-module inflated from S and $ext(W)$ is an extension of W to $F^\beta H$ with $\beta = \alpha inf(\omega)$. Because

$$dim_F S = dim_F inf(S) \quad \text{and} \quad dim_F W = dim_F ext(W),$$

we have

$$dim_F V = (dim_F S)(dim_F W)(G : H)$$

In case (i), $dim_F S$ divides $(H : N)$, by Proposition 10.12. Hence $dim_F V$ divides $(G : N)dim_F W$. In case (ii), $F^{\omega^{-1}}(H/N) \cong F(H/N)$ by Corollary 10.5 and Lemma 10.3. Hence $F^{\omega^{-1}}(H/N)$ is a local ring and therefore $dim_F S = 1$. Thus $dim_F V$ again divides $(G : N)dim_F W$.

Turning to the general case, we use induction on $|G|$. So assume that the result is true for groups of lower order than G. By the above, we may assume that $char F = p > 0$ divides $(G : N)$, in which case G/N is p-solvable, by hypothesis. Because G/N is p-solvable, there exists a proper normal subgroup M of G containing N such that G/M is either p or p'-group.

Let W denote a simple submodule of V_M and let S be a simple submodule of W_N. Then S is clearly a simple submodule of V_N. By the foregoing,

$$dim_F V \quad \text{divides} \quad (G : M) dim_F W$$

and, by induction hypothesis,

$$dim_F W \quad \text{divides} \quad (M : N) dim_F S$$

Thus $dim_F V$ divides

$$(G : M)(M : N) dim_F S = (G : N) dim_F S$$

as required. ∎

10.14. Corollary.   *Let A be a abelian normal subgroup of a finite group G, let F be an algebraically closed field and let $\alpha \in Z^2(G, F^*)$ be such that the restriction of α to $A \times A$ is a coboundary. If $char F = p > 0$ divides $(G : A)$, assume that G is p-solvable. Then the dimensions of simple $F^\alpha G$-modules divide $(G : A)$.*

Proof.   Let V be a simple $F^\alpha G$-module and let d be the dimension of a simple submodule of V_A. By Corollary 10.5 and Lemma 10.3,

$$F^\alpha A \cong FA$$

Since A is abelian, we obviously have $d = 1$. The desired conclusion is now a consequence of Theorem 10.13. ∎

10.15. Corollary.   *Let A be a cyclic normal subgroup of G, let F be an algebraically closed field and let $\alpha \in Z^2(G, F^*)$. If $char F = p > 0$ divides $(G : A)$, assume that G is p-solvable. Then the dimensions of simple $F^\alpha G$-modules divide $(G : A)$.*

Proof. Apply Corollaries 10.5 and 10.14 together with Lemma 10.3. ∎

10.16. Corollary. *(Dade (1968), Swan (1963)). Let A be an abelian normal subgroup of a finite group G, let F be an algebraically closed field of characteristic $p > 0$ and let G be p-solvable. Then the dimensions of simple FG-modules divide $(G : A)$.*

Proof. Direct consequence of Corollary 10.14. ∎

11.   Külshammer's theorems

Throughout this section, A denotes a finite-dimensional algebra over a field F. We define $[A, A]$. the *commutator subspace* of A, to be the F-linear span of all *Lie* products:

$$[x, y] = xy - yx$$

with $x, y \in A$. As usual, if (A, ψ) is a Frobenius algebra and X is a subset of A, we put

$$^{\perp}X = \{a \in A | \psi(aX) = 0\}$$

and

$$X^{\perp} = \{a \in A | \psi(Xa) = 0\}$$

Our aim is to establish certain general results pertaining to symmetric algebras. These results will have a number of applications in block theory.

11.1. Lemma. *(i) If (A, ψ) is a Frobenius algebra, then for any subset X of A, $l(X)$ is the largest left ideal of A contained in $^{\perp}X$.*
(ii) If (A, ψ) is a symmetric algebra, then

$$Z(A) = [A, A]^{\perp}$$

Proof. (i) It is a consequence of the definition of $^{\perp}X$, then $l(X) \subseteq {}^{\perp}X$. Denote by L any left ideal of A contained in $^{\perp}X$. Then

$$\psi(ALX) = \psi(LX) = 0$$

so $LX = 0$ and thus $L \subseteq l(X)$, as asserted.

(ii) If $z \in A$, then $z \in [A, A]^{\perp}$ if and only if for all $x, y \in A$

$$\psi((xy - yx)z) = \psi(xyz) - \psi(yxz) = \psi(xyz) - \psi(xzy) = 0$$

or if and only if

$$\psi(A(yz - zy)) = 0 \quad \text{for all} \quad y \in A$$

Since the latter is equivalent to $yz - zy = 0$ for all $y \in A$, the result is verified. ∎

From now on, we assume that $char F = p > 0$ and put

$$T(A) = \{a \in A | a^{p^n} \in [A, a] \quad \text{for some} \quad n \geq 1\}$$

It is clear that $T(A)$ is an F-subspace of A. Furthermore, by Lemma 11.2(ii) below, we have

$$T(A) \supseteq [A, A]$$

11.2. Lemma. *Let A be an algebra over a field F of characteristic $p > 0$.*

(i) If $x_1, x_2, \ldots, x_m \in A$ and if $n > 0$ is a given integer, then

$$\left(\sum_{i=1}^{m} x_i \right)^{p^n} \equiv \sum_{i=1}^{m} x_i^{p^n} (mod[A, A])$$

(ii) If $x \in [A, A]$, then $x^p \in [A, A]$.

Proof. By induction, to prove (i) it clearly suffices to treat the case $m = 2$. We now claim that, given $x, y \in A$,

$$(x + y)^p = x^p + y^p + z \quad \text{for some} \quad z \in [A, A] \tag{1}$$

Indeed, expanding by the distribution law,

$$(x + y)^p = x^p + y^p + \sum x_1 x_2 \ldots x_p \tag{2}$$

where the sum is over all products $x_1 x_2 \ldots x_p$ of p terms, $x_i \in \{x, y\}$ not all equal to x or y. With each word $x_1 x_2 \ldots x_p$ associate its cyclic permutations

$$x_1 x_2 \ldots x_p, x_2 x_3 \ldots x_p, \ldots, x_p x_1 x_2 \ldots x_{p-1}$$

All these products are congruent modulo $[A, A]$. This is so since for

$$a = x_{i_1} x_{i_2} \ldots x_{i_p}, b = x_{i_j} x_{i_{j+1}} \ldots x_{i_p} x_{i_1} \ldots x_{i_{j-1}}$$

we have $a - b = [c, d] \in [A, A]$, where

$$c = x_{i_1} x_{i_2} \ldots x_{i_{j-1}} \quad \text{and} \quad d = x_{i_j} x_{i_{j+1}} \ldots x_{i_p}$$

It follows that the sum of these cyclic permutations is $p x_1 \ldots x_p$ modulo $[A, A]$, and hence it belongs to $[A, A]$. Applying (2), (1) is therefore established.

By (1), we have

$$
\begin{aligned}
(xy - yx)^p &\equiv (xy)^p - (yx)^p (mod[A, A]) \\
&\equiv [x, (yx)^{p-1} y] (mod[A, A])
\end{aligned}
$$

proving (ii). Property (i) now follows from (1) and (ii) by induction on n. ∎

We are now ready to prove the following important result.

11.3. Theorem. *(Külshammer (1981)). Let A be a finite-dimensional algebra over a field F of characteristic $p > 0$.*
  (i) $T(A) = [A, A] + J(A)$ and $J(A)$ is the largest left ideal of A contained in $T(A)$.
  (ii) If A is a Frobenius algebra, then $Soc(A) = T(A)^\perp A$.
  (iii) If A is a symmetric algebra, then
  (a) $T(A)^\perp = Z(A) \cap Soc(A)$ and, in particular, $T(A)^\perp$ is an ideal of $Z(A)$.
  (b) $Soc(A) = (Z(A)) \cap Soc(A))A$.
  (c) $(T(A)^\perp)^2 = Z(B_1) \oplus \cdots \oplus Z(B_r)$, where $B_1, \ldots, B_r$ are all blocks of A which are simple algebras.

Proof. (i) We first treat the case where A is simple. Because for any $z \in Z(A)$ and $x, y \in A$, we have

$$(xy - yx)z = x(yz) - (yz)x \in [A, A],$$

both $[A, A]$ and $T(A)$ may be viewed as algebras over the field $Z(A)$. Thus we may assume that $Z(A) = F$, i.e. that A is a central simple F-algebra. Denote by E the algebraic closure of F. Then

$$A_E = E \otimes_F A \cong M_n(E)$$

for some $n \geq 1$. It is clear that

$$T(A_E) = [A_E, A_E]$$

and

$$dim_E T(A_E) = dim_E A_E - 1 \qquad (3)$$

Taking into account that

$$dim_F[A, A] = dim_E[A_E, A_E]$$

and

$$dim_F A = dim_E A_E,$$

we deduce that $dim_F[A, A] = dim_F A - 1$. If it were true that $A = T(A)$, then since $T(A_E)$ is an E-subspace of A_E, we would have $A_E = T(A_E)$, contrary to (3). Thus $A \neq T(A)$ and, since $T(A) \supseteq [A, A]$ by Lemma 11.2(ii), we must have $T(A) = [A, A]$. Since A is simple, (A, ψ) is a symmetric algebra for some $\psi \in Hom_F(A, F)$ (Proposition 1.9). But then

$$[A, A] \subseteq Ker\psi$$

and therefore $T(A)$ contains no nonzero left ideals of A. This proves the case where A is a simple algebra.

Turning to the general case, put $\bar{A} = A/J(A)$ and write

$$\bar{A} = A_1 \oplus \cdots \oplus A_m$$

where the A_i are pairwise orthogonal simple F-algebras. By the special case proved above, we have $T(A_i) = [A_i, A_i], 1 \leq i \leq m$. This implies that

$$
\begin{aligned}
([A, A] + J(A))/J(A) &= [\bar{A}, \bar{A}] \\
&= [A_1, A_1] \oplus \cdots \oplus [A_m, A_m] \\
&= T(A_1) \oplus \cdots \oplus T(A_m) \\
&= T(\bar{A})
\end{aligned}
$$

Since $[A, A] + J(A) \subseteq T(A)$ and $T(A)/J(A) \subseteq T(\bar{A})$, we have $T(\bar{A}) = T(A)/J(A)$ and $T(A) = [A, A] + J(A)$.

Finally, let L be a left ideal of A contained in $T(A)$. Then $(L + J(A))/J(A)$ is a left ideal of $\bar{A}$ contained in $T(\bar{A})$ and hence in $[\bar{A}, \bar{A}]$. But $\bar{A}$ is semisimple, so $\bar{A}$ is symmetric (Proposition 1.9) and thus, by the previous argument, $[\bar{A}, \bar{A}]$ contains no nonzero left ideals of $\bar{A}$. Consequently, $L \subseteq J(A)$ and the required assertion follows.

(ii) Assume that A is a Frobenius algebra. Then, by Theorem 6.1(i), we have

$$^{\perp}(T(A)^{\perp}) = T(A) \tag{4}$$

while, by Theorem 6.1(iii),

$$r(l(T(A)^{\perp}A)) = T(A)^{\perp}A \tag{5}$$

Owing to (i) and (4), $J(A)$ is the largest left ideal of A contained in $^{\perp}(T(A)^{\perp})$. Applying Lemma 11.1(i), we conclude that

$$J(A) = l(T(A)^{\perp}) = l(T(A)^{\perp}A) \tag{6}$$

Hence, by (5) and (6), we must have

$$Soc(A) = r(J(A)) = T(A)^{\perp}A,$$

as desired.

(iii) Assume that A is a symmetric algebra. Owing to (ii) and Lemma 11.1(ii), to prove (a), we need only show that

$$T(A)^{\perp} = [A, A]^{\perp} \cap T(A)^{\perp}A$$

Now $l(T(A)^{\perp}A)$ is a left ideal of A and so

$$(l(T(A)^{\perp}A))^{\perp} = r(l(T(A)^{\perp}A)) \tag{7}$$

by Theorem 6.1(ii). Thus

$$
\begin{aligned}
T(A)^{\perp} &= [A, A]^{\perp} \cap J(A)^{\perp} \quad \text{(by (i))} \\
&= [A, A]^{\perp} \cap [l(T(A)^{\perp}A)]^{\perp} \quad \text{(by (6))} \\
&= [A, A]^{\perp} \cap r(l(T(A)^{\perp}A)) \quad \text{(by (7))} \\
&= [A, A]^{\perp} \cap T(A)^{\perp}A, \quad \text{(by (5))}
\end{aligned}
$$

proving (a). Property (b) being a consequence of (a) and (ii), we are left to verify (c).

To prove (c), we may harmlessly assume that A is indecomposable, in which case it suffices to show that

$$(T(A)^\perp)^2 = \begin{cases} 0 & \text{if } J(A) \neq 0 \\ Z(A) & \text{if } J(A) = 0 \end{cases}$$

Assume that $J(A) = 0$. Then $1 \in Soc(A) \cap Z(A) = T(A)^\perp$. Because $T(A)^\perp$ is an ideal of $Z(A)$, we have $T(A)^\perp = Z(A)$ and hence

$$(T(A)^\perp)^2 = Z(A)$$

Now suppose that $J(A) \neq 0$. Then $1 \notin Soc(A)$ and hence $1 \notin T(A)^\perp$, by virtue of (a). Taking into account that $Z(A)$ is a local algebra (Proposition 7.7(iv)), we have

$$T(A)^\perp \subseteq J(Z(A)) \subseteq J(A)$$

However, by (a), $T(A)^\perp \subseteq Soc(A)$, so

$$(T(A)^\perp)^2 \subseteq J(A)Soc(A) = 0,$$

as we wished to show. ∎

11.4. Corollary. *Let A be a symmetric algebra over a field F of characteristic $p > 0$. If F is a splitting field for A, then the number of blocks of A that are simple algebras is equal to $dim_F(T(A)^\perp)^2$.*

Proof. Since F is a splitting field for A, if B is a block of A which is a simple algebra, then $dim_F Z(B) = 1$. Now apply Theorem 11.3(iii)(c). ∎

Assume that A is an algebra over a field F of characteristic $p > 0$. Then the maps

$$\begin{cases} A/[A, A] & \to & A/[A, A] \\ a + [A, A] & \mapsto & a^p + [A, A] \end{cases}$$

and

$$\begin{cases} Z(A) & \to & Z(A) \\ a & \mapsto & a^p \end{cases}$$

preserve addition, but need not be F-linear. However, if F is perfect, then these maps are "almost linear" in the sense defined below.

Let V and W be finite-dimensional vector spaces over a field F and let θ be an automorphism of F. A map

$$\rho : V \to W$$

is said to be F-*semilinear* with respect to θ if

$$\rho(x + y) = \rho(x) + \rho(y)$$

and

$$\rho(\lambda x) = \theta(\lambda)\rho(x)$$

for all $x, y \in V$ and $\lambda \in F$. Hence any F-linear map is an F-semilinear map with respect to the identity automorphism of F. Note that, as in the case of linear maps, $Ker\rho$ and $Im\rho$ are subspaces of V and W, respectively, and

$$dim_F V = dim_F Ker\rho + dim_F Im\rho$$

11.5. Lemma. *Let V, W, V' and W' be vector spaces over F, let $\rho : V \to W$ be an F-semilinear map with respect to $\theta \in AutF$, and let*

$$f : V \times V' \to F \quad g : W \times W' \to F$$

be nonsingular F-bilinear forms.

(i) For any $x \in W'$, there exists a unique $\rho^*(x) \in V'$ such that

$$g(\rho(y), x) = \theta f(y, \rho^*(x)) \quad \text{for all} \quad y \in V$$

(ii) The map $\rho^* : W' \to V'$, called the dual of ρ, is semilinear with respect to θ^{-1}. Moreover,

$$Ker\rho^* = \{x \in W' | g(Im\rho, x) = 0\}$$

and

$$Im\rho^* = \{y \in V' | f(Ker\rho, y) = 0\}$$

(iii) For any F-semilinear map $\pi : U \to V$ with respect to $\eta \in AutF$, the map $\rho \circ \pi$ is F-semilinear with respect to $\theta \circ \eta$. Moreover, if $U \times U' \to F$ is a nonsingular bilinear form, then $(\rho \circ \pi)^* = \pi^* \circ \rho^*$.

Proof. (i) Put $V^* = \mathrm{Hom}_F(V, F)$ and, for any given $x \in W'$, define $\psi_x : V \to F$ by

$$\psi_x(y) = \theta^{-1} g(\rho(y), x) \qquad \text{for all} \quad y \in V$$

We claim that $\psi_x \in V^*$. Indeed, given $v_1, v_2 \in V$, we have

$$
\begin{aligned}
\psi_x(v_1 + v_2) &= \theta^{-1} g(\rho(v_1 + v_2), x) \\
&= \theta^{-1}[g(\rho(v_1), x) + g(\rho(v_2), x)] \\
&= \theta^{-1} g(\rho(v_1), x) + \theta^{-1} g(\rho(v_2), x) \\
&= \psi_x(v_1) + \psi_x(v_2)
\end{aligned}
$$

Also, if $\lambda \in F$ and $v \in V$, we have

$$
\begin{aligned}
\psi_x(\lambda v) &= \theta^{-1} g(\rho(\lambda v), x) = \theta^{-1} g(\theta(\lambda)\rho(v), x) \\
&= \theta^{-1}[\theta(\lambda) g(\rho(v), x)] = \lambda \theta^{-1} g(\rho(v), x) \\
&= \lambda \psi_x(v),
\end{aligned}
$$

as claimed.

For any given $v \in V'$, we now define $\varphi_v(x) = f(x, v)$ for all $x \in V$. Then the map

$$\begin{cases} V' & \to & V^* \\ v & \mapsto & \varphi_v \end{cases}$$

is an F-isomorphism. This shows that $\psi_x = \varphi_{\rho^*(x)}$ for a unique $\rho^*(x) \in V'$. Hence, for all $y \in V$,

$$\psi_x(y) = \theta^{-1} g(\rho(y), x) = \varphi_{\rho^*(x)}(y) = f(y, \rho^*(x)),$$

proving (i).

(ii) The map ρ^* is clearly additive. If $x \in V'$ and $\lambda \in F$, then for all $y \in V$,

$$
\begin{aligned}
\theta f(y, \rho^*(\lambda x)) &= g(\rho(y), \lambda x) = g(\lambda \rho(y), x) = g(\rho(\theta^{-1}(\lambda)y, x) \\
&= \theta f(\theta^{-1}(\lambda)y, \rho^*(x)) \\
&= \theta f(y, \theta^{-1}(\lambda)\rho^*(x)),
\end{aligned}
$$

which shows that $\rho^*(\lambda x) = \theta^{-1}(\lambda)\rho^*(x)$. Consequently, ρ^* is semilinear with respect to θ^{-1}.

Since $\rho^*(x) = 0$ if and only if $g(\rho(y), x) = 0$ for all $y \in V$, the assertion regarding $Ker\rho^*$ follows.

It is clear that if $z \in Im\rho^*$, then $f(Ker\rho, z) = 0$. Conversely, suppose that $z \in V'$ is such that

$$f(Ker\rho, z) = 0$$

Then φ_z annihilates $Ker\rho$ and, by the definition of ψ_x, each ψ_x also annihilates $Ker\rho$.

Consider the F-linear map $\tau : W' \to V^*$ defined by $\tau(x) = \psi_x$. We claim that $Im\tau$ contains all elements of V^* annihilating $Ker\rho$; if sustained, it will follow that $\varphi_z = \psi_x$ for some $x \in W'$, hence the result. Consequently, it suffices to show that $dim_F Im\tau = dim_F(V/Ker\rho)^*$ or that

$$dim_F Im\tau = dim_F Im\rho$$

By the nature of $Ker\rho^*$, we have $Ker\tau = Ker\rho^*$ and

$$dim_F Ker\rho^* = dim_F W - dim_F Im\rho$$

Hence

$$
\begin{aligned}
dim_F Im\tau &= dim_F W' - dim_F Ker\rho^* \\
&= dim_F W - dim_F Ker\rho^* \\
&= dim_F Im\rho,
\end{aligned}
$$

as required.

(iii) This is straightforward. ∎

Again, assume that A is a finite-dimensional algebra over a field F of characteristic $p > 0$. For any given subset X of A, we denote by FX the F-linear span of X. Given a nonnegative integer n, put

$$
\begin{aligned}
T_n(A) &= \{a \in A \,|\, a^{p^n} \in [A, A]\} \\
A^{p^n} &= \{a^{p^n} \,|\, a \in A\} \\
P_n(A) &= FA^{p^n} + [A, A]
\end{aligned}
$$

Then $T_n(A)$ is an F-subspace of A containing $[A, A]$ and

$$T(A) = \cup_{n \geq 0} T_n(A)$$

From now on, we assume that F is a *perfect field* of characteristic $p > 0$. Since F is perfect, for any $\lambda \in F$, there exists a unique $\lambda^{p^{-n}} \in F$ such that

$$(\lambda^{p^{-n}})^{p^n} = \lambda$$

Moreover, the maps $\lambda \mapsto \lambda^{p^n}$ and $\lambda \mapsto \lambda^{p^{-n}}$ are automorphisms of F inverse to each other. We now provide an important example of a semilinear map.

11.6. Lemma. *The map*

$$\begin{cases} A/[A, A] & \xrightarrow{\rho} & A/[A, A] \\ a + [A, A] & \mapsto & a^{p^n} + [A, A] \end{cases}$$

is F-semilinear with respect to the automorphism $\lambda \mapsto \lambda^{p^n}$ of F. Moreover,

$$Ker\rho = T_n(A)/[A, A]$$

and

$$Im\rho = P_n(A)/[A, A]$$

Proof. Suppose that $a_1 + [A, A] = a_2 + [A, A]$ for some $a_1 a_2 \in A$. Then $a_1 - a_2 \in [A, A]$ and so, by Lemma 11.2(ii), $(a_1 - a_2)^{p^n} \in [A, A]$. Since, by Lemma 11.2(i),

$$(a_1 - a_2)^{p^n} \equiv a_1^{p^n} - a_2^{p^n} (mod[A, A]),$$

it follows that ρ is well defined. It is clear that ρ preserves addition. If $a \in A$ and $\lambda \in F$, then

$$\begin{aligned} \rho(\lambda a + [A, A]) &= (\lambda a)^{p^n} + [A, A] \\ &= \lambda^{p^n}(a^{p^n} + [A, A]) \\ &= \lambda^{p^n} \rho(a + [A, A]) \end{aligned}$$

which shows that ρ is F-semilinear with respect to $\lambda \mapsto \lambda^{p^n}$. The remaining assertion is a consequence of the definition of ρ. ∎

We have now come to the demonstration of the following result.

11.7. Theorem.   *(Külshammer (1985)). Let (A, ψ) be a symmetric algebra over a perfect field F of characteristic $p > 0$. Then, for any integer $n \geq 0$, there exists an F-semilinear map $\varphi_n : Z(A) \to Z(A)$ with respect to the automorphism $\lambda \mapsto \lambda^{p^{-n}}$ of F. The maps φ_n satisfy the following properties for all integers $n, m \geq 0$:*
  (i) $\psi(a^{p^n} z) = [\psi(a\varphi_n(z))]^{p^n}$ for all $a \in A, z \in Z(A)$.
  (ii) $Ker\varphi_n = P_n(A)^{\perp}$ and $Im\varphi_n = T_n(A)^{\perp}$.
  (iii) $\varphi_n \circ \varphi_m = \varphi_{n+m}$.
  (iv) $\varphi_n(y^{p^n} z) = y\varphi_n(z)$ for all $y, z \in Z(A)$.

Proof.   We know, from Corollary 1.2, that the map

$$A \times A \to F, (x, y) \mapsto \psi(xy)$$

is a nonsingular bilinear form on A. By Lemma 11.1(ii), $Z(A) = [A, A]^{\perp}$. It therefore follows that ψ induces a nonsingular bilinear form

$$\begin{cases} A/[A, A] \times Z(A) \overset{\psi_*}{\to} F \\ (a + [A, A], z) \mapsto \psi(az) \end{cases}$$

Now, by Lemma 11.6, the map

$$\begin{cases} A/[A, A] \overset{\rho}{\to} A/[A, A] \\ a + [A, A] \mapsto a^{p^n} + [A, A] \end{cases}$$

is F-semilinear with respect to the automorphism $\lambda \mapsto \lambda^{p^n}$ of F and

$$Ker\rho = T_n(A)/[A, A], Im\rho = P_n(A)/[A, A]$$

Applying Lemma 11.5 (with $V = W = A/[A, A], V' = W' = Z(A)$, $f = g = \psi_*$ and $\rho^* = \psi_n$) we conclude that φ_n satisfies all properties except for possibly (iv).

To establish (iv), we apply (i) twice to conclude that for all $a \in A$,

$$[\psi(a\varphi_n(y^{p^n} z))]^{p^n} = \psi(a^{p^n} y^{p^n} z) = \psi((ay)^{p^n} z) = [\psi(ay\varphi_n(z))]^{p^n}$$

Hence, for all $a \in A$,

$$\psi[a(\varphi_n(y^{p^n} z) - y\varphi_n(z))] = 0$$

and therefore $\varphi_n(y^{p^n}z) = y\varphi_n(z)$, as required. ∎

11.8. Theorem. *(Külshammer (1985)). Let (A, ψ) be a symmetric algebra over a perfect field F of characteristic $p > 0$. Then, for any integer $n \geq 0$, there exists an F-semilinear map*

$$\psi_n : A/[A, A] \to A/[A, A]$$

with respect to the automorphism $\lambda \mapsto \lambda^{p^{-n}}$ of F. The map ψ_n satisfies the following properties for all integers $n, m \geq 0$:
   (i) $\psi(az^{p^n}) = [f(\psi_n(a + [A, A]), z)]^{p^n}$ for all $a \in A, z \in Z(A)$
where $f : A/[A, A] \times Z(A) \to F$ is the associated F-bilinear form.
   (ii) $Ker\psi_n = P_n(Z(A))^{\perp}/[A, A]$.
   (iii) $Im\psi_n = T_n(Z(A))^{\perp}/[A, A]$.
   (iv) $\psi_n \circ \psi_m = \psi_{n+m}$
   (v) $\psi_n(az^{p^n} + [A, A]) = \psi_n(az + [A, A])$ for all $a \in A, z \in Z(A)$.
   (vi) $\psi_n(a^{p^n}z + [A, A]) = a\varphi_n(z) + [A, A]$ for all $a \in A, z \in Z(A)$.

Proof. For any integer $n \geq 0$, the map $Z(A) \to Z(A), z \mapsto z^{p^n}$ is F-semilinear with respect to the automorphism $\lambda \mapsto \lambda^{p^n}$ of F. Applying Lemma 11.5, we infer that there exists an F-semilinear map

$$\psi_n : A/[A, A] \to A/[A, A]$$

with respect to the automorphism $\lambda \mapsto \lambda^{p^{-n}}$ of F such that ψ_n satisfies
(i) - (iv).
   Given $a \in A$ and $y, z \in Z(A)$, we have

$$
\begin{aligned}
[f(\psi_n(az^{p^n} + [A, A]), y]^{p^n} &= \psi(az^{p^n}y^{p^n}) \quad \text{(by (i))} \\
&= [f(\psi_n(a + [A, A]), zy)]^{p^n} \quad \text{(by (i))} \\
&= [f(\psi_n(az + [A, A]), y)]^{p^n}
\end{aligned}
$$

where the last equality follows from associativity of f, proving (v).
   Finally, for all $a \in A$ and $y, z \in Z(A)$, we also have

$$
\begin{aligned}
[f(\psi_n(a^{p^n}z + [A, A]), y]^{p^n} &= \psi(a^{p^n}zy^{p^n}) \quad \text{(by (i))} \\
&= \psi((ya)^{p^n}z) \\
&= [\psi(ya\varphi_n(z))]^{p^n} \quad \text{(by Theorem 11.7(i))} \\
&= [f(a\varphi_n(z) + [A, A], y)]^{p^n},
\end{aligned}
$$

proving (vi). This completes the proof of the theorem. ∎

Our final aim is to provide an explicit description of φ_n and ψ_n in the case where $A = FG$, G a finite group. We need some preliminary information, in which G denotes a finite group and F a field of characteristic $p > 0$.

Let H be a subgroup of G. Then the natural projection

$$\pi : FG \to FH$$

defined by

$$\pi \left(\sum_{g \in G} x_g g \right) = \sum_{h \in H} x_h h$$

is obviously an F-linear map. The following result of Brauer plays an important role in modular representation theory. In what follows, for each subset X of G, we define $X^+ \in FG$ by

$$X^+ = \sum_{x \in X} x$$

11.9. Theorem. *Let P be a p-subgroup of G, let $S = C_G(P)$ and let H be a subgroup of G for which $S \subseteq H \subseteq N_G(P)$. Then the natural projection*

$$\pi : FG \to FS$$

induces a ring homomorphism of $Z(FG)$ into $Z(FH)$ whose kernel is the F-linear span of all C^+, where C is a conjugacy class of G such that $C \cap S = \emptyset$.

Proof. Let $C_1, \ldots, C_r$ be all conjugacy classes of G, let $X_i = C_i \cap S$ and suppose that $X_k \neq \emptyset$ for $k \in \{1, \ldots, t\}$ and $X_k = \emptyset$ for $k \in \{t+1, \ldots, r\}$. Note that the elements $C_1^+, C_2^+, \ldots, C_r^+$ form an F-basis for $Z(FG)$. By definition of π, we also have

$$\pi \left(\sum_{i=1}^{r} \lambda_i C_i^+ \right) = \sum_{i=1}^{t} \lambda_i X_i^+$$

where $X_1, \ldots, X_t$ are mutually disjoint and hence linearly independent. Since each X_i, $1 \leq i \leq t$, is a union of conjugacy classes of H, we deduce that π restricts to an F-linear map from $Z(FG)$ to $Z(FH)$ whose kernel is of the required form.

By the foregoing, it suffices to show that

$$\pi(C_i^+ C_j^+) = \pi(C_i^+)\pi(C_j^+)$$

for all $i,j \in \{1,\ldots,r\}$. Given $s \in S$, put

$$T_s = \{(x,y)|x \in C_i, y \in C_j \quad \text{and} \quad xy = s\}$$

and let $T = \cup_{s \in S} T_s$. If $T = \emptyset$, then either $X_i = \emptyset$ or $X_j = \emptyset$, in which case

$$\pi(C_i^+ C_j^+) = 0 = \pi(C_i^+)\pi(C_j^+)$$

Suppose that $T \neq \emptyset$. Then

$$\pi(C_i^+ C_j^+) = \sum_{(x,y)\in T} xy = \sum_{s\in S}\sum_{(x,y)\in T_s} xy$$

Because for all $h \in P$ and $s \in S$, $h^{-1}sh = s$ it follows that P acts as a permutation group on T_s via

$$(x,y)^h = (h^{-1}xh, h^{-1}yh)$$

Since $char F = p$, the sum of all elements xy, where (x,y) ranges all orbits of T_s of size $\neq 1$, is equal to 0. On the other hand, an orbit of T_s has size 1 if and only if it is of the form $\{(x,y)\}$ with $x,y \in C_G(P) = S$. The conclusion is that

$$\pi(C_i^+ C_j^+) = \sum_{(x,y)\in X_i \times X_j} xy = \pi(C_i^+)\pi(C_j^+),$$

as asserted. ∎

We shall refer to the homomorphism

$$Z(FG) \to Z(FH)$$

constructed in Theorem 11.9 as the *Brauer homomorphism*.

Let G denote a finite group and F a field. Define a map

$$tr : FG \to F$$

called the *trace*, by

$$tr(\sum_{g \in G} x_g g) = x_1$$

Note that, by Lemma 10.11, (FG, tr) is a symmetric F-algebra.

For any subset X of G and any integer $n \geq 0$, put

$$X^{p^{-n}} = \{g \in G | g^{p^n} \in X\}$$

where p is a prime. If $g \in G$, we say that g is a *p-element* if g has order equal to a power of p; we say that g is a *p'-element* (or is *p-regular*) if its order is prime to p. Each $g \in G$ can be written in a unique way $g = g_p g_{p'}$, where g_p is a p-element, $g_{p'}$ is a p'-element, and g_p and $g_{p'}$ commute. The elements g_p and $g_{p'}$ are called the p and p'-parts of g, respectively. If $x \in G$ is p-regular, we write $x^{p^{-n}}$ for a unique p-regular element $g \in G$ with $g^{p^n} = x$. Finally, in what follows φ_n and ψ_n are the maps given by Theorems 11.7 and 11.8, respectively, with respect to $A = FG$ and $\psi = tr$.

11.10. Theorem. *(Külshammer (1985)). Let G be a finite group and let F be a perfect field of characteristic $p > 0$. Then, for any integer $n \geq 0$, $g \in G$ and any conjugacy class C of G, we have*

(i) $\varphi_n(C^+) = (C^{p^{-n}})^+$.

(ii) $\psi_n(g + [FG, FG]) = [\{g_p\}^{p^{-n}}]^+ g_{p'}^{p^{-n}} + [FG, FG]$

Proof. Let $\pi_g : FG \to FC_G(g_p)$ be the natural projection and let

$$\varphi'_n : Z(FC_G(g_p)) \to Z(FC_G(g_p))$$

be the counterpart of φ_n with respect to the group algebra $FC_G(g_p)$. We first show that for all $c \in Z(FG)$,

$$\psi_n(gc + [FG, FG]) = \varphi'_n(g_p \pi_p(c) g_{p'}^{p^{-n}} + [FG, FG] \tag{8}$$

To this end, we first note that π_g is a homomorphism of left $FC_G(g_p)$-modules and that π_g restricts to the Brauer homomorphism:

$$Z(FG) \to Z(FC_G(g_p))$$

Let f be as in Theorem 11.8(i) with respect to $\psi = tr$. Taking into account that $g \in C_G(g_p)$, it follows from Theorems 11.7 and 11.8 that for all $z \in Z(FG)$,

$$
\begin{aligned}
[f(\psi_n(gc + [FG, FG]), z)]^{p^n} &= tr(gcz^{p^n}) = tr(\pi_g(gcz^{p^n})) \\
&= tr(g\pi_g(c)\pi_g(z)^{p^n}) \\
&= tr(g_p g_{p'} \pi_g(c)\pi_g(z)^{p^n}) \\
&= tr([g_{p'}^{p^{-n}} \pi_g(z)]^{p^n}(g_p \pi_g(c))] \\
&= \{tr[(g_p^{p^{-n}} \pi_g(z)\varphi'_n(g_p \pi_g(c)))]\}^{p^n} \\
&= \{tr[\varphi'_n(g_p \pi_g(c)g_{p'}^{p^{-n}} z]\}^{p^n} \\
&= [f(\varphi'_n(g_p \pi_g(c)g_{p'}^{p^{-n}} + [FG, FG]), z]^{p^n}
\end{aligned}
$$

Since ψ_n is uniquely determined by the equality (i) of Theorem 11.8, (8) is established.

Owing to Theorem 11.7(i), we have

$$
\begin{aligned}
[tr(g\varphi_n(C^+))]^{p^n} &= tr(g^{p^n} C^+) \\
&= tr(g(C^{p^{-n}})^+) \\
&= [tr(gC^{p^{-n}})^+]^{p^n}
\end{aligned}
$$

which clearly implies (i).

To prove (ii), we put $c = 1$ in (8) and apply (i) for $C_G(g_p)$. Since

$$
\{g_p\}^{p^{-n}} \subseteq C_G(g_p)
$$

we have

$$
\begin{aligned}
\psi_n(g + [FG, FG]) &= \varphi'_n(g_p)g_{p'}^{p^{-n}} + [FG, FG] \\
&= [\{g_p\}^{p^{-n}}]^+ g_{p'}^{p^{-n}} + [FG, FG],
\end{aligned}
$$

as desired. ∎

12. Applications

Our aim here is to provide a number of applications of the results proved in the previous section. In what follows, G denotes a finite group, $Cl(G)$

the set of all conjugacy classes of G and F a field of characteristic $p > 0$. For any subset X of G, we put

$$X^{-1} = \{x^{-1} | x \in X\}$$

and

$$X^+ = \sum_{x \in X} x \in FG$$

and denote by FX the F-linear span of X.

Let $C_1 = \{1\}, C_2, \ldots, C_r$ be all p-regular classes of G. By a p-regular section of G associated with C_i, we understand the set

$$S_i = \{g \in G | g_{p'} \in C_i\}$$

It is an easy consequence of the definition that

(i) $G = \cup_{i=1}^{r} S_i$ with $S_i \cap S_j = \emptyset$ for $i \neq j$.

(ii) Each S_i is a union of conjugacy classes of G and S_1 is a union of conjugacy classes of p-elements of G.

We begin by describing the commutator subspace of FG.

12.1. Lemma. *We have*

$$[FG, FG] = \{\sum x_g g \in FG | \sum_{g \in C} x_g = 0 \quad for\ all \quad C \in Cl(G)\}$$

Proof.   Given $C = \{g_1, \ldots, g_n\} \in Cl(G)$, assume that $x = \sum_{i=1}^{n} \lambda_i g_i$, where $\lambda_i \in F$ and $\sum_{i=1}^{n} \lambda_i = 0$. Then $x = \sum_{i=1}^{n-1} \lambda_i (g_i - g_n)$ and so in view of the identity

$$g - tgt^{-1} = (gt^{-1})t - t(gt^{-1}) \qquad (t, g \in G)$$

we have $x \in [FG, FG]$. Because $[FG, FG]$ is spanned by all $xy - yx$ with $x, y \in G$ and

$$xy - yx = xy - x^{-1}(xy)x,$$

the result follows. ∎

As a preliminary to the next auxiliary result, recall that the trace map

$$tr : FG \to F$$

is defined by

$$tr(\sum x_g g) = x_1 \qquad (x_g \in F, g \in G)$$

By Lemma 10.11, (FG, tr) is a symmetric F-algebra.

12.2. Lemma. *Let X be a subset of G and let $u = \sum u_g g \in FG$.*
(i) $tr(uX^+) = \sum_{g \in X^{-1}} u_g$ and, in particular, $tr(uS_i^+) = \sum_{g \in S_i^{-1}} u_g$.
(ii) $l(X^+) = \{\sum_{g \in G} u_g g \in FG | \sum_{g \in X^{-1}} u_{tg} = 0 \text{ for all } t \in G\}$

Proof. (i) By the definition of the product of elements in FG, we have

$$uX^+ = \sum_{t \in G} (\sum_{g \in X} u_{tg^{-1}}) t$$

It follows, from the definition of the trace map, that

$$tr(uX^+) = \sum_{g \in X} u_{g^{-1}} = \sum_{g \in X^{-1}} u_g,$$

proving (i).

(ii) Note that $uX^+ = 0$ if and only if for all $t \in G$,

$$\sum_{g \in X} u_{tg^{-1}} = \sum_{g \in X^{-1}} u_{tg} = 0,$$

as required. ∎

Recall that $T(FG)$ is defined by

$$T(FG) = \{a \in FG | a^{p^n} \in [FG, FG] \quad \text{for some} \quad n \geq 1\}$$

A description of $T(FG)$ and $T(FG)^{\perp}$ is provided by the following lemma.

12.3. Lemma. *Let $S_1, S_2, \ldots, S_r$ be all p-regular sections of G. Then*
(i) $T(FG) = \{\sum_{g \in G} x_g g \in FG | \sum_{g \in S_i} x_g = 0 \text{ for all } i \in \{1, \ldots, r\}\}$.
(ii) $T(FG)^{\perp} = \sum_{i=1}^{r} FS_i^+$.

Proof. (i) Let us show first that there exists $n \geq 1$ such that

$$g^{p^m} = g_{p'} \quad \text{for all} \quad m \equiv 0 (mod\, n) \quad \text{and all} \quad g \in G \qquad (1)$$

Indeed, write $|G| = p^a k$ with $(p, k) = 1$. Since $(p, k) = 1$, there exists $n \geq 1$ such that $p^n \equiv 1 (mod\, k)$. Replacing n by its multiple, if necessary, we may assume that $n \geq a$. Assume that m is divisible by n and let $g \in G$. Then $g_p^{p^m} = 1$ and, since $p^m \equiv 1 (mod\, k), g_{p'}^{p^m} = g_{p'}$. Hence

$$g^{p^m} = g_p^{p^m} g_{p'}^{p^m} = g_{p'},$$

proving (1).

Now fix $x = \sum_{g \in G} x_g g \in FG$. Then, by (1), we may choose $m \geq 1$ such that $g^{p^m} = g_{p'}$ for all $g \in G$ and such that $x \in T(FG)$ if and only if $x^{p^m} \in [FG, FG]$. Applying Lemma 11.2(i), we have

$$x^{p^m} \equiv \sum_{g \in G} x_g^{p^m} g^{p^m} (mod[FG, FG])$$

and therefore $x \in T(FG)$ if and only if

$$\sum_{g \in G} x_g^{p^m} g_{p'} \in [FG, FG]$$

It follows, from Lemma 12.1, that $x \in T(FG)$ if and only if

$$\sum_{g \in S_i} x_g^{p^m} = 0 \quad \text{for all} \quad i \in \{1, \ldots, r\}$$

Bearing in mind that

$$\sum_{g \in S_i} x_g^{p^m} = \left(\sum_{g \in S_i} x_g \right)^{p^m},$$

the required assertion follows.

(ii) Owing to (i) and Lemma 12.2(i), $\sum_{i=1}^r FS_i^+ \subseteq T(FG)^\perp$. Conversely, assume that $x = \sum_{g \in G} x_g g \in T(FG)^\perp$ and let $a, b \in S_i$. Then a^{-1} and b^{-1} belong to the p-regular section S_i^{-1} and thus, by (i), $a^{-1} - b^{-1} \in T(FG)$. But then

$$tr(xa^{-1} - xb^{-1}) = 0$$

which implies that

$$x_a = tr(xa^{-1}) = tr(xb^{-1}) = x_b,$$

proving (ii). ∎

We are now ready to prove the following result.

12.4. Theorem. *(Brauer (1955), Tsushima (1978)). Let F be a field of characteristic $p > 0$ and let $S_1, S_2, \ldots, S_r$ be all p-regular sections of G. Then*

$$Soc(FG) = \sum_{i=1}^{r} FG \cdot S_i^+$$

and

$$J(FG) = \cap_{i=1}^{r} l(S_i^+)$$

Proof. By Theorem 11.3,

$$Soc(FG) = FG \cdot T(FG)^{\perp}$$

Applying Lemma 12.3(ii), it follows that $Soc(FG) = \sum_{i=1}^{r} FG \cdot S_i^+$. Hence, by Proposition 1.4.22 and Theorem 6.3(iii),

$$
\begin{aligned}
J(FG) &= l(Soc(FG)) \\
&= l(\sum_{i=1}^{r} FG \cdot S_i^+) \\
&= \cap_{i=1}^{r} l(FG \cdot S_i^+) \\
&= \cap_{i=1}^{r} l(S_i^+),
\end{aligned}
$$

as required. ∎

12.5. Corollary. *Let G be a finite group, let F be a field of characteristic $p > 0$ and let $S_1, S_2, \ldots, S_r$ be all p-regular sections of G. Then*

$$J(FG) = \{\sum_{g \in G} x_g g \in FG \mid \sum_{s \in S_i} x_{gs} = 0 \quad \text{for all} \quad g \in G, i \in \{1, \ldots, r\}\}$$

Proof. This is a direct consequence of Theorem 12.4, Lemma 12.2(ii) and the fact that $S_i \mapsto S_i^{-1}$ is a permutation of the set $\{S_1, \ldots, S_r\}$. ∎

Our next aim is to provide a characterization of $J(Z(FG))$. We need a number of preliminary results.

12.6. Proposition. *(Osima (1955)).* *Let G be a finite group, let F be a field of characteristic $p > 0$ and let e be a nonzero central idempotent of FG. Then $Supp\, e$ consists of p'-elements.*

Proof. (Passman (1969)). Assume by way of contradiction that $z \in Supp\, e$ with $z = xy = yx$ where $x \neq 1$ is a p-element and y is a p'-element. Denote by P the subgroup of G generated by x. Then, in the notation of Theorem 11.9, $\pi(e)$ is a central idempotent of FS where $S = C_G(P)$. Moreover, since $z \in S$, we also have $z \in Supp\, \pi(e)$. Hence we may harmlessly assume that $x \in Z(G)$. Applying Lemma 12.1, we deduce that

$$x \notin Supp\, v \quad \text{for all} \quad v \in [FG, FG] \tag{2}$$

Let q be the order of y and choose an integer n with $p^n \geq |G|$ and with $p^n \equiv 1 (mod\, q)$. Put $t = y^{-1}e$ and write $t = \sum_{g \in G} t_g g$, $t_g \in F$. Then, by Lemma 11.2(i),

$$t^{p^n} \equiv \sum t_g^{p^n} g^{p^n} \quad (\text{mod } [FG, FG])$$

Since g^{p^n} is a p'-element, it follows from (2) that $x \notin Supp\, t^{p^n}$. On the other hand, because e is a central idempotent and $p^n \equiv 1 (mod\, q)$, we have

$$t^{p^n} = y^{-p^n} e = y^{-1} e = t$$

However, by the definition of t, $x \in Supp\, t$, a contradiction. ∎

12.7. Lemma. *Let e be any idempotent in an ideal I of a finite-dimensional algebra A over a field. Then the decompositions of e into primitive idempotents of A and of I are the same.*

Proof. Let $e = \sum_{i=1}^n e_i$, where the e_i are orthogonal primitive idempotents of A. Then, for all $j \in \{1, \ldots, n\}$,

$$ee_j = \sum_{i=1}^n e_i e_j = e_j^2 = e_j$$

so that $e_j \in I$, as required. ∎

12.8. Lemma. *Let I be a nilpotent left ideal of a ring R, and let $x \in R$ be a nonnilpotent element such that $x^2 - x \in I$. Then the left ideal Rx contains a nonzero idempotent y such that $y - x \in I$.*

Proof. Choose $k \geq 1$ with $I^k = 0$ and put $z_1 = x^2 - x$. If $z_1 = 0$, choose $y = x$, and we are done. If $z_1 \neq 0$, let $x_1 = x + z_1 - 2xz_1 \in Rx$. Then x, x_1 and z_1 commute with each other, and hence if x_1 is nilpotent, then so is $x = x_1 - z_1 + 2xz_1$, a contradiction. Thus x_1 is a nonnilpotent element of Rx, and a direct calculation shows that

$$x_1^2 - x_1 = 4z_1^3 - 3z_1^2$$

The element $z_2 = x_1^2 - x_1$ is nilpotent, contains z_1^2 as factor, and commutes with x_1. Continiung in this manner, we may construct a sequence $\{x_i\}$ of nonnilpotent elements of Rx such that $z_1^{2^i}$ occurs as a factor in $x_i^2 - x_i$. If we choose i such that $2^i \geq k$, we then have $x_i^2 - x_i = 0$. Furthermore, $x_i \neq 0$ since x_i is nonnilpotent. Hence $y = x_i$ is the desired idempotent. ∎

12.9. Lemma. *Let I be a nonnilpotent left ideal in a (left) artinian ring R. Then I contains a nonzero idempotent.*

Proof. Let X be a minimal member of the nonempty set of nonnilpotent left ideals contained in I. Then $X^2 \subseteq X$ and so $X^2 = X$. Let Y be minimal in the set of the left ideals contained in X such that $XY \neq 0$ (this set is nonempty since it contains X). Choose $a \in Y$ so that $Xa \neq 0$. Then $X \cdot Xa \neq 0$ and $Xa \subseteq Y$, whence $Y = Xa$. Thus $a = xa$ for some $x \in X$ and so $a = x^k a$ for all $k \geq 0$. It follows that X contains a nonnilpotent element x. Setting

$$L = \{b \in X \,|\, ba = 0\}$$

we have $x^2 - x \in L$. Moreover, $Xa = Y \neq 0$ so $L \subset X$ and thus L is nilpotent. Applying Lemma 12.8, we see that Rx contains a nonzero idempotent, hence so does I, since $I \supseteq X \supseteq Rx$. ∎

12.10. Lemma. *Let A be a finite-dimensional algebra over a field. Then*

$$J(Z(A)) = Z(A) \cap J(A)$$

Proof. If $x \in J(Z(A))$, then z is nilpotent and hence $I = Az$ is a nilpotent ideal of A. Thus $z \in I \subseteq J(A)$ and therefore

$$J(Z(A)) \subseteq Z(A) \cap J(A)$$

Conversely, if $z \in Z(A) \cap J(A)$, then z is nilpotent and hence $z \in J(Z(A))$, as required. ∎

We have now accumulated all the information necessary to prove the following result.

12.11. Theorem. *(Iizuka-Watanabe (1973)). Let G be a finite group, let F be a field of characteristic $p > 0$ and let c be the sum of all p-elements of G (including 1). Then*

$$J(Z(FG)) = Z(FG) \cap l(c)$$

Proof. (Külshammer (1981)). Let $R(FG)$ be the F-subspace of FG defined by

$$R(FG) = \{a \in FG \,|\, tr(a^{p^n}) = 0 \quad \text{for all sufficiently large} \quad n \geq 1\}$$

Denote by X and Y the sets of all p-regular elements (excluding 1) and all p-elements (including 1) of G, respectively. Because for all $y \in Y$, $y - 1$ is nilpotent, we have

$$R(FG) \subseteq \sum_{x \in X} Fx + \sum_{y \in Y} F(y - 1)$$

We claim that the reverse containment also holds. To verify the claim, it suffices to show that for $a \in R(FG)$ with $Supp\, a \subseteq Y$, we have $a \in \sum_{y \in Y} F(y - 1)$. But if $Supp\, a \subseteq Y$ and $a \in R(FG)$, then there exists $a_y \in F$ such that

$$a = \sum_{y \in Y} a_y(y - 1) + \sum_{y \in Y} a_y$$

and $\sum_{y \in Y} a_y \in R(FG)$, which is possible only in the case $a = \sum_{y \in Y} a_y(y - 1)$. Thus we have

$$R(FG) = \sum_{x \in X} Fx + \sum_{y \in Y} F(y - 1) \qquad (3)$$

It follows from (3) that $Fc \subseteq R(FG)^\perp$. Again, we claim that the reverse containment also holds. To substantiate the claim, suppose that

$$tr(aR(FG)) = 0$$

for some $a = \sum_{g \in G} a_g g \in FG$. Then, by (3), for all $x \in X, y \in Y$, we have $tr(ax) = 0$ and $tr(ay) = tra$. Hence, for all $x \in X, y \in Y, a_{x^{-1}} = 0$ and $a_1 = a_y$. This shows that $a = a_1 c$ and therefore

$$Fc = R(FG)^\perp \qquad (4)$$

We now claim that $R(FG)$ does not contain blocks of FG. Indeed, let $e = \sum e_g g$ be a block idempotent of FG and assume, by way of contradiction, that $FGe \subseteq R(FG)$. Owing to Proposition 12.6, there exists a p-regular element $g \in G$ such that $e_g \neq 0$. By assumption, $tr(eg^{-p^n}) = 0$ for all sufficiently large n. We may choose n so that $g^{-p^n} = g^{-1}$, in which case

$$0 = tr(eg^{-p^n}) = tr(eg^{-1}) = e_g,$$

a contradiction.

It is now an easy matter to complete the proof. By Lemma 12.10, $J(Z(FG)) = Z(FG) \cap J(FG)$. Thus, by Theorem 12.4,

$$J(Z(FG)) \subseteq l(c) \cap Z(FG)$$

To prove the reverse containment, it suffices to show that $l(c) \cap Z(FG)$ is a nilpotent ideal of $Z(FG)$.

Assume the contrary. Then, by Lemma 12.9, $l(c) \cap Z(FG)$ contains a primitive idempotent, say e. By Lemma 12.7, e is a block idempotent of FG and so $FGe \subseteq l(c)$. However, by (4) and Theorem 6.1(i),

$$l(c) = l(R(FG)^\perp) \subseteq {}^\perp(R(FG)^\perp) = R(FG)$$

which implies that $R(FG)$ contains the block FGe, a contradiction. ∎

12.12. Corollary. *Let G be a finite group and let F be a field of characteristic $p > 0$. Then*

$$J(Z(FG)) = \{\sum x_g g \in Z(FG)|\ \sum_{s \in G_p} x_{gs} = 0 \quad \text{for all} \quad g \in G\}$$

where G_p is the set of all p-elements of G including 1.

Proof. This is a direct consequence of Theorem 12.11 and Lemma 12.2(ii). ∎

As a preliminary to the next result, we record the following property.

12.13. Lemma. *Let A be a finite-dimensional algebra over a field F of characteristic $p > 0$, let F be a splitting field for A and let*

$$T(A) = \{a \in A | a^{p^n} \in [A, A] \quad \text{for some} \quad n \geq 1\}$$

Then the number of nonisomorphic irreducible A-modules is equal to $dim_F(A/T(A))$.

Proof. It is clear that $T(A)$ is an F-subspace of A. By Lemma 11.2(ii), we have $T(A) \supseteq [A, A]$. Since $J(A)$ is nilpotent, it follows from the definition of $T(A)$ that $J(A) \subseteq T(A)$.

Now put $\bar{A} = A/J(A)$ and $\bar{T} = T(A)/J(A)$. Denote by r the number of nonisomorphic irreducible A-modules. Since F is a splitting field for A, we have

$$\bar{A} \cong \prod_{i=1}^{r} M_{n_i}(F)$$

for some positive integers $n_1, \ldots, n_r$. Next put $\bar{S} = ([A, A]+J(A))/J(A)$. Then clearly $\bar{S} = [\bar{A}, \bar{A}]$ and $\bar{T} = T(\bar{A})$. Let $S_i = [A_i, A_i]$ and $T_i = T(A_i)$, $1 \leq i \leq r$, where $A_i = M_{n_i}(F)$. Then

$$dim_F(A/T(A)) = dim_F(\bar{A}/\bar{T}) = \sum_{i=1}^{r} dim_F(A_i/T_i)$$

and so we are left to show that

$$dim_F(A_i/T_i) = 1 \qquad (1 \leq i \leq r)$$

Let e_{st} denote the $n_i \times n_i$ matrix with (s,t)-th entry equal to 1 and all other entries 0. Then, for any $s \neq t$,

$$e_{st} = e_{sj}e_{jt} - e_{jt}e_{st} \in S_i$$

and

$$e_{ss} - e_{tt} = e_{st}e_{ts} - e_{ts}e_{st} \in S_i$$

Hence S_i contains the $n_i^2 - 1$ linearly independent elements $e_{st}, 1 \leq s, t \leq n_i, s \neq t$, and $e_{ss} - e_{11}(s = 2, \ldots, n_i)$. Since $S_i \subseteq T_i$, this implies that

$$dim_F(A_i/T_i) = dim_F A_i - dim_F T_i \leq n_i^2 - (n_i^2 - 1) = 1$$

On the other hand, the elements of S_i are all $n_i \times n_i$ matrices with trace 0. In particular, $e_{11} \notin S_i$. But $e_{11}^{p^m} = e_{11}$ for all $m \geq 1$, so $e_{11} \notin T_i$. Hence $T_i \neq A_i$ and therefore $dim_F(A_i/T_i) = 1$, as required. ∎

12.14. Theorem. *(Brauer (1935)). Let G be a finite group and let F be a field of characteristic $p > 0$ such that F is a splitting field for FG. Then the number of nonisomorphic irreducible FG-modules is equal to the number of p-regular classes of G.*

Proof. Let $S_1, \ldots, S_r$ be all p-regular sections of G. Then, by Lemma 12.3(i),

$$T(FG) = \{\sum x_g g \in FG | \sum_{g \in S_i} x_g = 0 \quad \text{for all} \quad i \in \{1, \ldots, r\}\} \qquad (5)$$

Define $f : FG \to F \times F \times \cdots \times F$ (r copies) as the F-linear map given by

$$f(\sum x_g g) = (\alpha_1, \alpha_2, \ldots, \alpha_r)$$

where

$$\alpha_i = \sum_{g \in S_i} x_g \qquad (1 \leq i \leq r)$$

Then f is clearly surjective and, by (5), $\ker f = T(FG)$. Hence

$$dim_F(FG/T(FG)) = r$$

and the result follows by virtue of Lemma 12.13. ∎

Let A be an algebra over a field and let $Soc(A)$ be the left socle of A. We shall refer to the ideal $Z(A) \cap Soc(A)$ of $Z(A)$ as the *Reynolds ideal* of A. The name is partially justified by the fact that Reynolds (1972) introduced this ideal for the special case where A is a group algebra. In what follows, we put $Rey(A) = Z(A) \cap Soc(A)$.

12.15. Theorem. *Let G be a finite group, let F be a field of characteristic $p > 0$ and let $S_1, \ldots, S_r$ be all p-regular sections of G.*
(i) $Rey(FG) = \sum_{i=1}^r FS_i^+$.
(ii) If F is algebraically closed, $V_1, \ldots, V_r$ are all nonisomorphic irreducible FG-modules and χ_i is the character of G afforded by $V_i, 1 \leq i \leq r$, then the elements

$$z_i = \sum_{g \in G} \chi_i(g^{-1})g \qquad (1 \leq i \leq r)$$

form an F-basis of $Rey(FG)$.

Proof. (i) Apply Lemma 12.3 and Theorem 11.3(iii).
(ii) By Theorem 12.14, there are precisely, r nonisomorphic irreducible FG-modules. Let χ be an F-character of G. We first show that

$$\chi(g) = \chi(g_{p'}) \qquad \text{for all} \quad g \in G \tag{6}$$

Fix $g \in G$. We may choose an integer $m \geq 1$ such that g^{p^m} is p-regular. In a representation of G with character χ, the linear transformation corresponding to g has eigenvalues $\varepsilon_1, \varepsilon_2, \ldots, \varepsilon_n$, say, in F and $\chi(g) = \varepsilon_1 + \cdots + \varepsilon_n$. Similarly, $\chi(g^{p^m}) = \varepsilon_1^{p^m} + \cdots + \varepsilon_n^{p^m}$. Since F has characteristic p, the binomial theorem shows that

$$\chi(g)^{p^m} = \varepsilon_1^{p^m} + \cdots + \varepsilon_n^{p^m} = \chi(g^{p^m})$$

Taking into account that $g^{p^m} = (g_{p'})^{p^m}$, we have

$$\chi(g)^{p^m} = \chi(g^{p^m}) = \chi((g_{p'})^{p^m}) = \chi(g_{p'})^{p^m}$$

and therefore

$$[\chi(g) - \chi(g_{p'})]^{p^m} = \chi(g)^{p^m} - \chi(g_{p'})^{p^m} = 0,$$

proving (6).

Owing to (6), each χ_i is constant on p-regular sections. Hence each z_i lies in $Rey(FG)$, by virtue of (i). Note also that, by (i),

$$dim_F Rey(FG) = r$$

and so we need only verify that $z_1, \ldots, z_r$ are F-linearly independent. If $\sum_{i=1}^{r} \lambda_i z_i = 0$ for some $\lambda_i \in F$, then

$$0 = \sum_{i=1}^{r} \sum_{g \in G} \lambda_i \chi_i(g^{-1})g = \sum_{g \in G} \left(\sum_{i=1}^{r} \lambda_i \chi_i(g^{-1}) \right) g$$

and therefore

$$\sum_{i=1}^{r} \lambda_i \chi_i(g) = 0 \quad \text{for all} \quad g \in G$$

Since $\chi_1, \ldots, \chi_r$ are F-linearly independent, all $\lambda_i = 0$ and the result follows. ∎

To illustrate significance of the Reynolds ideal, we need some auxiliary results.

12.16. Lemma. *Let A be a finite-dimensional algebra over a field F and let $e_1, \ldots, e_n$ be all block idempotents of A. Let $F_i = Z(A)e_i/J(Z(A)e_i)$, let $\theta_i : Z(A) \to F_i$ be the natural homomorphism and let $f_i : F_i \to M_{n_i}(F)$ be the regular representation of F_i. Define $\gamma_i : Z(A) \to M_{n_i}(F)$ by $\gamma_i = f_i \circ \theta_i$. Then*

$$Ker\gamma_i = Ker\theta_i = Z(A)(1 - e_i) + J(Z(A))e_i \qquad (1 \le i \le n)$$

and $\{\gamma_1, \ldots, \gamma_n\}$ is a complete set of nonequivalent irreducible matrix representations of $Z(A)$. Each γ_i satisfies $\gamma_i(e_i) = 1$ and $\gamma_i(e_j) = 0$ for $j \ne i$. Furthermore, if F is a splitting field for $Z(A)$, then each $n_i = 1$.

Proof. It is obvious that $Ker\gamma_i = Ker\theta_i$. Assume that $z \in Ker\theta_i$. Then $ze_i = xe_i$ for some $x \in J(Z(A))$, so

$$z = z(1 - e_i) + xe_i \in Z(A)(1 - e_i) + J(Z(A))e_i$$

and therefore $Ker\theta_i \subseteq Z(A)(1-e_i)+J(Z(A))e_i$. The opposite inclusion being obvious, we deduce that $Ker\theta_i = Z(A)(1 - e_i) + J(Z(A))e_i$.

It follows from Proposition 7.7(iii), (iv) that

$$Z(A)/J(Z(A)) = \oplus_{i=1}^{n} F_i$$

and that each F_i is a field. Furthermore, if F is a splitting field for $Z(A)$, then each F_i is isomorphic to F and hence $n_i = 1$. Since f_i is the only irreducible representation of F_i, the result follows. ∎

The representation $\gamma_i : Z(A) \to M_{n_i}(F)$ described in Lemma 12.16 is uniquely determined, up to equivalence, by e_i. We call γ_i the *irreducible representation of $Z(A)$ associated with e_i*.

12.17. Lemma. *Let A be a finite-dimensional algebra over a field, let I be an ideal of $Z(A)$ and let γ be the irreducible representation of $Z(A)$ associated with the block idempotents e of A. Then $e \in I$ if and only if $\gamma(I) \neq 0$.*

Proof. If $e \in I$, then $\gamma(e) = 1 \neq 0$ and hence $\gamma(I) \neq 0$. Conversely, assume that $\gamma(I) \neq 0$. Since $\gamma(I)$ is a nonzero ideal of the field $\gamma(Z(A))$, we have

$$\gamma(I) = \gamma(Z(A))$$

Thus $\gamma(e) \in \gamma(I)$. It therefore follows, from Lemma 12.16, that there are elements $x \in I$, $y \in J(Z(A))$ and $z \in Z(A)$ such that

$$x = e + z(1 - e) + y$$

Multiplying both sides by e gives $e = xe - ye$. Hence, by raising both sides to the k-th power, we find

$$e = u + (-1)^k (ye)^k$$

with $u \in I$. Since ye is nilpotent, the result follows. ∎

Let G be a finite group, let p be a prime and let C be a conjugacy class of G. If $g \in C$, then a Sylow p-subgroup of $C_G(g)$ is called a *defect group* of C (with respect to p). Thus all defect groups of C are conjugate and so have a common order, say p^d. The integer d is called the *defect* of C. We denote by $\delta(C)$ any defect group of C. If H_1 and H_2 are subgroups of G, we write $H_1 \subseteq_G H_2$ (respectively, $H_1 \subset_G H_2$) to indicate that H_1 is conjugate in G to a subgroup of H_2 (respectively, to a proper subgroup of H_2), while $H_1 =_G H_2$ will mean that H_1 and H_2 are conjugate in G. In what follows, F denotes a field of characteristic $p > 0$.

Let e be a block idempotent of FG and let $B = B(e)$. Since $e \in Z(FG)$, we have

$$Supp\, e = C_1 \cup C_2 \cup \ldots \cup C_t$$

for some conjugacy classes $C_1, \ldots, C_t$ of G. The largest of the defect groups of C_i, $1 \leq i \leq t$, denoted by $\delta(e)$ (or $\delta(B)$), is called a *defect group* of e (or of B). It will be shown (Proposition 12.19) that all defect groups of e are conjugate, say of order p^d. The integer d is called the *defect* of e (or of B).

12.18. Lemma. *Let D be a p-subgroup of G and let $C_1, C_2, \ldots, C_t$ be all conjugacy classes of G with $\delta(C_i) \subseteq_G D$. Then the F-linear span $I_D(G)$ of $C_1^+, C_2^+, \ldots, C_t^+$ is an ideal of $Z(FG)$.*

Proof. Fix $i \in \{1, \ldots, t\}$ and denote by C_j any conjugacy class of G. It clearly suffices to show that $C_i^+ C_j^+ \in I_D(G)$.

We may assume that $C_i^+ C_j^+ \neq 0$, in which case we may choose $g \in Supp\, C_i^+ C_j^+$. Bearing in mind that

$$C_i^+ C_j^+ = \sum_{(x,y) \in C_i \times C_j} xy$$

we have $g = uv$ for some $u \in C_i$ and $v \in C_j$. Let P be a defect group of the conjugacy class containing g and put

$$X = \{(x,y) | xy = g \quad \text{and} \quad (x,y) \in C_i \times C_j\}$$

The P acts on X by conjugation, so by the argument employed in the proof of Theorem 11.9, we have $C_i \cap C_G(P) \neq \emptyset$. Therefore $P \subseteq_G \delta(C_i)$ and thus $P \subseteq_G D$. Hence $C_i^+ C_j^+ \in I_D(G)$, as required. ∎

Let C be a conjugacy class of G. If D is a defect group of C, we shall write $I[C]$ instead of $I_D(G)$. The fact that, by Lemma 12.17, $I[C]$

is an ideal of $Z(FG)$ will be exployted in the proof of the following result.

12.19. Proposition. *Let e be a block idempotent of FG and let γ be the irreducible representation of $Z(FG)$ associated with e. Then there exists p-regular classes $C_1, C_2, \ldots, C_r$ of G and $t \leq r$ such that*

(i) $\operatorname{Supp} e = C_1 \cup C_2 \cup \ldots \cup C_t \cup C_{t+1} \cup \ldots \cup C_r$.

(ii) For all $i \in \{1, \ldots, t\}$, $e \in I[C_i]$ and $\delta(C_i) =_G \delta(C_j)$ for all $i, j \in \{1, \ldots, t\}$.

(iii) $\delta(C_k) \subset_G \delta(C_i)$ for all $k \in \{t+1, \ldots, r\}$ and $i \in \{1, \ldots, t\}$.

(iv) If C is a conjugacy class of G with $\delta(C) \subset_G \delta(C_i)$ for some $i \in \{1, \ldots, t\}$, then $\gamma(C^+) = 0$.

(v) There exists $i \in \{1, \ldots, t\}$ such that $\gamma(C_i^+) \neq 0$.

(vi) If C is a conjugacy class of G with $\gamma(C^+) \neq 0$, then $\delta(e) \subseteq_G \delta(C)$.

Proof. (i) - (iii): By Proposition 12.6, $e = \sum_{i=1}^r \lambda_i C_i^+$ for some nonzero λ_i in F and some p-regular classes $C_1, \ldots, C_r$ of G. Because $\gamma(1) = \gamma(e) \neq 0$, there exists $k \in \{1, \ldots, r\}$ such that $\gamma(C_k^+) \neq 0$. But $C_k^+ \in I[C_k]$, so we must have $\gamma(I[C_k]) \neq 0$. Hence, by Lemma 12.17, we have $e \in I[C_k]$. By renumbering $C_1, \ldots, C_r$ in such a way that $\{C_1, \ldots, C_t\}$ is the subset of $\{C_1, \ldots, C_r\}$ consisting of all C_i with $\delta(C_i) =_G \delta(C_k)$, we deduce that (i), (ii) and (iii) hold.

(iv) By the foregoing, $\delta(C) \subset_G \delta(C_k)$. Assume that $\gamma(C^+) \neq 0$. Since $C^+ \in I[C]$ we have $\gamma(I[C]) \neq 0$ and so, by Lemma 12.17, $e \in I[C]$. Hence $\delta(C_k) \subseteq_G \delta(C)$, a contradiction.

(v) This was established in (i).

(vi) By the argument of (iv), $e \in I[C]$. The desired assertion is now a consequence of the definition of $I[C]$. ∎

12.20. Lemma. *Let F be a field, let $n \geq 1$ and let $A = M_n(F)$. If $\varphi, \psi \in A^*$ are such that (A, φ) and (A, ψ) are symmetric algebras, then there exists nonzero λ in F such that $\varphi(a) = \lambda\psi(a)$ for all $a \in A$.*

Proof. Owing to Lemma 1.5, the maps

$$f : A \to A^* \quad \text{and} \quad g : A \to A^*$$

given by $f(a)(x) = \varphi(xa)$ and $g(a)(x) = \psi(xa)$, $x, a \in A$, are iso-morphisms of (A, A)-bimodules. Hence $g^{-1}f : A \to A$ is an (A, A)-bimodule automorphism. Let A° be the opposite ring to A so that A is an $A \otimes_F A^\circ$-module via

$$(a_1 \otimes a_2)a = a_1 a a_2 \qquad (a_1, a \in A, a_2 \in A^\circ)$$

Then $g^{-1}f$ is in fact an automorphism of $A \otimes_F A^\circ$-modules. Since $A \otimes_F A^\circ$ is a simple algebra with F as a splitting field and A is a simple $(A \otimes_F A^\circ)$-module, we deduce that there exists $0 \neq \lambda \in F$ such that $(g^{-1}f)(a) = \lambda a$ for all $a \in A$. Thus $f(a) = \lambda g(a)$ for all $a \in A$. In particular, for all $a \in A$, we have

$$\varphi(a) = f(a)(1) = \lambda g(a)(1) = \lambda \psi(a),$$

as required. ∎

Let p be a prime. A conjugacy class C of G is called *p-singular* if all elements of C have order divisible by p.

12.21. Lemma. *Let G be a finite group, let F be a field of char-acteristic $p > 0$ and let $e = \sum_{g \in G} e_g g$, $e_g \in F$, be an idempotent of FG.*
(i) For any p-singular conjugacy class C of G,

$$\sum_{g \in C} e_g = tr(eC^+) = 0$$

Furthermore, if $eFGe$ is simple, then $eC^+ = 0$.
(ii) If e is primitive and F is a splitting field for FG, then $\sum_{g \in C} e_g \neq 0$ for some p-regular conjugacy class C of G.

Proof. (i) Choose a positive integer n such that for all $g \in G$, g^{p^n} is a p'-element. Owing to Lemma 11.2(i), we have

$$e = e^{p^n} \equiv \sum_{g \in G} e_g^{p^n} g^{p^n} \quad (\text{mod } [FG, FG])$$

By our choice of n, for all $g \in C$, e_g is the coefficient of g in

$$e - \sum_{g \in G} e_g^{p^n} g^{p^n} \in [FG, FG]$$

Hence, by Lemma 12.1, we have $\sum_{g \in C} e_g = 0$. Since C^{-1} is p-singular, we also have $tr(e(C^+)) = \sum_{g \in C^{-1}} e_g = 0$.

Suppose that the algebra $eFGe$ is simple, say $eFGe \cong M_n(D)$ for a suitable division ring D. If E is a maximal subfield of D, then

$$eEGe \cong E \otimes_F M_n(D) \cong M_n(E)$$

We may thus assume that

$$eFGe = \sum_{1 \leq i,j \leq n} F e_{ij}$$

with matrix units e_{ij}. The map

$$\left\{ \begin{array}{ccc} eFGe & \overset{\psi}{\to} & F \\ \sum_{i,j} \lambda_{ij} e_{ij} & \mapsto & \sum_i \lambda_{ii} \end{array} \right.$$

is such that $(eFGe, \psi)$ is a symmetric algebra. On the other hand, by Lemma 4.2, $(eFGe, tr)$ is also a symmetric algebra. Thus, by Lemma 12.20, there exists $\lambda \in F$ such that for $c = \lambda e$, $tr(ac) = \psi(a)$ for all $a \in eFGe$. Since $eC^+ = eC^+e \in Z(eFGe)$, we have $eC^+ = \mu e$ for some $\mu \in F$. It follows that

$$\begin{aligned} \mu &= \psi(\mu e e_{11}) = \psi(eC^+ e_{11}) \\ &= tr(C^+ e_{11} c) = tr(\lambda C^+ e_{11}) \\ &= \lambda tr(e_{11} C^+) \end{aligned}$$

But e_{11} is an idempotent of FG, so $tr(e_{11} C^+) = 0$ and therefore $eC^+ = 0$, as required.

(ii) Assume that e is primitive and F is a splitting field for FG. By (i) and Lemma 12.1, it suffices to verify that $e \notin [FG, FG]$. Assume by way of contradiction that $e \in [FG, FG]$. Since the image of e in $FG/J(FG)$ is a primitive idempotent contained in the commutator subspace of $FG/J(FG)$, we may harmlessly assume that $J(FG) = 0$. Because F is a splitting field for FG, we have

$$FG \cong \prod_{i=1}^{k} M_{n_i}(F)$$

for some positive integers k and n_i. Write $e = (e_1, \ldots, e_k)$ with $e_i \in M_{n_i}(F)$, $1 \leq i \leq k$. Then, each $e_i \in [M_{n_i}(F), M_{n_i}(F)]$ and, for some $j \in \{1, \ldots, k\}$, e_j is a primitive idempotent of $M_{n_j}(F)$. But all primitive idempotents in $M_n(F)$ have trace 1, whereas the elements of $[M_{n_j}(F), M_{n_j}(F)]$ have trace zero, a contradiction. ■

We are now ready to provide our final result extracted from Okuyama (1980), Reynolds (1972) and Tsushima (1978).

12.22. Theorem. *Let G be a finite group and let F be a field of characteristic $p > 0$.*

(i) The primitive idempotents of $Rey(FG)$ are the same as the block idempotents of FG of defect zero. In particular, a block B of FG is of zero defect if and only if B is a simple F-algebra.

(ii) $Rey(FG)^2 = Z(B_1) \oplus \cdots \oplus Z(B_r)$.

where $B_1, B_2, \ldots, B_r$ are all blocks of FG of defect zero. In particular, if F is a splitting field for FG, then $\dim_F Rey(FG)^2$ is equal to the number of blocks of FG of defect zero.

(iii) If F is a splitting field for FG, then the number of nonisomorphic simple FG-modules in a block $B = B(e)$ of FG is equal to $\dim_{Fe} Rey(FG)$.

Proof. (i) Assume that e is a primitive idempotent of $Rey(FG)$. Then, by Lemma 12.7, e is a block idempotent of FG. Since $Supp\, e$ consists of p'-elements, it follows that each p-regular section in $Supp\, e$ is a conjugacy class of G. Thus $Supp\, e$ is a union of conjugacy classes of p-defect zero and therefore e is a block idempotent of defect zero.

Conversely, suppose that e is a block idempotent of defect zero. Then $Supp\, e$ is a union of conjugacy classes of p-defect zero. Since the latter classes are also p-regular sections, we conclude from Theorem 12.15(i) that $e \in Rey(FG)$.

Let $B = B(e)$ be a block of FG. Since $J(B) = J(FG)e$, we have $J(B) = 0$ if and only if $e \in Soc(FG) \cap Z(FG) = Rey(FG)$. Since $J(B) = 0$ if and only if B is a simple F-algebra, the desired assertion follows from the first statement.

(ii) Owing to Theorem 11.3(iii)

$$Rey(FG)^2 = Z(B_1) \oplus \cdots \oplus Z(B_r)$$

where $B_1, \ldots, B_r$ are all blocks of FG which are simple F-algebras. By (i), $B_1, \ldots, B_r$ are all blocks of FG of defect zero. Suppose that F is a splitting field for FG. Since each B_i is a block of $FG/J(FG)$, it follows that $B_i \cong M_{n_i}(F)$ for some $n_i \geq 1, 1 \leq i \leq r$. Hence $dim_F Z(B_i) = 1$, $1 \leq i \leq r$, and the required assertion follows.

(iii) Let F be a splitting field for FG and let f be a primitive idempotent of FG. We first show that, for any given $z \in Rey(FG)$,

$$fz = 0 \quad \text{if and only if} \quad tr(fz) = 0 \tag{7}$$

Indeed, let $z \in Rey(FG)$ be such that $tr(fz) = 0$. Given $y \in FG$, we have

$$fy = fyf + (ffy - fyf)$$

and so

$$fzFG \subseteq fzFGf + [FG, FG]$$

Since F is a splitting field for FG, we have

$$fFGf/fJ(FG)f \cong F$$

and thus

$$fFGf = Ff + fJ(FG)f$$

Applying Theorem 12.4, it follows that

$$\begin{aligned} fzFG &\subseteq Ffz + fJ(FG)zf + [FG, FG] \\ &= Ffz + [FG, FG] \end{aligned}$$

Thus $tr(fzFG) = 0$ and therefore $fz = 0$, proving (7).

Let $\bar{e}$ denote the image of e in $B/J(B)$ and write

$$\bar{e} = \varepsilon_1 + \varepsilon_2 + \cdots + \varepsilon_n$$

as a sum of block idempotents of $B/J(B)$. Then we may write

$$e = e_1 + e_2 + \cdots + e_n$$

where $e_1, \ldots, e_n$ are orthogonal idempotents of B such that $\bar{e}_i = \varepsilon_i$, $1 \le i \le n$. Since F is a splitting field for FG, B has exactly n nonisomorphic simple modules. We are therefore left to verify that

$$eRey(FG) = \oplus_{i=1}^n e_i Rey(FG) \tag{8}$$

and

$$dim_F e_i Rey(FG) = 1 \quad \text{for all} \quad i \in \{1, \ldots, n\} \tag{9}$$

To prove (8), we must show that $e_i Rey(FG) \subseteq eRey(FG)$, $1 \le i \le n$; hence it suffices to show that $e_i Rey(FG) \subseteq Rey(FG)$ for all $i \in \{1, \ldots, n\}$. So fix $u \in \{e_1, \ldots, e_n\}$ and a p-regular section S. Because $u + J(FG)$ is a central element of $FG/J(FG)$, we have $ux - xu \in J(FG)$ for all $x \in FG$. Hence, by Theorem 12.4, $S^+(ux - xu) = 0$ and so $S^+ u \in Z(FG)$. Write

$$S^+ u = \lambda_1 C_1^+ + \cdots + \lambda_k C_k^+ + \lambda_{k+1} C_{k+1}^+ + \cdots + \lambda_t C_t^+$$

where $\lambda_i \in F, C_1, \ldots, C_k$ are p-regular and $C_{k+1}, \ldots, C_t$ are p-singular classes. Then

$$S^+ u = \lambda_1 C_1^+ + \cdots + \lambda_k C_k^+ + \lambda_1 (C_1^+ - S_1^+) + \cdots + \lambda_k (C_H^+ - S_k^+)$$

$$+ \lambda_{k+1} C_{k+1}^+ + \cdots + \lambda_t C_t^+$$

and $Supp(C_i^+ - S_i^+)$ consists of p-singular classes for all $i \in \{1, \ldots, k\}$. Hence

$$S^+ u = x + y$$

for some $x \in Rey(FG)$ and some $y \in Z(FG)$ such that $Supp \, y$ consists of p-singular classes. By Lemma 12.21(i), for any primitive idempotent f in FG, $tr(fy) = 0$. Since f is primitive, we have either $fuS^+ = 0$ or $fuS^+ = fS^+$. Hence either $fy = -fx$ or $fy = f(S^+ - x)$ and thus, by (7), $fy = 0$. Since f is an arbitrary primitive idempotent of FG, we conclude that $y = 0$, proving (8).

To prove (9), let v be an idempotent of FG such that $v + J(FG)$ is a block idempotent of $FG/J(FG)$. Write

$$v = v_1 + v_2 + \cdots + v_n$$

where the v_i are mutually orthogonal primitive idempotents of FG. Then

$$v_1 FG \cong v_i FG \quad \text{for all} \quad i \in \{1,\ldots,n\}$$

It follows that $v_1 = x_i y_i$ and $v_i = y_i x_i$ for some $x_i \in v_1 FG$ and $y_i \in v_i FG$. Thus

$$v_i - v_1 \in [FG, FG] \quad \text{for all} \quad i \in \{1,\ldots,n\}$$

Consider the F-linear maps

$$\left\{ \begin{array}{ccc} Rey(FG) & \overset{\varphi}{\to} & v_1 Rey(FG) \\ x & \mapsto & v_1 x \end{array} \right.$$

and

$$\left\{ \begin{array}{ccc} Rey(FG) & \overset{\psi}{\to} & F \\ x & \mapsto & tr(v_1 x) \end{array} \right.$$

By (7), $Ker\varphi = Ker\psi$ and, by Lemma 12.21, $\psi(S^+) \neq 0$ for some p-regular section S. Thus

$$v_1 Rey(FG) = \{\lambda v_1 S^+ | \lambda \in F\}$$

Since $v_i - v_1 \in [FG, FG]$, we have

$$(v_i - v_1)x \in [FG, FG] \quad \text{for all} \quad x \in Rey(FG)$$

Hence $tr(v_i x) = tr(v_1 x)$ for all $i \in \{1,\ldots,n\}$ and $x \in Rey(FG)$. It follows that

$$v_i Rey(FG) = \{\lambda v_i S^+ | \lambda \in F\}$$

with $tr(v_i S^+) \neq 0$, $1 \leq i \leq n$. Given $x \in Rey(FG)$, write

$$v_i x = \lambda_i v_i S^+$$

with $\lambda_i \in F$. Then

$$\begin{aligned} \lambda_i tr(v_i S^+) &= tr(v_i x) = tr(v_1 x) \\ &= \lambda_1 tr(v_1 S^+) = \lambda_1 tr(v_i S^+), \end{aligned}$$

proving that $\lambda_i = \lambda_1$, $1 \leq i \leq n$. The conclusion is that

$$vx = \lambda_1 v S^+$$

and hence that $dim_F v Rey(FG) = 1$, proving (9). ∎

Chapter 3

Symmetric local algebras

The aim of this chapter is twofold: first to investigate symmetric local algebras A over an algebraically closed field F with $dim_F Z(A) \leq 5$ and second to apply the results obtained to modular representation theory of finite groups. More specifically, we first prove that if A is a symmetric local algebra with $dim_F Z(A) \leq 4$, then A is commutative. Second, we show that if A is a symmetric local algebra with $dim_F Z(A) = 5$, then $dim_F A = 5$ or $dim_F A = 8$. Now assume that $char F = p > 0$ and that B is a block of a group algebra FG with exactly one irreducible FG-module. Denote by $k(B)$ the number of irreducible complex characters in B and by D a defect group of B. As an application of our results on symmetric local algebras, we demonstrate that

    (i) If $k(B) \leq 4$, then $k(B) = |D|$.

    (ii) If $k(B) = 5$, then either $|D| = 5$ or D is nonabelian of order 8.

The results above are relevant to the Brauer's conjecture which states that for any B, $k(B) \leq |D|$.

1. Symmetric local algebras A with $dim_F Z(A) \leq 4$

Throughout A denotes a finite-dimensional algebra over a field F. If V is an A-module, we put

$$S_0(V) = 0, S_1(V) = Soc(V)$$

and define the i-th socle $S_i(V)$ of V by

$$S_i(V)/S_{i-1}(V) = Soc(V/S_{i-1}(V)) \quad (i \geq 2)$$

If $V =_A A$, then we put $S_i(A) = S_i(V)$ and refer to $S_i(A)$ as the i-th socle of A. For any ideal I of A, we also put $I^0 = A$. If (A, ψ) is a symmetric algebra over a field F and X is a subset of A, we put

$$X^\perp = \{a \in A | \psi(Xa) = 0\}$$

Recall from Theorem 2.6.1 that if X is a subspace of A, then

$$dim_F X^\perp = dim_F(A/X)$$

while if X is an ideal of A, then

$$X^\perp = l(X) = r(X)$$

Finally, if X is a subset of A, then FX denotes the F-linear span of X.

1.1. Lemma.   *(i) For all $i \geq 0$, $S_i(A) = r(J(A)^i)$.*
(ii) If A is a Frobenius algebra, then for all $i \geq 0$

$$dim_F(S_{i+1}(A)/S_i(A)) = dim_F J(A)^i / J(A)^{i+1}$$

Proof.   (i) Owing to Proposition 1.4.21, $Soc(A) = r(J(A))$. Assume by induction that

$$S_{i-1}(A) = r(J(A)^{i-1})$$

Since $S_i(A)/S_{i-1}(A)$ is a semisimple A-module, we have

$$J(A)S_i(A) \subseteq S_{i-1}(A)$$

and hence

$$J(A)^i S_i(A) \subseteq J(A)^{i-1} S_{i-1}(A) = 0$$

Thus $S_i(A) \subseteq r(J(A)^i)$.

Conversely, suppose that $J(A)^i x = 0$ for some $x \in A$. Then

$$J(A)x \subseteq r(J(A)^{i-1}) = S_{i-1}(A)$$

and therefore $(Ax + S_{i-1}(A))/S_{i-1}(A)$ is a semisimple A-module. Thus

$$Ax + S_{i-1}(A) \subseteq S_i(A),$$

proving that $x \in S_i(A)$, as required.

(ii) By Theorem 2.6.1, $dim_F r(J(A)^i) = dim_F A/J(A)^i$. Hence the desired assertion follows from (i). ∎

1.2. Theorem. *(Külshammer (1984a)). Let (A, ψ) be a symmetric algebra over an arbitrary field F. Assume that*

$$dim_F(A/J(A)) = 1 \quad and \quad dim_F Z(A) \leq 4$$

Then A is commutative.

Proof. By Lemma 1.1(ii), $dim_F Soc(A) = dim_F A/J(A)$ and therefore $dim_F Soc(A) = 1$. Bearing in mind that $Ker\psi$ (and hence $[A, A]$) contains no nonzero ideals of A, we deduce that

$$[A, A] \cap Soc(A) = 0 \tag{1}$$

By Lemma 2.11.1(ii), we also have

$$dim_F A = dim_F[A, A] + dim_F Z(A) \tag{2}$$

Owing to Theorem 2.11.3(iii),

$$(Z(A) \cap Soc(A))A = Soc(A)$$

But $dim_F Soc(A) = 1$, hence $Soc(A) \subseteq Z(A)$. Bearing in mind that $J(A)$ is a unique maximal left ideal of A, we infer that

$$Soc(A) \subseteq Z(A) \cap J(A) = J(Z(A)) \tag{3}$$

Define the higher commutator subspaces of A as follows:

$$K_1(A) = A, K_2(A) = [A, A], K_{i+1}(A) = [K_i(A), A]$$

By hypothesis, we may write $A = F + J(A)$, so

$$[A, A] = [J(A), J(A)] \subseteq J(A)^2$$

and

$$[A, A] \cap Z(A) \subseteq J(A) \cap Z(A) = J(Z(A)) \tag{4}$$

Moreover, by induction we have

$$K_n(A) = 0 \quad \text{for sufficiently large} \quad n \tag{5}$$

For the sake of clarity, we divide the rest of the proof into five steps.

Step 1. Assume that A is noncommutative. We wish to show that

$$2 \le \dim_F([A, A] \cap Z(A)) \le \dim_F Z(A) - 2 \tag{6}$$

Assume by way of contradiction that $\dim_F([A, A] \cap Z(A)) \le 1$. Then, by (2),

$$\dim_F([A, A] + Z(A)) \ge \dim_F A - 1$$

and we may write

$$A = [A, A] + Z(A) + Fx$$

for some $x \in A$. But then, by (5), we have

$$
\begin{aligned}
[A, A] &= [[A, A] + Fx, [A, A] + Fx] \\
&= [[A, A], [A, A]] + [[A, A], Fx] \\
&\subseteq K_3(A) \subseteq K_4(A) \subseteq \cdots \subseteq 0,
\end{aligned}
$$

a contradiction. Thus

$$\dim_F([A, A] \cap Z(A)) \ge 2$$

Applying (3) and (4), we have

$$Soc(A) + ([A, A] \cap Z(A)) \subseteq J(Z(A))$$

But $\dim_F Soc(A) = 1$, hence by (1) we have

$$
\begin{aligned}
1 + \dim_F([A, A] \cap Z(A)) &\le \dim_F J(Z(A)) \\
&\le \dim_F Z(A) - 1,
\end{aligned}
$$

proving (6).

Step 2. Given a subset X of A, let FX be the F-space spanned by X. Let I be an ideal of A, and let m, n be natural numbers with $m \le n$. Assume that

$$I^n = F\{x_{i1} \cdots x_{in} | 1 \le i \le d\} + I^{n+1}$$

with elements $x_{ij} \in I$. Our aim is to prove that

$$I^{n+m} = F\{x_{j1} \cdots x_{jm} x_{i1} \cdots x_{in} | 1 \leq i, j \leq d\} + I^{n+m+1}$$

In particular,

$$dim_F(I^{n+m}/I^{n+m+1}) \leq \left[dim_F(I^n/I^{n+1})\right]^2$$

For any element $y \in I^m$ and any number $j \in \{1, \ldots, d\}$, we have

$$x_{j,m+1} \cdots x_{jn} y \in I^n = F\{x_{i1} \cdots x_{in} | 1 \leq i \leq d\} + I^{n+1}$$

which implies

$$x_{j1} \cdots x_{jn} y \in F\{x_{j1} \cdots x_{jm} x_{i1} \cdots x_{in} | 1 \leq i \leq d\} + I^{n+m+1},$$

as required.

Step 3. Our aim here is to show that if $dim_F Z(A) = 4$, then

$$dim_F J(A)^2 / J(A)^3 = 1$$

By Step 2, any counter-example A must satisfy the following conditions

$$dim_F A/J(A) = 1, \quad dim_F J(A)/J(A)^2 \geq 2$$

$$dim_F J(A)^2/J(A)^3 \geq 2, \quad dim_F J(A)^3/J(A)^4 \geq 1$$

In particular, A is not commutative. Owing to (6), we have

$$dim_F([A, A] \cap Z(A)) = 2$$

and we may write

$$A = [A, A] + Z(A) + Fx + Fy$$

for some $x, y \in A$. Then

$$
\begin{aligned}
[A, A] &= [[A, A] + Fx + Fy, [A, A] + Fx + Fy] \\
&\subseteq K_3(A) + F[x, y]
\end{aligned}
$$

and

$$dim_F A/K_3(A) \le 5$$

But $K_3(A)$ is contained in $J(A)^3$, so from $dim_F A/J(A)^3 \ge 5$, we deduce that $J(A)^3 = K_3(A)$. It follows that $J(A)^3$ is an ideal of A contained in $[A, A]$, and we get the contradiction $J(A)^3 = 0$.

Step 4.   Here we wish to show that if

$$dim_F J(A)^n/J(A)^{n+1} = 1$$

for some $n \ge 1$, then

$$J(A)^{n-1}A \subseteq Z(A)$$

To this end, we first note that by Step 2

$$
\begin{aligned}
dim_F J(A)^n/J(A)^{n+1} &= \cdots = dim_F J(A)^m/J(A)^{m+1} \\
&= dim_F J(A)^m = 1
\end{aligned}
$$

for some $m \ge n$. Applying Lemma 1.1(ii), we deduce that

$$
\begin{aligned}
dim_F(S_{n+1}(A)/S_n(A)) &= \cdots = dim_F(S_{m+1}(A)/S_m(A)) \\
&= dim_F A/J(A) = 1
\end{aligned}
$$

In particular, $A/S_n(A)$ has exactly one composition series and

$$dim_F J(A/S_n(A))/J(A/S_n(A))^2 \le 1$$

Again, by Step 2,

$$A = F\{1, x, x^2, \ldots\} + S_n(A)$$

for some $x \in J(A)$ and

$$
\begin{aligned}
[A, A] &= [F\{x, x^2, \ldots\} + S_n(A), F\{x, x^2, \ldots\} + S_n(A)] \\
&\subseteq S_{n-1}(A)
\end{aligned}
$$

Now because A is symmetric, it follows from Lemma 1.1(i) and Theorem 2.6.1 that

$$S_i(A)^\perp = J(A)^i \qquad \text{for all} \quad i$$

Thus

$$J(A)^{n-1}A = S_{n-1}(A)^{\perp} \subseteq [A, A]^{\perp} = Z(A),$$

as desired.

Step 5. Here we complete the proof by applying the information above. The case where $dim_F Z(A) \leq 3$ is a consequence of (6). So suppose that $dim_F Z(A) = 4$. Then by Step 3,

$$dim_F J(A)^2 / J(A)^3 = 1$$

Applying Step 4, we therefore conclude that $J(A) \subseteq Z(A)$. Because $A = F + J(A)$, it follows that $Z(A) = A$, thus completing the proof. ■

2. Some technical lemmas

Our aim here is to record a number of preliminary results required for subsequent investigations.

2.1. Lemma. *Let (A, ψ) be a symmetric algebra over a field F such that $dim_F A/J(A) = 1$. Then*
   (i) $[A, A] = [J(A), J(A)] \subseteq J(A)^2$.
   (ii) $Z(A)^{\perp} = [A, A]$ and, in particular,

$$dim_F Z(A) = dim_F A/[A, A]$$

   (iii) $dim_F J(A)^{\perp} = 1$.
   (iv) *If $J(A)^n = 0$ for some $n \geq 1$, then $J(A)^{n-1} \subseteq J(A)^{\perp}$ and, in particular,*

$$dim_F J(A)^{n-1} \leq 1$$

   Proof. (i) Since $A = F + J(A)$, we have $[A, A] = [J(A), J(A)] \subseteq J(A)^2$.
   (ii) Apply Lemma 2.11.1(ii) and Theorem 2.6.1(i).
   (iii) Apply Theorem 2.6.1(i).
   (iv) If $J(A)^n = 0$, then $J(A)^{n-1} \subseteq r(J(A)) = J(A)^{\perp}$ by Theorem 2.6.1(ii). Hence, by (iii), $dim_F J(A)^{n-1} \leq 1$, as required. ■

2.2. Lemma. *Let (A, ψ) be a symmetric algebra over a field F such that $dim_F A/J(A) = 1$ and $dim_F Z(A) = 5$. Then*

(i) $J(A)^2 \neq 0$, $J(A)^2 \neq J(A)^3, J(A) \neq J(A)^2$, $J(A)^2 \nsubseteq [A, A]$.

(ii) $dim_F J(A)/J(A)^3 \in \{1, 2, 3\}$.

(iii) If $J(A)^3 = 0$, then $dim_F A = 5$ and $dim_F J(A)/J(A)^2 = 3$.

(iv) If $dim_F J(A)/J(A)^2 = 1$ or $dim_F J(A)^2/J(A)^3 = 1$, then

$$dim_F A = 5$$

(v) If $dim_F J(A)/J(A)^2 = 2$, then $dim_F J(A)^2/J(A)^3 \in \{1, 2\}$ and

$$dim_F([A, A] + J(A)^3)/J(A)^3 \leq 1$$

(vi) If $dim_F J(A)/J(A)^2 = 3$, then

$$dim_F J(A)^2/J(A)^3 \in \{1, 2, 3\}$$

and

$$dim_F([A, A] + J(A)^3)/J(A)^3 \leq 3$$

Proof. (i) If $J(A)^2 = 0$, then by Lemma 2.1(iv), $dim_F J(A) \leq 1$ and hence $dim_F A \leq 2$, a contradiction. By Nakayama's lemma, we have $J(A)^2 \neq J(A)^3$. In particular, $J(A) \neq J(A)^2$. Since A is symmetric, $J(A)^2 \neq 0$ implies that $J(A)^2 \nsubseteq [A, A]$, as required.

(ii) By Lemma 2.1(i), $[A, A] \subseteq J(A)^2$ and, by Lemma 2.1(ii),

$$dim_F A/[A, A] = 5$$

Hence $dim_F A/J(A)^2 \leq dim_F A/[A, A] = 5$. Assume that

$$dim_F A/J(A)^2 = 5$$

Then $J(A)^2 = [A, A]$, contrary to (i). Hence

$$dim_F A/J(A)^2 \leq 4$$

and, since $dim_F A/J(A) = 1$, we have

$$dim_F J(A)/J(A)^2 \leq 3$$

Since, by (i), $J(A) \neq J(A)^2$ the required assertion follows.

(iii) Assume that $J(A)^3 = 0$. Then, by Lemma 2.1(iv),

$$dim_F J(A)^2 \leq 1$$

Hence, by (ii),

$$5 = dim_F Z(A) \leq dim_F A \leq 5$$

and so $dim_F A = 5$. By (ii), we have $dim_F J(A)/J(A)^2 = 3$.

(iv) First assume that $dim_F J(A)^2/J(A)^3 = 1$. Then $J(A) \subseteq Z(A)$ by Step 4 in the proof of Theorem 1.2. Hence $A = F + J(A) \subseteq Z(A)$, so $A = Z(A)$ and $dim_F A = 5$. The argument for the case $dim_F J(A)/J(A)^2 = 1$ is similar.

(v) Assume that $dim_F J(A)/J(A)^2 = 2$ and write

$$J(A) = Fa + Fb + J(A)^2$$

for some $a, b \in J(A)$. Then $A = F\{1, a, b\} + J(A)^2$ and $[A, A] \subseteq F[a, b] + J(A)^3$; in particular, $dim_F([A, A] + J(A)^3)/J(A)^3 \leq 1$. By (iii), $J(A)^3 \neq 0$ and, since A is symmetric, we must have $J(A)^3 \not\subseteq [A, A]$. Hence

$$dim_F(A/[A, A] + J(A)^3) < dim_F A/[A, A] = 5$$

and therefore $dim_F A/J(A)^3 \leq 5$. In particular, $dim_F J(A)^2/J(A)^3 \leq 2$. Because $J(A)^2 \neq J(A)^3$, the required assertion follows.

(vi) Assume that $dim_F J(A)/J(A)^2 = 3$ and write

$$J(A) = F\{a, b, c\} + J(A)^2$$

for some $a, b, c \in J(A)$. Then

$$[A, A] \subseteq F\{[a, b], [a, c], [b, c]\} + J(A)^3$$

and, in particular, $dim_F([A, A] + J(A)^3)/J(A)^3 \leq 3$. By (i), $J(A)^2 \neq J(A)^3$ and so we are left to verify that

$$dim_F J(A)^2/J(A)^3 \leq 3$$

Assume by way of contradiction that $dim_F J(A)^2/J(A)^3 \geq 4$. Then $dim_F A/J(A)^3 \geq 8$ and

$$dim_F A/([A, A] + J(A)^3) \geq 5 = dim_F A/[A, A]$$

Thus $[A, A] + J(A)^3 = [A, A]$ and $J(A)^3 \subseteq [A, A]$. Since A is symmetric, $J(A)^3 = 0$. Hence, by (iii), $dim_F A = 5$, a contradiction. ∎

2.3. Lemma. *Let (A, ψ) be a symmetric algebra over a field F such that $dim_F A/J(A) = 1$ and $dim_F Z(A) = 5$.*

(i) If $dim_F A \leq 8$, then $dim_F A = 5$ or $dim_F A = 8$.

(ii) If $x^2 \in J(A)^3$ for all $x \in J(A)$, then $dim_F A = 5$ or $dim_F A = 8$. If, furthermore, $dim_F A = 8$, then

$$dim_F J(A)/J(A)^2 = dim_F J(A)^2/J(A)^3 = 3$$

(iii) If $dim_F J(A)^3/J(A)^4 = 1$, then $dim_F A = 5$ or $dim_F A = 8$. If, furthermore, $dim_F A = 8$, then

$$dim_F J(A)/J(A)^2 = dim_F J(A)^2/J(A)^3 = 3$$

Proof. (i) Assume that $dim_F A \leq 7$. Because $dim_F Z(A) = 5$, we may find elements $a, b \in A$ such that

$$A = Fa + Fb + Z(A)$$

Then $[A, A] = F[a, b]$ and, in particular,

$$dim_F [A, A] \cap Z(A) \leq dim_F [A, A] \leq 1$$

It follows from Step 1 in the proof of Theorem 1.2 that A is commutative. Thus $dim_F A = dim_F Z(A) = 5$ and the required assertion is proved.

(ii) By Lemma 2.2(ii), we have $dim_F J(A)/J(A)^2 \leq 3$. Hence we may write

$$J(A) = F\{a, b, c\} + J(A)^2$$

for some $a, b, c \in J(A)$. Then

$$a^2, b^2, (a + b)^2 \in J(A)^3$$

so

$$ab + ba = (a + b)^2 - a^2 - b^2 \in J(A)^3$$

and thus

$$ba \equiv -ab \pmod{J(A)^3}$$

A similar argument shows that

$$ca \equiv -ac \pmod{J(A)^3}$$

and

$$cb \equiv -bc \pmod{J(A)^3}$$

We conclude therefore that

$$\begin{aligned} J(A)^2 &= F\{a^2, ab, ac, ba, b^2, bc, ca, cb, c^2\} + J(A)^3 \\ &= F\{ab, ac, bc\} + J(A)^3 \end{aligned}$$

and, in particular, $dim_F J(A)^2/J(A)^3 \leq 3$. Invoking Step 2 in the proof of Theorem 1.2, we deduce that

$$\begin{aligned} J(A)^3 &= F\{a^2 b, a^2 c, abc, bab, bac, b^2 c\} + J(A)^4 \\ &= Fabc + J(A)^4 \end{aligned}$$

which implies $dim_F J(A)^3/J(A)^4 \leq 1$. Again, applying Step 2 in the proof of Theorem 1.2, we obtain

$$J(A)^4 = Fa^2 bc + J(A)^5 = J(A)^5$$

and hence, by Nakayama's lemma, $J(A)^4 = 0$. It follows that $dim_F A \leq 8$ and so, by (i), $dim_F A = 5$ or $dim_F A = 8$. Furthermore, if $dim_F A = 8$, then

$$dim_F J(A)/J(A)^2 = 3 = dim_F J(A)^2/J(A)^3,$$

as required.

(iii) Assume that $dim_F J(A)^3/J(A)^4 = 1$. Then, by Step 4 in the proof of Theorem 1.2, we have

$$J(A)^2 \subseteq Z(A) \cap J(A)$$

and, in particular, $dim_F J(A)^2 \leq dim_F Z(A) - 1 = 4$. On the other hand, by Lemma 2.2(ii), $dim_F J(A)/J(A)^2 \leq 3$ and thus $dim_F A \leq 8$.

Owing to (i), $dim_F A = 5$ or $dim_F A = 8$. Assume that $dim_F A = 8$. Then we must have $dim_F J(A)/J(A)^2 = 3$ and $dim_F J(A)^2 = 4$. Since $dim_F J(A)^3/J(A)^4 = 1$, we have $dim_F J(A)^2/J(A)^3 \leq 3$.

Assume by way of contradiction that $dim_F J(A)^2/J(A)^3 \leq 2$. Bearing in mind that, by Lemma 2.2(i), $J(A)^2 \neq J(A)^3$, it follows from

Lemma 2.2(iv) that $dim_F J(A)^2/J(A)^3 = 2$. Thus $dim_F J(A)^4 = 1$. Because A is symmetric, $\psi(J(A)^4) \neq 0$, so $J(A)^2 \not\subseteq [J(A)^2]^\perp$ and therefore

$$dim_F[J(A)^2 + (J(A)^2)^\perp] > dim_F[J(A)^2]^\perp$$
$$= dim_F A/J(A)^2$$
$$= 4$$

On the other hand, $[A, A] \subseteq J(A)^2$ implies

$$[J(A)^2]^\perp \subseteq [A, A]^\perp = Z(A)$$

and hence $J(A)^2 + [J(A)^2]^\perp \subseteq Z(A)$. Comparing dimensions we obtain

$$Z(A) = J(A)^2 + [J(A)^2]^\perp$$

However $1 \notin J(A)^2$ and $1 \notin [J(A)^2]^\perp$ and hence

$$Z(A) = J(A)^2 + [J(A)^2]^\perp \subseteq J(A),$$

a contradiction. ∎

2.4. Lemma.   *Let (A, ψ) be a symmetric algebra over a field F such that $dim_F A/J(A) = 1$ and $dim_F Z(A) = 5$. Then the configuration*

$$dim_F J(A)/J(A)^2 = 3, \quad dim_F J(A)^2/J(A)^3 = 2$$

is impossible.

Proof.   Assume by way of contradiction that the above equalities hold. Then $dim_F A \geq 6$ and, by Lemma 2.3(i), $dim_F A \geq 8$; in particular, $J(A)^3 \neq 0$. By Nakayama's lemma, it follows that $J(A)^3 \neq J(A)^4$. Hence, by Lemma 2.3(iii),

$$dim_F J(A)^3/J(A)^4 \geq 2 \tag{1}$$

In what follows, we shall prove that this leads to a contradiction. For the sake of clarity, we divide the proof into three steps.

Step 1.   Here we prove the existence of $a \in J(A)$ such that

$$a^2 \notin J(A)^3 \quad \text{but} \quad J(A)^2 = aJ(A) \quad \text{or} \quad J(A)^2 = J(A)a$$

Assume that for all $x \in J(A)$, $x^2 \in J(A)^3$. Then, by Lemma 2.3(ii), $dim_F A \in \{5, 8\}$ and, since $dim_F A \geq 8$, we must have $dim_F A = 8$. It follows from Lemma 2.3(ii) that $dim_F J(A)^2/J(A)^3 = 3$, a contradiction. Hence there is an element $a \in J(A)$ such that $a^2 \notin J(A)^3$. In particular, $a \notin J(A)^2$ and since $dim_F J(A)/J(A)^2 = 3$, there are elements $b, c \in J(A)$ such that

$$J(A) = F\{a, b, c\} + J(A)^2$$

If $ab \notin Fa^2 + J(A)^3$, then

$$J(A)^2 = Fa^2 + Fab + J(A)^3$$

since $dim_F J(A)^2/J(A)^3 = 2$. Hence $J(A)^2 = aJ(A) + J(A)^3$, so

$$J(A)^2 = aJ(A)$$

by Nakayama's lemma, as required. We may thus assume that $ab \in Fa^2 + J(A)^3$ and similarly $ba, ac, ca \in Fa^2 + J(A)^3$.

We next consider the case where $b^2 \notin Fa^2 + J(A)^3$ and, in particular, $b^2 \notin J(A)^3$. Then we can interchange the roles of a and b and may thus assume that

$$ab, ba, bc, cb \in Fb^2 + J(A)^3$$

Taking into account that $dim_F J(A)^2/J(A)^3 = 2$, we see that the elements $a^2 + J(A)^3$, $b^2 + J(A)^3$ form a basis of $J(A)^2/J(A)^3$. It follows that

$$ab, ba \in J(A)^3$$

whence

$$(a + b)^2 + J(A)^3 = a^2 + b^2 + J(A)^3$$

and

$$(a + b)b + J(A)^3 = b^2 + J(A)^3$$

are linearly independent. The conclusion is that

$$\begin{aligned} J(A)^2 &= F(a+b)^2 + F(a+b)b + J(A)^3 \\ &= (a+b)J(A) + J(A)^3 \\ &= (a+b)J(A) \end{aligned}$$

and the required assertion follows in this case.

By the foregoing, we may assume that $b^2 \in Fa^2 + J(A)^3$ and, similarly, $c^2 \in Fa^2 + J(A)^3$. Then we have

$$
\begin{aligned}
J(A)^2 &= F\{a^2, ab, ac, ba, b^2, bc, ca, cb, c^2\} + J(A)^3 \\
&= F\{a^2, bc, cb\} + J(A)^3
\end{aligned}
$$

Bearing in mind that $dim_F J(A)^2/J(A)^3 = 2$ and $a^2 \notin J(A)^3$, we obtain

$$
J(A)^2 = F\{a^2, bc\} + J(A)^3
$$

or

$$
J(A)^2 = F\{a^2, cb\} + J(A)^3
$$

By interchanging b and c, if necessary, we may assume that

$$
J(A)^2 = F\{a^2, bc\} + J(A)^3
$$

It therefore follows, from Step 2 in the proof of Theorem 1.2, that

$$
\begin{aligned}
J(A)^3 &= F\{a^3, abc, ba^2, b^2c\} + J(A)^4 \\
&= F\{a^3, a^2c\} + J(A)^4 \\
&= Fa^3 + J(A)^4
\end{aligned}
$$

Hence $dim_F J(A)^3/J(A)^4 \leq 1$, a contradiction to (1).

Step 2.   Choose $a \in J(A)$ as in Step 1. We claim that

$$
a^3 + J(A)^4, a^2b + J(A)^4
$$

form a basis of $J(A)^3/J(A)^4$. Indeed, by symmetry, we may assume that $J(A)^2 = aJ(A)$. Because $dim_F J(A)^2/J(A)^3 = 2$, there exists $b \in J(A)$ such that $ab \notin Fa^2 + J(A)^3$, so $J(A)^2 = Fa^2 + Fab + J(A)^3$ and $b \notin Fa + J(A)^2$. In view of $a \notin J(A)^2$ and $dim_F J(A)/J(A)^2 = 3$, we may choose an element $c \in J(A)$ such that $c \notin Fa + Fb + J(A)^2$, so $J(A) = F\{a, b, c\} + J(A)^2$. Furthermore,

$$
J(A)^3 = aJ(A)^2 = a^2J(A) = Fa^3 + Fa^2b + J(A)^4
$$

and, in particular, $dim_F J(A)^3 / J(A)^4 \leq 2$. It follows from (1) that $dim_F J(A)^3 / J(A)^4 = 2$ and hence $a^3 + J(A)^4, a^2 b + J(A)^4$ form a basis of $J(A)^3 / J(A)^4$.

Step 3. Here we complete the proof by deriving a final contradiction. By Step 2, we may choose elements $\lambda, \mu \in F$ such that

$$ac \equiv \lambda a^2 + \mu ab \pmod{J(A)^3}$$

Setting $c' = c - \lambda a - \mu b$, we then have

$$J(A) = F\{a, b, c'\} + J(A)^2$$

and

$$ac' \equiv ac - \lambda a^2 - \mu ab \equiv 0 \pmod{J(A)^3}$$

We may therefore replace c by c' and assume that $ac \in J(A)^3$.
Choose elements $\lambda_i, \mu_i \in F$, $1 \leq i \leq 4$, such that

$$
\begin{aligned}
bc &\equiv \lambda_1 a^2 + \mu_1 ab \pmod{J(A)^3} \\
ca &\equiv \lambda_2 a^2 + \mu_2 ab \pmod{J(A)^3} \\
cb &\equiv \lambda_3 a^2 + \mu_3 ab \pmod{J(A)^3} \\
c^2 &\equiv \lambda_4 a^2 + \mu_4 ab \pmod{J(A)^3}
\end{aligned}
$$

Then

$$
\begin{aligned}
0 &\equiv (ac)a \equiv a(ca) \equiv \lambda_2 a^3 + \mu_2 a^2 b \pmod{J(A)^4} \\
0 &\equiv (ac)b \equiv a(cb) \equiv \lambda_3 a^3 + \mu_3 a^2 b \pmod{J(A)^4} \\
0 &\equiv (ac)a \equiv a(c^2) \equiv \lambda_4 a^3 + \mu_4 a^2 b \pmod{J(A)^4}
\end{aligned}
$$

which implies that

$$\lambda_2 = \mu_2 = \lambda_3 = \mu_3 = \lambda_4 = \mu_4 = 0$$

In particular, $ca, cb, c^2 \in J(A)^3$ and thus

$$
\begin{aligned}
0 &\equiv b(c^2) \equiv (bc)c \equiv \lambda_1 a^2 c + \mu_1 abc \\
&\equiv \lambda_1 \mu_1 a^3 + \mu_1^2 a^2 b \pmod{J(A)^4}
\end{aligned}
$$

which implies that $\mu_1 = 0$. Hence

$$0 \equiv b(cb) \equiv (bc)b \equiv \lambda_1 a^2 b \quad (\text{mod } J(A)^4)$$

and therefore $\lambda_1 = 0$. In particular, $bc \in J(A)^3$. Accordingly,

$$[a, c], [b, c] \in J(A)^3$$

and

$$\begin{aligned}
[A, A] &\subseteq F\{[a, b], [a, c], [b, c]\} + J(A)^3 \\
&\subseteq F[a, b] + J(A)^3
\end{aligned}$$

In particular $dim_F([A, A] + J(A)^3)/J(A)^3 \leq 1$ and hence

$$dim_F A/([A, A] + J(A)^3) \geq 5 = dim_F A/[A, A]$$

It follows that $[A, A] + J(A)^3 = [A, A]$ and, in particular, $J(A)^3 \subseteq [A, A]$. But A is symmetric, hence $J(A)^3 = 0$ which contradicts (1). ■

2.5. Lemma. *Let (A, ψ) be a symmetric algebra over a field F such that $dim_F A/J(A) = 1$, $dim_F Z(A) = 5$, $dim_F J(A)/J(A)^2 = 2 = dim_F J(A)^2/J(A)^3$. Then $dim_F A = 8$.*

Proof. For the sake of clarity, we divide the proof into five steps.

Step 1. Here we demonstrate that $dim_F A \geq 8$ and

$$\begin{aligned}
dim_F A/([A, A] + J(A)^3) &= dim_F A/([A, A] + J(A)^4) \\
&= 4
\end{aligned}$$

We first note that, by Lemma 2.2(iii), $J(A)^3 \neq 0$. Hence $dim_F A > 5$ and therefore, by Lemma 2.3(i), $dim_F A \geq 8$. By Nakayama's lemma, we have $J(A)^3 \neq J(A)^4$. Hence, by Lemma 2.3(iii), we may assume that

$$dim_F J(A)^3/J(A)^4 \geq 2 \tag{2}$$

In particular, we have $J(A)^4 \neq 0$ for otherwise $dim_F J(A)^3 \leq 1$, a contradiction to (2). Hence, by Nakayama's lemma

$$J(A)^4 \neq J(A)^5$$

Because A is symmetric, $J(A)^3 \neq 0$ implies that $J(A)^3 \not\subseteq [A, A]$. Hence

$$dim_F(J(A)^3 + [A, A])/[A, A] \geq 1$$

and taking into account that $dim_F A/[A, A] = 5$, we deduce that

$$dim_F A/(J(A)^3 + [A, A]) \leq 4$$

On the other hand, $dim_F([A, A] + J(A)^3)/J(A)^3 \leq 1$ by Lemma 2.2(v), so $dim_F A/J(A)^3 = 5$ implies that

$$dim_F A/(J(A)^3 + [A, A]) \geq 4$$

Thus $dim_F A/([A, A] + J(A)^3) = 4$.

Similarly, $J(A)^4 \neq 0$ implies $J(A)^4 \not\subseteq [A, A]$, so

$$dim_F(J(A)^4 + [A, A])/[A, A] \geq 1$$

and

$$4 \geq dim_F A/(J(A)^4 + [A, A]) \geq dim_F A/(J(A)^3 + [A, A]) = 4$$

Thus $dim_F A/(J(A)^4 + [A, A]) = 4$, as required.

Step 2. We now prove that there exists $a \in J(A)$ such that

$$J(A)^2 = aJ(A) \quad \text{or} \quad J(A)^2 = J(A)a \tag{3}$$

Owing to Lemma 2.3(ii), we may assume that there is an element $a \in J(A)$ such that $a^2 \notin J(A)^3$; in particular, $a \notin J(A)^2$. Because $dim_F J(A)/J(A)^2 = 2$, there exists an element $b \in J(A)$ such that

$$J(A) = Fa + Fb + J(A)^2$$

If $ab \notin Fa^2 + J(A)^3$, then

$$J(A)^2 = Fa^2 + Fab + J(A)^3 = aJ(A) + J(A)^3$$

since $dim_F J(A)^2/J(A)^3 = 2$. It follows from Nakayama's lemma that $J(A)^2 = aJ(A)$. Thus we may assume that

$$ab \in Fa^2 + J(A)^3$$

and, similarly,

$$ba \in Fa^2 + J(A)^3$$

Then we have

$$
\begin{aligned}
J(A)^2 &= F\{a^2, ab, ba, b^2\} + J(A)^3 \\
&= F\{a^2, b^2\} + J(A)^3
\end{aligned}
$$

Because $dim_F J(A)^2/J(A)^3 = 2$, the elements $a^2 + J(A)^3, b^2 + J(A)^3$ form a basis of $J(A)^2/J(A)^3$. In particular, $b^2 \notin J(A)^3$ and, interchanging the roles of a and b, we may assume that

$$ab, ba \in Fb^2 + J(A)^3$$

It follows that $ab, ba \in J(A)^3$ and, in particular, $[a, b] \in J(A)^3$. Taking into account that $A = F\{1, a, b\} + J(A)^2$, we conclude that

$$[A, A] \subseteq F[a, b] + J(A)^3 \subseteq J(A)^3$$

Hence $dim_F A/([A, A] + J(A)^3) = dim_F A/J(A)^3 = 5$, contradicting Step 1.

Step 3. Choose $a \in J(A)$ such that (3) holds. By symmetry, we may assume that $J(A)^2 = aJ(A)$. Because $J(A)^2 \neq J(A)^3$, it follows that $a \notin J(A)^2$ and so

$$J(A) = Fa + Fb + J(A)^2$$

for some $b \in J(A)$. Hence

$$J(A)^2 = aJ(A) = Fa^2 + Fab + J(A)^3 \tag{4}$$

and

$$
\begin{aligned}
J(A)^3 &= aJ(A)^2 = a^2 J(A) \\
&= Fa^3 + Fa^2 b + J(A)^4, \tag{5}
\end{aligned}
$$

whence $dim_F J(A)^3/J(A)^4 \leq 2$. Since $dim_F J(A)^3/J(A)^4 \geq 2$, we deduce that

$$dim_F J(A)^3/J(A)^4 = 2$$

which implies that $a^3 + J(A)^4$ and $a^2b + J(A)^4$ form a basis of $J(A)^3/J(A)^4$. Furthermore, we also have

$$A = F\{1, a, b, a^2, ab, a^3, a^2b\} + J(A)^4$$

which implies that

$$[A, A] \subseteq F\{[a, b], [a, ab], [a, a^2b], [b, a^2], [b, ab],$$
$$[b, a^3], [b, a^2b], [a^2, ab]\} + J(A)^5$$

Now, by (4) and (5), we have

$$J(A)^2 = F\{a^2, ab, a^3, a^2b\} + J(A)^4$$

and therefore we may find elements $\lambda_i, \mu_i, t_i, s_i \in F, i = 1, 2$, such that

$$ba \equiv \lambda_1 a^2 + \mu_1 ab + t_1 a^3 + s_1 a^2b \pmod{J(A)^4}$$
$$b^2 \equiv \lambda_2 a^2 + \mu_2 ab + t_2 a^3 + s_2 a^2b \pmod{J(A)^4}$$

Step 4. Here we prove that $dim_F A = 8$ under the assumption that $\mu_1 \neq 1$. To this end, we put

$$\alpha = \lambda_1(1 - \mu_1)^{-1} \quad \text{and} \quad b' = b - \alpha a$$

Then

$$J(A) = Fa + Fb' + J(A)^2$$

and

$$b'a \equiv ba - \alpha a^2 \equiv (\lambda_1 - \alpha)a^2 + \mu_1 ab$$
$$\equiv (\lambda_1 - \alpha + \mu_1 \alpha)a^2 + \mu_1 ab'$$
$$\equiv \mu_1 ab' \pmod{J(A)^3}$$

We may thus assume that $\lambda_1 = 0$, by replacing b by b'. Then

$$0 \equiv (b^2)a - b(ba) \equiv \lambda_2 a^3 + \mu_2 aba - \mu_1 bab$$
$$\equiv \lambda_2 a^3 + \mu_1 \mu_2 a^2b - \mu_1^2 ab^2$$
$$\equiv (\lambda_2 - \lambda_2 \mu_1^2)a^3 + (\mu_1 \mu_2 - \mu_1^2 \mu_2)a^2b \pmod{J(A)^4}$$

and similarly

$$0 \equiv (b^2)b - b(b^2)$$
$$\equiv (\lambda_2\mu_2 - \lambda_2\mu_1\mu_2)a^3 + (\lambda_2 + \mu_2^2 - \lambda_2\mu_1^2 - \mu_1\mu_2^2)a^2b \quad (\bmod\ J(A)^4)$$

Taking into account that $a^3 + J(A)^4$ and $a^2b + J(A)^4$ form a basis of $J(A)^3/J(A)^4$, we derive

$$\lambda_2 - \lambda_2\mu_1^2 = 0 \tag{6}$$

$$\mu_1\mu_2 - \mu_1^2\mu_2 = 0 \tag{7}$$

$$\lambda_2\mu_2 - \lambda_2\mu_1\mu_2 = 0 \tag{8}$$

$$\lambda_2 + \mu_2^2 - \lambda_2\mu_1^2 - \mu_1\mu_2^2 = 0 \tag{9}$$

Subtracting (6) from (9), we have $\mu_2^2 = \mu_1\mu_2^2$. Because $\mu_1 \neq 1$, this implies that $\mu_2 = 0$. From (6) we also deduce that $\lambda_2 = 0$ or $\mu_1^2 = 1$.

Assume by way of contradiction that $\lambda_2 = 0$. Then

$$[a, ab] \equiv a^2b - aba \equiv (1 - \mu_1)a^2b \quad (\bmod\ J(A)^4)$$
$$\left[b, a^2\right] \equiv ba^2 - a^2b \equiv \mu_1 aba - a^2b \equiv (\mu_1^2 - 1)a^2b \quad (\bmod\ J(A)^4)$$
$$[b, ab] \equiv bab - ab^2 \equiv \mu_1 ab^2 \equiv 0 \quad (\bmod\ J(A)^4)$$

It therefore follows that

$$[A, A] \subseteq F[a, b] + Fa^2b + J(A)^4$$

and, in particular, $dim_F([A, A] + J(A)^4)/J(A)^4 \leq 2$. But then

$$dim_F A/([A, A] + J(A)^4) \geq 5,$$

contrary to Step 1. Thus we may assume that $\lambda_2 \neq 0$ and $\mu_1^2 = 1$.

Because $\mu_1 \neq 1$, the above implies that $\mu_1 = -1$ and $charF \neq 2$. It now easily follows that

$$[a, a^2b] \equiv 2a^3b \quad (\bmod\ J(A)^5)$$
$$\left[b, a^2\right] \equiv -2s_1a^3b \quad (\bmod\ J(A)^5)$$
$$\left[b, a^3\right] \equiv -2a^3b \quad (\bmod\ J(A)^5)$$
$$\left[b, a^2b\right] \equiv [a^2, ab] \equiv 0 \quad (\bmod\ J(A)^5)$$

and therefore

$$[A, A] \subseteq F\{[a, b], [a, ab], [b, ab], a^3b\} + J(A)^5,$$

proving that $dim_F([A, A] + J(A)^5)/J(A)^5 \leq 4$.

On the other hand,

$$dim_F A/([A, A] + J(A)^5) \leq dim_F A/[A, A] = 5$$

and so $dim_F A/J(A)^5 \leq 9$. Suppose that $dim_F A/J(A)^5 = 9$. Then

$$dim_F A/([A, A] + J(A)^5) = dim_F A/[A, A]$$

and so $J(A)^5 \subseteq [A, A]$. Because A is symmetric, we conclude that $J(A)^5 = 0$, so $dim_F J(A)^4 \leq 1$ and $dim_F A \leq 8$. Hence we may assume that $dim_F A/J(A)^5 \leq 8$, in which case $dim_F J(A)^4/J(A)^5 \leq 1$.

Because $J(A)^4 \neq J(A)^5$, the latter implies that

$$dim_F J(A)^4/J(A)^5 = 1 \quad \text{and} \quad dim_F A/J(A)^5 = 8$$

By Step 4 in the proof of Theorem 1.2, this also forces $J(A)^3 \subseteq Z(A)$ and, in particular, $a^2b \in Z(A)$. Hence

$$a^3b \equiv a^2ba \equiv -a^3b \pmod{J(A)^5}$$

and, since $char F \neq 2$, the above forces $a^3b \in J(A)^5$. Thus

$$[A, A] \subseteq F\{[a, b], [a, ab], [b, ab]\} + J(A)^5$$

and, in particular, $dim_F([A, A] + J(A)^5)/J(A)^5 \leq 3$. Consequently,

$$dim_F A/([A, A] + J(A)^5) \geq 5 = dim_F A/[A, A]$$

which forces $J(A)^5 \subseteq [A, A]$. But A is symmetric, so $J(A)^5 = 0$ and hence $dim_F A = 8$.

Step 5. Here we complete the proof by treating the case where $\mu_1 = 1$. Assume by way of contradiction that $\lambda_1 = 0$. Then $[a, b] \in J(A)^3$ and $[A, A] \subseteq J(A)^3$, so

$$dim_F A/([A, A] + J(A)^3) = dim_F A/J(A)^3 = 5,$$

contrary to Step 1. Hence we must have $\lambda_1 \neq 0$.

Setting $a' = \lambda_1 a$, we then have

$$J(A)^2 = a'J(A) \quad \text{and} \quad J(A) = Fa' + Fb + J(A)^2$$

and

$$\begin{aligned}
ba' &\equiv \lambda_1 ba \equiv \lambda_1^2 a^2 + \lambda_1 ab \\
&\equiv (a')^2 + a'b \pmod{J(A)^3}
\end{aligned}$$

Replacing a by a', we may therefore assume that $\lambda_1 = 1$. As in the previous step, we compute

$$0 \equiv (b^2)a - b(ba) \equiv (\mu_2 - 2)a^3 - 2a^2 b \pmod{J(A)^4}$$

Because $a^3 + J(A)^4$ and $a^2 b + J(A)^4$ form a basis of $J(A)^3/J(A)^4$, this forces $char F = 2$ and $\mu_2 = 0$. Thus

$$\begin{aligned}
[a, a^2 b] &\equiv a^4 \pmod{J(A)^5} \\
\left[b, a^2\right] &\equiv s_1 a^4 \pmod{J(A)^5} \\
\left[b, a^3\right] &\equiv a^4 \pmod{J(A)^5} \\
\left[b, a^2 b\right] &\equiv [a^2, ab] \equiv 0 \pmod{J(A)^5}
\end{aligned}$$

and therefore

$$[A, A] \subseteq F\{[a, b], [a, ab], [b, ab], a^4\} + J(A)^5$$

which shows that $dim_F([A, A] + J(A)^5)/J(A)^5 \leq 4$.
On the other hand,

$$dim_F A/([A, A] + J(A)^5) \leq dim_F A/[A, A] = 5$$

and so $dim_F A/J(A)^5 \leq 9$. Suppose that $dim_F A/J(A)^5 = 9$. Then

$$dim_F A/([A, A] + J(A)^5) = dim_F A/[A, A]$$

and hence $J(A)^5 \subseteq [A, A]$. But A is symmetric, so $J(A)^5 = 0$. It follows that $dim_F J(A)^4 \leq 1$ and $dim_F A \leq 8$. Thus we may assume that

$dim_F A/J(A)^5 \leq 8$. Then $dim_F J(A)^4/J(A)^5 \leq 1$ and, since $J(A)^4 \neq J(A)^5$, this implies

$$dim_F J(A)^4/J(A)^5 = 1 \quad \text{and} \quad dim_F A/J(A)^5 = 8$$

It follows from Step 4 in the proof of Theorem 1.2 that $J(A)^3 \subseteq Z(A)$ and, in particular, $a^2 b \in Z(A)$. Hence

$$a^3 b \equiv a^2 ba \equiv a^4 + a^3 b \pmod{J(A)^5}$$

whence $a^4 \in J(A)^5$ and $J(A)^5 = a^2 J(A)^3 = a^4 J(A) = J(A)^6$. Applying Nakayama's lemma, we conclude that $J(A)^5 = 0$. Thus $dim_F A = 8$ and the result follows. ■

2.6. Lemma. *Let (A, ψ) be a symmetric local algebra over an algebraically closed field F such that*

$$dim_F Z(A) = 5 \quad \text{and}$$

$$dim_F J(A)/J(A)^2 = dim_F J(A)^2/J(A)^3 = 3$$

Then
(i) $dim_F A \geq 8$, $J(A)^3 \neq 0$ and $J(A)^3 \neq J(A)^4$.
(ii) $J(A)^2 = [A, A] + J(A)^3$.
(iii) If for all $x \in J(A), xJ(A) \neq J(A)^2$ and $J(A)^2 \neq J(A)x$, then $dim_F A = 8$

Proof. (i) Our assumptions imply that $dim_F A \geq 7$. Hence, by Lemma 2.3(i), $dim_F A \geq 8$ and, in particular, $J(A)^3 \neq 0$. By Nakayama's lemma, $J(A)^3 \neq J(A)^4$.

(ii) Owing to Lemma 2.2(vi), $dim_F([A, A] + J(A)^3)/J(A)^3 \leq 3$. Assume by way of contradiction that $dim_F([A, A] + J(A)^3)/J(A)^3 \leq 2$. Because $dim_F A/J(A)^3 = 7$, we must have

$$5 \leq dim_F A/([A, A] + J(A)^3) \leq dim_F A/[A, A] = 5$$

and therefore
$$J(A)^3 \subseteq [A, A] + J(A)^3 = [A, A]$$

Since A is symmetric, we obtain the contradiction $J(A)^3 = 0$, by applying (i). Thus $dim_F([A, A] + J(A)^3)/J(A)^3 = 3$. Since

$$[A, A] + J(A)^3 \subseteq J(A)^2 \quad \text{and} \quad dim_F J(A)^2/J(A)^3 = 3,$$

we conclude that $J(A)^2 = [A, A] + J(A)^3$, as required.

(iii) By Lemma 2.3(ii), we may assume that $a^2 \notin J(A)^3$ for some $a \in J(A)$. Assume by way of contradiction that $\{ay, ya\} \subseteq Fa^2 + J(A)^3$ for all $y \in J(A)$. SInce $a \notin J(A)^2$ and $dim_F J(A)/J(A)^2 = 3$, there exist elements $b, c \in J(A)$ such that

$$J(A) = F\{a, b, c\} + J(A)^2$$

and hence $ab, ba, ac, ca \in Fa^2 + J(A)^3$. Taking into account that

$$A = F\{1, a, b, c\} + J(A)^2,$$

it follows from (ii) that

$$
\begin{aligned}
J(A)^2 &= [A, A] + J(A)^3 \\
&= F\{[a, b], [a, c], [b, a]\} + J(A)^3 \\
&\subseteq F\{a^2, [b, c]\} + J(A)^3
\end{aligned}
$$

contrary to the fact that $dim_F J(A)^2/J(A)^3 = 3$. Thus there exists $b \in J(A)$ such that $\{ab, ba\} \not\subseteq Fa^2 + J(A)^3$. By symmetry, we may assume that $ab \notin Fa^2 + J(A)^3$ and, in particular, $b \notin Fa + J(A)^3$.

Because $dim_F J(A)/J(A)^2 = 3$, we may find $c \in J(A)$ such that

$$J(A) = F\{a, b, c\} + J(A)^2$$

in which case

$$F\{a^2, ab, ac, ba, b^2, bc, ca, cb, c^2\} + J(A)^3 = J(A)^2 = [A, A] + J(A)^3$$

$$= F\{[a, b], [a, c], [b, c]\} + J(A)^3$$

by (ii). Taking into account that $dim_F J(A)^2/J(A)^3 = 3$, there exists

$$d \in \{ac, ba, b^2, bc, ca, cb, c^2\}$$

such that
$$J(A)^2 = F\{a^2, ab, d\} + J(A)^3$$

The rest of the proof will be divided into seven steps treating the corresponding values of d.

Step 1. Assume that $d = ac$. Then $J(A)^2 = aJ(A) + J(A)^3$ and hence, by Nakayama's lemma $J(A)^2 = aJ(A)$, contrary to our assumption.

Step 2. Assume that $d = ba$. Then there exist $\lambda_i, \mu_i, t_i \in F$, $1 \le i \le 6$, such that

$$
\begin{aligned}
ac &\equiv \lambda_1 a^2 + \mu_1 ab + t_1 ba \pmod{J(A)^3} \\
b^2 &\equiv \lambda_2 a^2 + \mu_2 ab + t_2 ba \pmod{J(A)^3} \\
bc &\equiv \lambda_3 a^2 + \mu_3 ab + t_3 ba \pmod{J(A)^3} \\
ca &\equiv \lambda_4 a^2 + \mu_4 ab + t_4 ba \pmod{J(A)^3} \\
cb &\equiv \lambda_5 a^2 + \mu_5 ab + t_5 ba \pmod{J(A)^3} \\
c^2 &\equiv \lambda_6 a^2 + \mu_6 ab + t_6 ba \pmod{J(A)^3}
\end{aligned}
$$

If $t_1 \neq 0$, then by Step 1 we have a contradiction. Thus $t_1 = 0$ and, after replacing c by $c - \lambda_1 a - \mu_1 b$, we may assume that $\lambda_1 = \mu_1 = 0$. Similarly, we have $\mu_4 = 0$; for otherwise we are in Step 1 for the opposite algebra of A which leads to a contradiction. Furthermore, replacing b by $b - t_2 a$, we may also assume that $t_2 = 0$. By our assumption,

$$J(A)^2 \neq (\alpha a + \beta b + \gamma c)J(A) \quad \text{for all} \quad \alpha, \beta, \gamma \in F$$

Therefore, by Nakayama's lemma,

$$
\begin{aligned}
J(A)^2 \neq\ & (\alpha a + \beta b + \gamma c)J(A) + J(A)^3 \\
=\ & F\{(\alpha a + \beta b + \gamma c)a, (\alpha a + \beta b + \gamma c)b, (\alpha a + \beta b + \gamma c)c\} \\
& + J(A)^3 \\
=\ & F\{(\alpha + \lambda_4\gamma)a^2 + (\beta + t_4\gamma)ba, (\lambda_2\beta + \lambda_5\gamma)a^2 + \\
& + (\alpha + \mu_2\beta + \mu_5\gamma)ab + t_5\gamma ba, \\
& (\lambda_3\beta + \lambda_6\gamma)a^2 + (\mu_3\beta + \mu_6\gamma)ab + (t_3\beta + t_6\gamma)ba\} + J(A)^3
\end{aligned}
$$

Because $a^2 + J(A)^3$, $ab + J(A)^3$, $ba + J(A)^3$ form a basis of $J(A)^2/J(A)^3$, this implies that

$$
0 = \begin{vmatrix}
\alpha + \lambda_4\gamma & 0 & \beta + t_4\gamma \\
\lambda_2\beta + \lambda_5\gamma & \alpha + \mu_2\beta + \mu_5\gamma & t_5\gamma \\
\lambda_3\beta + \lambda_6\gamma & \mu_3\beta + \mu_6\gamma & t_3\beta + t_6\gamma
\end{vmatrix}
$$

$$
\begin{aligned}
= \ & t_3\alpha^2\beta + t_6\alpha^2\gamma + (\mu_2 t_3 - \lambda_3)\alpha\beta^2 \\
+ \ & (\mu_2 t_6 + \mu_5 t_3 + \lambda_4 t_3 - \mu_3 t_5 - \lambda_6 - \lambda_3 t_4)\alpha\beta\gamma \\
+ \ & (\mu_5 t_6 + \lambda_4 t_6 - \mu_6 t_5 - \lambda_6 t_4)\alpha\gamma^2 + (\lambda_2\mu_3 - \lambda_3\mu_2)\beta^3 \\
+ \ & (\lambda_4\mu_2 t_3 + \lambda_2\mu_6 + \lambda_5\mu_3 + \lambda_2\mu_3 t_4 - \lambda_3\mu_5 - \lambda_6\mu_2 - \lambda_3\mu_2 t_4)\beta^2\gamma \\
+ \ & (\lambda_4\mu_2 t_6 + \lambda_4\mu_5 t_3 - \lambda_4\mu_3 t_5 + \lambda_5\mu_6 + \lambda_2\mu_6 t_4 \\
+ \ & \lambda_5\mu_3 t_4 - \lambda_6\mu_5 - \lambda_3\mu_5 t_4 - \lambda_6\mu_2 t_4)\beta\gamma^2 \\
+ \ & (\lambda_4\mu_5 t_6 - \lambda_4\mu_6 t_5 + \lambda_5\mu_6 t_4 - \lambda_6\mu_5 t_4)\gamma^3
\end{aligned}
$$

for $\alpha, \beta, \gamma \in F$. Because F is infinite, this implies that all coefficients on the right hand side have to vanish and, in particular,

$$
0 = t_3 = t_6 = \lambda_3 = \lambda_2\mu_3 \quad \text{and} \quad \lambda_6 = -\mu_3 t_5.
$$

Then, similarly, we have

$$
\begin{aligned}
J(A)^2 \ \neq \ & F\{a(\alpha a + \beta b + \gamma c), b(\alpha a + \beta b + \gamma c), c(\alpha a + \beta b + \gamma c)\} \\
+ \ & J(A)^3 \\
= \ & F\{\alpha a^2 + \beta ab, \lambda_2\beta a^2 + (\mu_2\beta + \mu_3\gamma)ab + \alpha ba, \\
& (\lambda_4\alpha + \lambda_5\beta + \lambda_6\gamma)a^2 + (\mu_5\beta + \mu_6\gamma)ab + (t_4\alpha + t_5\beta)ba\} \\
+ \ & J(A)^3
\end{aligned}
$$

for $\alpha, \beta, \gamma \in F$. As before, we work out the corresponding determinant and obtain

$$
\begin{aligned}
0 \ = \ & (\mu_2 t_4 - \mu_5 + \lambda_4)\alpha^2\beta + (\mu_3 t_4 - \mu_6)\alpha^2\gamma \\
+ \ & (\mu_2 t_5 - \lambda_2 t_4 + \lambda_5)\alpha\beta^2 - \lambda_2 t_5\beta^3
\end{aligned}
$$

for $\alpha, \beta, \gamma \in F$. Again, this implies that

$$
\mu_5 = \mu_2 t_4 + \lambda_4, \ \mu_6 = \mu_3 t_4, \ \lambda_5 = \lambda_2 t_4 - \mu_2 t_5, \ \lambda_2 t_5 = 0
$$

On the other hand,

$$
\begin{aligned}
J(A)^2 &= F\{[a,b],[a,c],[b,c]\} + J(A)^3 \\
&= F\{ab - ba, \lambda_4 a^2 + t_4 ba, (\mu_2 t_5 - \lambda_2 t_4)a^2 \\
&\quad + (\mu_3 - \lambda_4 - \mu_2 t_4)ab - t_5 ba\} + J(A)^3
\end{aligned}
$$

Bearing in mind that $a^2 + J(A)^3, ab + J(A)^3$ and $ba + J(A)^3$ form a basis of $J(A)^2/J(A)^3$, a computation of the corresponding determinant gives

$$
\lambda_4 t_5 + \mu_2 t_4 t_5 - \lambda_2 t_4^2 - \lambda_4 \mu_3 + \lambda_4^2 + \lambda_4 \mu_2 t_4 \neq 0
$$

Furthermore, because $J(A)^2 = F\{a^2, ab, ba\} + J(A)^3$, it follows from Step 2 in the proof of Theorem 1.2 that

$$
\begin{aligned}
J(A)^3 &= F\{a^3, a^2 b, aba, ba^2, bab, b^2 a\} + J(A)^4 \\
&= F\{a^3, a^2 b, aba, ba^2, bab\} + J(A)^4
\end{aligned}
$$

From now on, we distinguish two cases, namely $\lambda_4 \neq 0$ and $\lambda_4 = 0$.

Assume that $\lambda_4 \neq 0$. Repalcing a by $\lambda_4 a$, if necessary, we may assume that $\lambda_4 = 1$. Then

$$
\begin{aligned}
0 &\equiv a(ca) - (ac)a \equiv a^3 + t_4 aba \pmod{J(A)^4} \\
0 &\equiv b(ca) - (bc)a \equiv (\mu_2 t_4 - \lambda_2 t_4^2 - \mu_3)aba + ba^2 \pmod{J(A)^4}
\end{aligned}
$$

Assume that $\mu_3 \neq 0$. Then we have $\lambda_2 = 0$ since $\lambda_2 \mu_3 = 0$. Furthermore,

$$
\begin{aligned}
0 &\equiv (b^2)c - b(bc) \equiv \mu_2 \mu_3 a^2 b - \mu_3 bab \pmod{J(A)^4} \\
0 &\equiv a(c^2) - (ac)c \equiv \mu_3 t_4 a^2 b + \mu_3 t_4 t_5 aba \pmod{J(A)^4}
\end{aligned}
$$

and, in particular, $J(A)^3 = F a^2 b + F aba + J(A)^4$. Suppose that $a^2 b + J(A)^4$ and $aba + J(A)^4$ are linearly independent. Then $t_4 = 0$ and we obtain the contradiction

$$
0 \equiv a(cb) - (ac)b \equiv a^2 b + t_5 aba \pmod{J(A)^4}
$$

Thus $a^2 b + J(A)^4$ and $aba + J(A)^4$ are linearly dependent. Because $J(A)^3 \neq J(A)^4$, this implies that $dim_F J(A)^3/J(A)^4 = 1$. Thus, by Lemma 2.3(iii), $dim_F A = 8$.

Assume that $\mu_3 = 0$ and $\mu_2 \neq 0$. Replacing b by $\mu_2^{-1}b$, we may assume that $\mu_2 = 1$. Then we have

$$0 \equiv (b^2)b - b(b^2) \equiv (1 + \lambda_2)a^2b - \lambda_2^2 t_4^2 aba - bab \pmod{J(A)^4}$$
$$0 \equiv a(cb) - (ac)b \equiv (1 + t_4)a^2b + (t_4 t_5 - \lambda_2 t_4^2 + t_5)aba \pmod{J(A)^4}$$

and, in particular, $J(A)^3 = Fa^2b + Faba + J(A)^4$. If $a^2b + J(A)^4$ and $aba + J(A)^4$ are linearly independent, then $t_4 = -1$ and $\lambda_2 = 0$, which leads to the contradiction

$$\lambda_4 t_5 + \mu_2 t_4 t_5 - \lambda_2 t_4^2 - \lambda_4 \mu_3 + \lambda_4^2 + \lambda_4 \mu_2 t_4 = 0$$

Thus $a^2b + J(A)^4$ and $aba + J(A)^4$ are linearly dependent. Because $J(A)^3 \neq J(A)^4$, we must have $dim_F J(A)^3/J(A)^4 = 1$. Hence, by Lemma 2.3(iii), $dim_F A = 8$.

Assume that $\mu_3 = 0$ and $\mu_2 = 0$. Then we have

$$0 \equiv a(cb) - (ac)b \equiv a^2b + (t_5 - \lambda_2 t_4^2)aba \pmod{J(A)^4}$$
$$0 \equiv b(cb) - (bc)b \equiv \lambda_2^2 t_4^3 aba + bab \pmod{J(A)^4}$$

and, in particular, $J(A)^3 = Faba + J(A)^4$. As in the previous case, we obtain $dim_F A = 8$.

Now consider the case $\lambda_4 = 0$. Then

$$0 \neq \lambda_4 t_5 + \mu_2 t_4 t_5 - \lambda_2 t_4^2 - \lambda_4 \mu_3 + \lambda_4^2 + \lambda_4 \mu_2 t_4$$
$$= \mu_2 t_4 t_5 - \lambda_2 t_4^2$$

and, in particular, $t_4 \neq 0$. Replacing b by $t_4 b$, we may assume that $t_4 = 1$. Consequently,

$$0 \equiv a(ca) - (ac)a \equiv aba \pmod{J(A)^4}$$

We distinguish two more cases, namely $\lambda_2 \neq 0$ and $\lambda_2 = 0$.

Assume that $\lambda_2 \neq 0$. Then $\mu_3 = t_5 = 0$, since $0 = \lambda_2 \mu_3 = \lambda_2 t_5$. Replacing a by $\sqrt{\lambda_2}\, a$, we may therefore assume that $\lambda_2 = 1$. Then

$$0 \equiv b(ca) - (bc)a \equiv a^3 \pmod{J(A)^4}$$
$$0 \equiv b(cb) - (bc)b \equiv ba^2 + \mu_2 bab \pmod{J(A)^4}$$
$$0 \equiv a(cb) - (ac)b \equiv \mu_2 a^2b \pmod{J(A)^4}$$
$$0 \equiv (b^2)b - b(b^2) \equiv a^2b \pmod{J(A)^4}$$

and, in particular, $J(A)^3 = Fbab + J(A)^4$. As before, we deduce that $dim_F A = 8$.

Assume that $\lambda_2 = 0$. Then we have

$$0 \neq \mu_2 t_4 t_5 - \lambda_2 t_4 = \mu_2 t_5$$

i.e. $\mu_2 \neq 0$ and $t_5 \neq 0$. Replacing a by $\mu_2 a$, we may assume that $\mu_2 = 1$. Then

$$
\begin{aligned}
0 &\equiv a(cb) - (ac)b \equiv a^2 b - t_5 a^3 \pmod{J(A)^4} \\
0 &\equiv (b^2)b - b(b^2) \equiv t_5 a^3 - bab \pmod{J(A)^4} \\
0 &\equiv (bc)c - b(c^2) \equiv (\mu_3^2 t_5 - \mu_3 t_5)a^3 + \mu_3 t_5 ba^2 \pmod{J(A)^4}
\end{aligned}
$$

and, in particular, $J(A)^3 = Fa^3 + Fba^2 + J(A)^4$. Assume by way of contradiction that $a^3 + J(A)^4$ and $ba^2 + J(A)^4$ are linearly independent. Then $\mu_3 = 0$ since $t_5 \neq 0$, which leads to the contradiction

$$0 \equiv b(cb) - (bc)b \equiv t_5 a^3 - t_5 ba^2 \pmod{J(A)^4}$$

Thus $a^3 + J(A)^4$ and $ba^2 + J(A)^4$ are linearly dependent. As in the previous cases, we deduce that $dim_F A = 8$.

Step 3. Assume that $d = b^2$. Then we may also assume that

$$ba \in Fa^2 + Fab + J(A)^3,$$

since otherwise we are in Step 2. Similarly, we may assume that

$$ba \in Fb^2 + Fab + J(A)^3$$

since otherwise we interchange a and b and are in Step 2 again. Thus $ba \in Fab + J(A)^3$ and so there exists $\lambda \in F$ such that

$$ba \equiv \lambda ab \pmod{J(A)^3}$$

Setting $b' = a + b$, we then have

$$ab' = a^2 + ab, (b')^2 \equiv a^2 + (1 + \lambda)ab + b^2 \pmod{J(A)^3}$$

and, in particular, $J(A) = F\{a, b, c\} + J(A)^2$ and

$$J(A)^2 = F\{a^2, ab', (b')^2\} + J(A)^3$$

We may thus similarly assume that $b'a \in Fab' + J(A)^3$.

Now write $b'a \equiv \mu ab'$ (mod $J(A)^3$) for some $\mu \in F$. Then

$$\begin{aligned} \mu a^2 + \mu ab &\equiv \mu ab' \equiv b'a \equiv (a+b)a \\ &\equiv a^2 + ba \equiv a^2 + \lambda ab \quad (\text{mod } J(A)^3) \end{aligned}$$

Because $a^2 + J(A)^3$ and $ab + J(A)^3$ are linearly independent, we deduce that $\lambda = \mu = 1$. In particular,

$$[a, b] \in J(A)^3 \quad \text{and} \quad J(A)^2 = F[a, c] + F[b, c] + J(A)^3,$$

contrary to the fact that $dim_F J(A)^2 / J(A)^3 = 3$.

Step 4.   Assume that $d = bc$. Then we may also assume that

$$ac, ba, b^2 \in Fa^2 + Fab + J(A)^3$$

since otherwise we are in Step 1, 2 or 3. Now write

$$\begin{aligned} ac &\equiv \lambda_1 a^2 + \mu_1 ab \quad (\text{mod } J(A)^3) \\ ba &\equiv \lambda_2 a^2 + \mu_2 ab \quad (\text{mod } J(A)^3) \\ b^2 &\equiv \lambda_3 a^2 + \mu_3 ab \quad (\text{mod } J(A)^3) \end{aligned}$$

for some $\lambda_i, \mu_i \in F, 1 \le i \le 3$. Replacing c by $c - \lambda_1 a - \mu_1 b$, we may therefore assume that $0 = \lambda_1 = \mu_1$.

Given $\alpha \in F$, we have $J(A)^2 \ne (a + \alpha b)J(A)$ by our hypothesis. Thus, by Nakayama's lemma

$$\begin{aligned} J(A)^2 \ne\ & (a + \alpha b) + J(A)^3 \\ =\ & F\{(a + \alpha b)a, (a + \alpha b)b, (a + \alpha b)c\} + J(A)^3 \\ =\ & F\{(1 + \lambda_2 \alpha)a^2 + \mu_2 \alpha ab, \lambda_3 \alpha a^2 + (1 + \mu_3)ab, \alpha bc\} + J(A)^3 \end{aligned}$$

Because $a^2 + J(A)^3, ab + J(A)^3, bc + J(A)^3$ form a basis of $J(A)^2 / J(A)^3$, this gives

$$0 = \begin{vmatrix} 1 + \lambda_2 \alpha & \mu_2 \alpha & 0 \\ \lambda_3 \alpha & 1 + \mu_3 \alpha & 0 \\ 0 & 0 & \alpha \end{vmatrix}$$

$$= \alpha + (\lambda_2 + \mu_3)\alpha^2 + (\lambda_2\mu_3 - \lambda_3\mu_2)\alpha^3$$

for all $\alpha \in F$. This is impossible, since F is infinite.

Step 5. Assume that $d = ca$. Then we can write

$$
\begin{aligned}
ac &\equiv \lambda_1 a^2 + \mu_1 ab + t_1 ca \quad (\text{mod } J(A)^3) \\
ba &\equiv \lambda_2 a^2 + \mu_2 ab + t_2 ca \quad (\text{mod } J(A)^3) \\
b^2 &\equiv \lambda_3 a^2 + \mu_3 ab + t_3 ca \quad (\text{mod } J(A)^3) \\
bc &\equiv \lambda_4 a^2 + \mu_4 ab + t_4 ca \quad (\text{mod } J(A)^3) \\
cb &\equiv \lambda_5 a^2 + \mu_5 ab + t_5 ca \quad (\text{mod } J(A)^3) \\
c^2 &\equiv \lambda_6 a^2 + \mu_6 ab + t_6 ca \quad (\text{mod } J(A)^3)
\end{aligned}
$$

for some $\lambda_i, \mu_i, t_i \in F, 1 \leq i \leq 6$. We may assume that $t_i = 0$ for $i \in \{1, 2, 3, 4\}$, since otherwise we are in Steps 1, 2, 3, 4, respectively. Replacing c by $c - \lambda_1 a - \mu_1 b$, we may also assume that $\lambda_1 = \mu_1 = 0$. Furthermore, $\mu_2 = 0$ since otherwise we are in Step 1 for the opposite algebra of A which leads to a contradiction. Similarly, we may assume that $\lambda_3 = 0$, since otherwise we interchange a and b and are then in Step 4 for the opposite algebra of A. Again, replacing b by $b - t_5 a$, we may then assume that $t_5 = 0$. Moreover, we may assume that $\lambda_6 = 0$, since otherwise we replace (a, b, c) by (b, c, a) and are then in Step 4 again. Finally, we may also assume that $\mu_6 = 0$, since otherwise we interchange b and c and are then in Step 3 for the opposite algebra of A.

As in Step 2, we now have

$$
\begin{aligned}
J(A)^2 \neq\ & F\{(\alpha a + \beta b + c)a, (\alpha a + \beta b + c)b, (\alpha a + \beta b + c)c\} + J(A)^3 \\
=\ & F\{(\alpha + \lambda_2\beta)a^2 + ca, \lambda_5 a^2 + (\alpha + \mu_3\beta + \mu_5)ab, \\
& \lambda_4\beta a^2 + \mu_4\beta ab + t_6 ca\} + J(A)^3
\end{aligned}
$$

for all $\alpha, \beta \in F$. Because $a^2 + J(A)^3, ab + J(A)^3, ca + J(A)^3$ form a basis of $J(A)^2/J(A)^3$, we may compute the corresponding determinant and obtain

$$
\begin{aligned}
0 =\ & t_6\alpha^2 + (\mu_3 t_6 + \lambda_2 t_6 - \lambda_4)\alpha\beta + \mu_5 t_6\alpha \\
& + (\lambda_2\mu_3 t_6 - \lambda_4\mu_3)\beta^2 + (\lambda_2\mu_5 t_6 + \lambda_5\mu_4 - \lambda_4\mu_5)\beta
\end{aligned}
$$

for all $\alpha, \beta \in F$.

Because F is infinite, this implies that all coefficients on the right hand side vanish and, in particular, $t_6 = \lambda_4 = 0$. Then, similarly,

$$
\begin{aligned}
J(A)^2 \;\neq\; & F\{a(a + \beta b + c), b(a + \beta b + c), c(a + \beta b + c)\} + J(A)^3 \\
= \; & F\{a^2 + \beta ab, \lambda_2 a^2 + (\mu_3 \beta + \mu_4)ab, \lambda_5 \beta a^2 + \mu_5 \beta ab + ca\} \\
& + J(A)^3
\end{aligned}
$$

for all $\beta \in F$. Computing the corresponding determinant, we have

$$
0 = (\mu_3 - \lambda_2)\beta + \mu_4
$$

for all $\beta \in F$, which implies $\mu_3 = \lambda_2$ and $\mu_4 = 0$.

Finally, we may assume that

$$
\begin{aligned}
J(A)^2 \;\neq\; & F\{(\alpha a + b + c)^2, (\alpha a + b + c)a, a(\alpha a + b + c)\} + J(A)^3 \\
= \; & F\{(\alpha^2 + \lambda_2 \alpha + \lambda_5)a^2 + (\alpha + \lambda_2 + \mu_4 + \mu_5)ab + \alpha ca, \\
& (\alpha + \lambda_2)a^2 + ca, \alpha a^2 + ab\} + J(A)^3
\end{aligned}
$$

for all $\alpha \in F$; indeed, otherwise we replace (a, b) by $(\alpha a + b + c, a)$ and are in Step 2 again. Computing the corresponding determinant, we obtain

$$
0 = \alpha^2 + (\lambda_2 + \mu_4 + \mu_5)\alpha - \lambda_5
$$

for all $\alpha \in F$, which is impossible.

Step 6.   Assume that $d = cb$. Then we may write

$$
\begin{aligned}
ac &\equiv \lambda_1 a^2 + \mu_1 ab + t_1 cb \pmod{J(A)^3} \\
ba &\equiv \lambda_2 a^2 + \mu_2 ab + t_2 cb \pmod{J(A)^3} \\
b^2 &\equiv \lambda_3 a^2 + \mu_3 ab + t_3 cb \pmod{J(A)^3} \\
bc &\equiv \lambda_4 a^2 + \mu_4 ab + t_4 cb \pmod{J(A)^3} \\
ca &\equiv \lambda_5 a^2 + \mu_5 ab + t_5 cb \pmod{J(A)^3} \\
c^2 &\equiv \lambda_6 a^2 + \mu_6 ab + t_6 cb \pmod{J(A)^3}
\end{aligned}
$$

for some $\lambda_i, \mu_i, t_i \in F$, $1 \leq i \leq 6$. We may assume that $t_i = 0$ for $i \in \{1, \ldots, 5\}$, since otherwise we are in Step 1, 2, $\ldots$, 5, respectively.

Replacing c by $c - \lambda_1 a - \mu_1 b$, we may then assume that $\lambda_1 = \mu_1 = 0$. We may also assume that $\mu_2 = 0$, since otherwise we are in Step 4 for the opposite algebra of A. Similarly, $\lambda_3 = 0$ since otherwise we interchange a and b and are then in Step 1 for the opposite algebra of A.

As in the previous cases, we obtain

$$
\begin{aligned}
J(A)^2 \neq\ & F\{(\alpha a + \beta b + c)a, (\alpha a + \beta b + c)b, (\alpha a + \beta b + c)c\} + J(A)^3 \\
=\ & F\{(\alpha + \lambda_2\beta + \lambda_5)a^2 + \mu_5 ab, (\alpha + \mu_3\beta)ab + cb, \\
& (\lambda_4\beta + \lambda_6)a^2 + (\mu_4\beta + \mu_6)ab + t_6 cb\} + J(A)^3
\end{aligned}
$$

for all $\alpha, \beta \in F$. Computing the corresponding determinant, we obtain

$$
\begin{aligned}
0 =\ & t_6\alpha^2 + (\mu_3 t_6 + \lambda_2 t_6 - \mu_4)\alpha\beta + (\lambda_5 t_6 - \mu_6)\alpha \\
& + (\lambda_2\mu_3 t_6 - \lambda_2\mu_4)\beta^2 \\
& + (\lambda_5\mu_3 t_6 - \lambda_2\mu_6 - \lambda_5\mu_4 + \lambda_4\mu_5)\beta + (\lambda_6\mu_5 - \lambda_5\mu_6)
\end{aligned}
$$

for all $\alpha, \beta \in F$. Thus the coefficients on the right hand side must vanish and, in particular, $0 = t_6 = \mu_4 = \mu_6$. Similarly, we have

$$
\begin{aligned}
J(A)^2 \neq\ & F\{a(\alpha a + b + c), b(\alpha a + b + c), c(\alpha a + b + c)\} + J(A)^3 \\
=\ & F\{\alpha a^2 + ab, (\lambda_2\alpha + \lambda_4)a^2 + \mu_3 ab, \\
& (\lambda_5\alpha + \lambda_6)a^2 + \mu_5\alpha ab + cb\} + J(A)^3
\end{aligned}
$$

for all $\alpha \in F$. We work out the corresponding determinant and obtain

$$
0 = (\mu_3 - \lambda_2)\alpha - \lambda_4
$$

for all $\alpha \in F$, which implies that $\mu_3 = \lambda_2$ and $\lambda_4 = 0$.

We may also assume that

$$
\begin{aligned}
J(A)^2 \neq\ & F\{(\alpha a + \beta b + c)^2, (\alpha a + \beta b + c)a, a(\alpha a + \beta b + c)\} + J(A)^3 \\
=\ & F\{(\alpha^2 + \lambda_2\alpha\beta + \lambda_5\alpha + \lambda_6)a^2 + (\alpha\beta + \mu_5\alpha + \lambda_2\beta^2)ab + \beta cb, \\
& (\alpha + \lambda_2\beta + \lambda_5)a^2 + \mu_5 ab, \alpha a^2 + \beta ab\} + J(A)^3
\end{aligned}
$$

for all $\alpha, \beta \in F$, since otherwise we replace (a, b, c) by $(\alpha a + \beta b + c, a, b)$ and are then in Step 2 again. Computing the corresponding determinant, we obtain

$$
0 = \alpha\beta^2 + \lambda_2\beta^3 - \mu_5\alpha\beta + \lambda_5\beta^2
$$

for all $\alpha, \beta \in F$, which is impossible.

Step 7. Assume that $d = c^2$. Then we may also assume that

$$ac, ba, b^2, bc, ca, cb \in Fa^2 + Fab + J(A)^3$$

since otherwise we are in Step 1, 2, ..., 6, respectively. Now write

$$\begin{aligned}
ac &\equiv \lambda_1 a^2 + \mu_1 ab \pmod{J(A)^3} \\
ca &\equiv \lambda_2 a^2 + \mu_2 ab \pmod{J(A)^3} \\
cb &\equiv \lambda_3 a^2 + \mu_3 ab \pmod{J(A)^3}
\end{aligned}$$

for some $\lambda_i, \mu_i \in F$, $i = 1, 2, 3$. Replacing c by $c - \lambda_1 a - \mu_1 b$, we may assume that $\lambda_1 = \mu_1 = 0$. As in the previous cases, we obtain

$$\begin{aligned}
J(A)^2 \ &\neq\ F\{(a + \alpha c)a, (a + \alpha c)b, (a + \alpha c)c\} + J(A)^3 \\
&=\ F\{(1 + \lambda_2\alpha)a^2 + \mu_2\alpha ab, \lambda_3\alpha a^2 + (1 + \mu_3\alpha)ab, \alpha c^2\} + J(A)^3
\end{aligned}$$

for all $\alpha \in F$. Computing the corresponding determinant, we obtain a contradiction as in Step 4. ■

3.   Symmetric local algebras A with $dim_F Z(A) = 5$

Our aim is to prove a theorem due to Chlebowitz and Külshammer (1989) which asserts that a symmetric local algebra over an algebraically closed field whose center is 5-dimensional has dimension 5 or 8. We can slightly relax the above assumptions, by virtue of the following observation.

3.1. Lemma. *Let A be a symmetric algebra over a field F such that $dim_F A/J(A) = 1$, let E be the algebraic closure of F and let $A_E = E \otimes_F A$. Then A_E is a symmetric E-algebra with*

$$dim_E A_E = dim_F A \quad and \quad dim_E Z(A_E) = dim_F Z(A)$$

Proof. It is clear that $dim_E A_E = dim_F A$ and that $dim_E Z(A_E) = dim_F Z(A)$. Since

$$(A/J(A))_E \cong A_E/J(A)_E$$

it follows that $J(A)_E = J(A_E)$ and $dim_E A_E/J(A_E) = 1$. Finally, by Proposition 2.1.11(ii), A_E is a symmetric E-algebra, which proves the result. ∎

3.2. Theorem. *(Chlebowitz and Külshammer (1989)). Let A be a symmetric algebra over a field F such that*

$$dim_F A/J(A) = 1 \quad and \quad dim_F Z(A) = 5$$

Then $dim_F A = 5$ or $dim_F A = 8$.

Proof. By Lemma 3.1, we may assume that F is algebraically closed. By lemma 2.2, it suffices to consider the following cases
(a) $dim_F J(A)/J(A)^2 = dim_F J(A)^2/J(A)^3 = 2$.
(b) $dim_F J(A)/J(A)^2 = 3, dim_F J(A)^2/J(A)^3 = 2$.
(c) $dim_F J(A)/J(A)^2 = dim_F J(A)^2/J(A)^3 = 3$.
By Lemma 2.4, the case (b) must be excluded. By Lemma 2.5, (a) implies that $dim_F A = 8$. Thus it remains to treat the case (c). By Lemma 2.6(iii), we may assume that there exists $a \in J(A)$ such that

$$J(A)^2 = aJ(A) \quad or \quad J(A)^2 = J(A)a$$

By symmetry, we may assume that $J(A)^2 = aJ(A)$. Since $J(A)^2 \neq J(A)^3$ (Lemma 2.2(i)), there exist $b, c \in J(A)$ such that

$$J(A) = F\{a, b, c\} + J(A)^2$$

Then Lemma 2.6(ii) implies that

$$\begin{aligned} F\{[a, b], [a, c], [b, c]\} + J(A)^3 &= [A, A] + J(A)^3 \\ &= J(A)^2 = aJ(A) \\ &= F\{a^2, ab, ac\} + J(A)^3 \end{aligned}$$

and

$$J(A)^3 = aJ(A)^2 = a^2 J(A) = F\{a^3, a^2 b, a^2 c\} + J(A)^4$$

In particular, we must have $dim_F J(A)^3/J(A)^4 \leq 3$. Owing to Lemma 2.3(iii), we may assume that $dim_F J(A)^3/J(A)^4 \in \{2, 3\}$. The rest of the proof will be devoted to demonstrating that this configuration is impossible, which will complete the proof. Note that $J(A)^4 \neq 0$, since otherwise $dim_F J(A)^3 \leq 1$ by Lemma 2.1(iv). By Nakayama's lemma, we have $J(A)^4 \neq J(A)^5$. Because

$$J(A)^4 = aJ(A)^3 = a^3 J(A),$$

we deduce that $a^3 \notin J(A)^4$. For the sake of clarity, we divide the rest of the proof into a number of steps.

Step 1. Here we prove that

$$J(A)^2 = [A, A] + J(A)^4$$

By Lemma 2.1(i), $[A, A] + J(A)^4 \subseteq J(A)^2$. Assume by way of contradiction that $[A, A] + J(A)^4 \neq J(A)^2$. Then

$$dim_F A/([A, A] + J(A)^4) \geq dim_F A/J(A)^2 = 4$$

and therefore

$$5 \leq dim_F A/([A, A] + J(A)^4) \leq dim_F A/[A, A] = 5$$

proving that $J(A)^4 \subseteq [A, A] + J(A)^4 = [A, A]$. But A is symmetric, hence $J(A)^4 = 0$, a contradiction.

Step 2. Let us write

$$
\begin{aligned}
ba &\equiv \lambda_1 a^2 + \mu_1 ab + t_1 ac \pmod{J(A)^3} \\
b^2 &\equiv \lambda_2 a^2 + \mu_2 ab + t_2 ac \pmod{J(A)^3} \\
bc &\equiv \lambda_3 a^2 + \mu_3 ab + t_3 ac \pmod{J(A)^3} \\
ca &\equiv \lambda_4 a^2 + \mu_4 ab + t_4 ac \pmod{J(A)^3} \\
cb &\equiv \lambda_5 a^2 + \mu_5 ab + t_5 ac \pmod{J(A)^3} \\
c^2 &\equiv \lambda_6 a^2 + \mu_6 ab + t_6 ac \pmod{J(A)^3}
\end{aligned}
$$

for some $\lambda_i, \mu_i, t_i \in F$, $1 \leq i \leq 6$. Our aim is to show that a, b, c can be chosen such that one of the following holds:

$$\lambda_1 = \mu_1 = \lambda_4 = 0, t_1 = 1, \lambda_5 = \lambda_3 - 1, \mu_4 + t_4 \neq 1 \qquad (1)$$

$$\lambda_1 = \mu_1 = 0, t_1 = \lambda_4 = 1, t_4 = 1 - \mu_4, \mu_5 - \mu_3 + t_5 - t_3 \neq 0 \quad (2)$$

$$\lambda_1 = t_1 = \lambda_4 = \mu_4 = 0, t_4 = \mu_1 \neq 1, \lambda_3 = 1 \neq \lambda_5 \quad (3)$$

We distinguish two cases, namely $t_1 \neq 0$ and $t_1 = 0$. Assume that $t_1 \neq 0$. Then, replacing c by $\lambda_1 a + \mu_1 b + t_1 c$, we may assume that $\lambda_1 = \mu_1 = 0$ and $t_1 = 1$. Now consider two subcases, namely $\mu_4 + t_4 \neq 1$ and $\mu_4 + t_4 = 1$.

Assume that $\mu_4 + t_4 \neq 1$. Then, setting $\alpha = \lambda_4(\mu_4 + t_4 - 1)^{-1}$ and replacing b by $b + \alpha a$ and c by $c + \alpha a$, we have $\lambda_4 = 0$. Hence

$$
\begin{aligned}
J(A)^2 &= F\{[a,b],[a,c],[b,c]\} + J(A)^3 \\
&= F\{ab - ac, \mu_4 ab + (t_4 - 1)ac, (\lambda_3 - \lambda_5)a^2 \\
&\quad + (\mu_3 - \mu_5)ab + (t_3 - t_5)ac\} + J(A)^3
\end{aligned}
$$

and, in particular, $\lambda_3 \neq \lambda_5$. Now we replace a by $(\lambda_3 - \lambda_5)^{1/2}a$ and may then assume that $\lambda_5 = \lambda_3 - 1$.

Assume that $\mu_4 + t_4 = 1$. Then we have

$$
\begin{aligned}
J(A)^2 &= F\{[a,b],[a,c],[b,c]\} + J(A)^3 \\
&= F\{ab - ac, \lambda_4 a^2 + \mu_4 ab - \mu_4 ac, \\
&\quad (\lambda_3 - \lambda_5)a^2 + (\mu_3 - \mu_5)ab + (t_3 - t_5)ac\} + J(A)^3
\end{aligned}
$$

Now $a^2 + J(A)^3, ab + J(A)^3, ac + J(A)^3$ form a basis of $J(A)^2/J(A)^3$. Computing the corresponding determinant, we then obtain $0 \neq \lambda_4(\mu_5 - \mu_3 + t_5 - t_3)$. Hence

$$\mu_5 - \mu_3 + t_5 - t_3 \neq 0 \neq \lambda_4$$

and, replacing a by $\lambda_4 a$, we may therefore assume that $\lambda_4 = 1$.

Assume that $t_1 = 0$. Then we may also assume that $\mu_4 = 0$, since otherwise we interchange b and c and are in case $t_1 \neq 0$ again. Similarly, we may assume that $t_4 = \mu_1$, since otherwise we replace b by $b + c$ and are then in case $t_1 \neq 0$ again. Thus

$$
\begin{aligned}
J(A)^2 &= F\{[a,b],[a,c],[b,c]\} + J(A)^3 \\
&= F\{\lambda_1 a^2 + (\mu_1 - 1)ab, \lambda_4 a^2 + (\mu_1 - 1)ac, \\
&\quad (\lambda_3 - \lambda_5)a^2 + (\mu_3 - \mu_5)ab + (t_3 - t_5)ac\} + J(A)^3
\end{aligned}
$$

Because $dim_F J(A)^2/J(A)^3 = 3$, this implies that $\mu_1 \neq 1$. Now we replace b by $b + \lambda_1(\mu_1 - 1)^{-1}a$ and c by $c + \lambda_4(\mu_1 - 1)^{-1}a$ and may then assume that $\lambda_1 = \lambda_4 = 0$. Then we have $\lambda_3 \neq 0$ or $\lambda_5 \neq 0$. Interchanging b and c, if necessary, we may then assume that $\lambda_3 \neq 0$. Finally, replacing b by $\lambda_3^{-1}b$, we may assume that $\lambda_3 = 1$.

Step 3. Here we prove that case (1) does not occur. We argue by contradiction and distinguish two cases, namely $dim_F J(A)^3/J(A)^4 = 3$ and $dim_F J(A)^3/J(A)^4 = 2$.

Assume that $dim_F J(A)^3/J(A)^4 = 3$. Then the elements $a^3 + J(A)^4, a^2b + J(A)^4, a^2c + J(A)^4$ form a basis of $J(A)^3/J(A)^4$. Since

$$
\begin{aligned}
0 &\equiv (b^2)a - b(ba) \equiv (\lambda_2 - \lambda_6)a^3 + (\mu_4t_2 - \mu_6)a^2b \\
&\quad + (\mu_2 + t_2t_4 - t_6)a^2c \pmod{J(A)^4}
\end{aligned}
$$

we deduce that $\lambda_6 = \lambda_2, \mu_6 = \mu_4t_2$ and $t_6 = \mu_2 + t_2t_4$. Similarly, using the fact that $0 = (bc)a - b(ca) + c(ba) - (cb)a + J(A)^4$, we obtain $\mu_4 = -1$, so $t_4 \neq 2$. This also demonstrates that $t_5 = \mu_3 - \mu_5 + t_3$ and $0 = (2 - t_4)(\mu_3 - \mu_5)$. Because $t_4 \neq 2$, it follows that $\mu_5 = \mu_3$ and $t_5 = t_3$. Then, using the fact that $0 = (b^2)b - b(b^2) + J(A)^4$ and $0 = (bc)b - b(cb) + J(A)^4$, we deduce that

$$
0 = (\lambda_2 - \lambda_3 + 1)(\mu_2 - t_2) = (\lambda_2 - \lambda_3 + 1)(\mu_3 - t_3)
$$

We now distinguish two cases $\lambda_3 \neq \lambda_2 + 1$ and $\lambda_3 = \lambda_2 + 1$.

Suppose that $\lambda_3 \neq \lambda_2 + 1$. Then $t_2 = \mu_2$ and $t_3 = \mu_3$. Furthermore, since $0 = (bc)a - b(ca) + J(A)^4$, we have $\mu_2t_4 = 0$. Assume that $t_4 \neq 0$, so $\mu_2 = 0$. Then, using the fact that $0 = (bc)b - b(cb) + J(A)^4$, we obtain $0 = t_4(1 - \lambda_3)$, so $\lambda_3 = 1$. This, however, is impossible since $0 = (bc)b - b(cb) + J(A)^4$. Assume that $t_4 = 0$. Then, using the fact that $0 = (bc)b - b(cb) + J(A)^4$, we obtain $2\lambda_3 = 1$ and, in particular, $char F \neq 2$. Then, using the fact that $0 = (bc)a - b(ca) + J(A)^4$, we deduce that $\mu_3 = 0$. Applying the fact that $0 = (c^2)a - c(ca) + J(A)^4$, we see that $\mu_2 = 0$. Finally, using the fact that $0 = (bc)c - b(c^2) + J(A)^4$, we also deduce that $\lambda_3 = 0$. This is, however, impossible since $0 = (bc)b - b(cb) + J(A)^4$.

Suppose that $\lambda_3 = \lambda_2 + 1$. Since $0 = (bc)a - b(ca) + J(A)$, we then have $2\lambda_2 + 1 - \lambda_2 t_4 = 0$. But $A = F\{1, a, b, c, a^2, ab, ac\} + J(A)^3$, so by Step 1,

$$
\begin{aligned}
J(A)^2 &= [A, A] + J(A)^4 \\
&= F\{[a, b], [a, c], [b, c], [a, ab], [a, ac], [b, a^2], \\
&\quad [b, ab], [b, ac], [c, a^2], [c, ab], [c, ac]\} + J(A)^4 \\
&\subseteq F\{[a, b], [a, c], [b, c], a^2b, a^2c\} + J(A)^4
\end{aligned}
$$

as is easily verified. This is a contradiction, since $\dim_F J(A)^2/J(A)^4 = 6$.

Assume that $\dim_F J(A)^3/J(A)^4 = 2$. We treat separately the cases $a^2b \in Fa^3 + J(A)^4$ and $a^2b \notin Fa^3 + J(A)^4$.

Assume that $a^2b \in Fa^3 + J(A)^4$. Then we have

$$
\begin{aligned}
J(A)^3 &= F\{a^3, a^2b, a^2c\} + J(A)^4 \\
&= F\{a^3, a^2c\} + J(A)^4
\end{aligned}
$$

and we may write

$$
a^2b \equiv \lambda a^3 \quad (\text{mod } J(A)^4)
$$

for some $\lambda \in F$. Then

$$
a^3c \equiv a^2ba \equiv \lambda a^3 \quad (\text{mod } J(A)^4)
$$

which implies that

$$
J(A)^4 = aJ(A)^3 = F\{a^4, a^3c\} + J(A)^5 = Fa^4 + J(A)^5
$$

Because $J(A)^4 \neq J(A)^5$, the latter forces $\dim_F J(A)^4/J(A)^5 = 1$. By Step 4 in the proof of Theorem 1.2, we have $J(A)^3 \subseteq Z(A)$ and, in particular, $a^2c \in Z(A)$. Thus

$$
\begin{aligned}
0 &\equiv (a^2c)a - a(a^2c) \equiv a^2(ca) - a^3c \\
&\equiv (\mu_4 + t_4 - 1)\lambda a^3 \quad (\text{mod } J(A)^4)
\end{aligned}
$$

Since $\mu_4 + t_4 \neq 1$ and $a^3 \notin J(A)^4$, we see that $\lambda = 0$, which implies

$$
\begin{aligned}
0 &\equiv a^2(bc) - (a^2b)c \equiv \lambda_3 a^4 \quad (\text{mod } J(A)^5) \\
0 &\equiv a^2(b^2) - (a^2b)b \equiv \lambda_2 a^4 \quad (\text{mod } J(A)^5) \\
0 &\equiv (b^2)a - b(ba) \equiv (\lambda_2 - \lambda_6)a^3 + (\mu_2 + t_2 t_4 - t_6)a^2c \quad (\text{mod } J(A)^4) \\
0 &\equiv (cb)a - c(ba) \equiv (\lambda_3 - 1 - \lambda_3\mu_4 - \lambda_6 t_4)a^3 \\
&\quad + (\mu_5 + t_4 t_5 - \mu_4 t_3 - t_4 t_6)a^2c \quad (\text{mod } J(A)^4)
\end{aligned}
$$

This forces the contradiction $0 = \lambda_3 = \lambda_2 = \lambda_6 = -1$.

Now assume that $a^2 b \notin Fa^3 + J(A)^4$. Because $a^3 \notin J(A)^4$ and $\dim_F J(A)^3/J(A)^4 = 2$, the elements $a^3 + J(A)^4$ and $a^2 b + J(A)^4$ form a basis of $J(A)^3/J(A)^4$. Now write $a^2 c \equiv \lambda a^3 + \mu a^2 b \pmod{J(A)^4}$ for some $\lambda, \mu \in F$. Because

$$J(A)^4 = aJ(A)^3 = Fa^4 + Fa^3 b + J(A)^5$$

and $J(A)^4 \neq J(A)^5$, we have $\dim_F J(A)^4/J(A)^5 \in \{1, 2\}$.

Assume that $\dim_F J(A)^4/J(A)^5 = 1$. Then, by Step 4 in the proof of Theorem 1.2, we have $J(A)^3 \subseteq Z(A)$. In particular, $a^2 b, a^2 c \in Z(A)$ and so

$$
\begin{aligned}
0 &\equiv (a^2 b)a - a(a^2 b) \equiv a^2(ba) - a^3 b \\
&\equiv a^3 c - a^3 b \equiv \lambda a^4 + (\mu - 1)a^3 b \pmod{J(A)^5} \\
0 &\equiv (a^2 c)a - a(a^2 c) \equiv a^2(ca) - a^3 c \\
&\equiv (t_4 - 1)\lambda a^4 + (\mu_4 + t_4\mu - \mu)a^3 b \\
&\equiv (\mu_4 + t_4 - 1)a^3 b \pmod{J(A)^5}
\end{aligned}
$$

Because $\mu_4 + t_4 \neq 1$, the latter implies that $a^3 b \in J(A)^5$, so $J(A)^4 = Fa^4 + J(A)^5$ and $\lambda a^4 \in J(A)^5$. But $\dim_F J(A)^4/J(A)^5 = 1$, so $\lambda = 0$ and therefore

$$
\begin{aligned}
0 &\equiv (b^2)a - b(ba) \equiv (\lambda_2 - \lambda_6)a^3 \\
&+ (\mu_4 t_2 - \mu_6 + \mu_2\mu + t_2 t_4\mu - t_6\mu)a^2 b \pmod{J(A)^4}
\end{aligned}
$$

In particular, $\lambda_6 = \lambda_2$. Similarly, applying the fact that

$$0 = (bc)a - b(ca) + c(ba) - (cb)a + J(A)^4$$

it follows that $\mu_4 = -1$ and, in particular, $t_4 \neq 2$. Thus

$$
\begin{aligned}
0 &\equiv a^2(cb) - (a^2 c)b \equiv (\lambda_3 - 1 - \lambda_2\mu)a^4 \pmod{J(A)^5} \\
0 &\equiv a^2(c^2) - (a^2 c)c \equiv (\lambda_2 - \lambda_3\mu)a^4 \pmod{J(A)^5}
\end{aligned}
$$

Because $a^4 \notin J(A)^5$, this forces $\lambda_2 = \lambda_3\mu$ and $\lambda_3 - \lambda_3\mu^2 = 1$; in particular, $\lambda_3 \neq 0$ and $\mu^2 \neq 1$. However, because $a^2 c \in Z(A)$, we have

$$
\begin{aligned}
0 &\equiv (a^2 c)b - b(a^2 c) \equiv (a^2 c)b - (ba)(ac) \\
&\equiv (\mu^2 + 1 - t_4\mu)\lambda_3 a^4 \pmod{J(A)^5} \\
0 &\equiv (a^2 c)b - b(a^2 c) \equiv (2 - t_4\mu)\lambda_3\mu^2 a^4 \pmod{J(A)^5}
\end{aligned}
$$

which forces $t_4\mu = 2$ and $\mu^2 = 1$, a contradiction.

Finally assume that $dim_F J(A)^4/J(A)^5 = 2$. Then

$$
\begin{aligned}
0 &\equiv (a^2 c)a - a^2(ca) \\
  &\equiv (\lambda + \lambda\mu - t_4\lambda)a^4 + (\mu^2 - \mu_4 - t_4\mu)a^3 b \quad (\text{mod } J(A)^5) \\
0 &\equiv (a^2 c)b - a^2(cb) \equiv (\lambda_2\mu + t_2\lambda\mu - \lambda_3 + 1 - t_5\lambda)a^4 \\
  &+ (\lambda + \mu_2\mu + t_2\mu^2 - \mu_5 - t_5\mu)a^3 b \quad (\text{mod } J(A)^5) \\
0 &\equiv (a^2 c)c - a^2 c^2 \equiv (\lambda_3\mu + t_3\lambda\mu - \lambda_6 - t_6\lambda + \lambda^2)a^4 \\
  &+ (\lambda\mu + \mu_3\mu + t_3\mu^3 - \mu_6 - t_6\mu)a^3 b \quad (\text{mod } J(A)^5)
\end{aligned}
$$

Taking into account that $a^4 + J(A)^5$ and $a^3 b + J(A)^5$ form a basis of $J(A)^4/J(A)^5$, the above implies that all coefficients on the right hand side vanish. In particular, $\lambda + \lambda\mu - t_4\lambda = 0$. If $\lambda \neq 0$, then $\mu = t_4 - 1$ and we obtain a contradiction $0 = \mu^2 - \mu_4 - t_4\mu = 1 - \mu_4 - t_4$. Thus we must have $\lambda = 0$, in which case

$$
\begin{aligned}
0 &\equiv (b^2)a - b(ba) \equiv (\lambda_2 - \lambda_6)a^3 \\
  &+ (\mu_4 t_2 - \mu_6 + \mu_2\mu + t_2 t_4\mu - t_6\mu)a^2 b \quad (\text{mod } J(A)^4)
\end{aligned}
$$

and, in particular, $\lambda_6 = \lambda_2$. A similar argument, by using the fact that $0 = (bc)a - b(ca) + c(ba) - (cb)a + J(A)^4$ shows that $\mu_4 = -1$. Therefore $\mu^2 - t_4\mu = -1$ and, in particular, $\mu \neq 0$. Thus

$$
0 = \lambda_2\mu + t_2\lambda\mu - \lambda_3 + 1 - t_5\lambda = \lambda_2\mu - \lambda_3 + 1
$$

and $\lambda_3 = \lambda_2\mu + 1$, which forces

$$
0 = \lambda_3\mu + t_3\lambda\mu - \lambda_6 - t_6\lambda + \lambda^2 = \lambda_2\mu^2 + \mu - \lambda_2
$$

In particular, $\lambda_2 \neq 0$ and $\mu^2 \neq 1$. This implies the contradiction

$$
\begin{aligned}
0 &\equiv b(a^2 c) - (ba)ac \equiv -\mu^2 a^4 \\
  &+ (\mu_5 t_4\mu - \mu_2\mu + t_4 t_5\mu^2 - t_2\mu^2 - \mu_6 t_4 + \mu_3 - t_4 t_6\mu + t_3\mu)a^3 b \\
  &\quad (\text{mod } J(A)^5)
\end{aligned}
$$

Step 4. Here we prove that case (2) does not occur. Again, we argue by contradiction and treat the cases $dim_F J(A)^3/J(A)^4 = 3$ and

$dim_F J(A)^3/J(A)^4 = 2$ separately.

Assume that $dim_F J(A)^3/J(A)^4 = 3$. Then we have

$$
\begin{aligned}
0 &\equiv (bc)a - b(ca) \\
&\equiv (\lambda_3 + t_3 - 1 - \lambda_5\mu_4 - \lambda_6\mu_4)a^3 \\
&+ (\mu_4 t_3 - \mu_4 - \mu_4\mu_5 - \mu_6 + \mu_4\mu_6)a^2 b \\
&= (\mu_3 + t_3 - \mu_4 t_3 - 1 + \mu_4 - \mu_4 t_5 - t_6 + \mu_4 t_6)a^2 c \pmod{J(A)^4}
\end{aligned}
$$

Taking into account that $a^3 + J(A)^4, a^2b + J(A)^4, a^2c + J(A)^4$ form a basis of $J(A)^3/J(A)^4$, we derive

$$0 = \lambda_3 + t_3 - 1 - \lambda_5\mu_4 - \lambda_6 + \lambda_6\mu_4 \tag{4}$$

$$0 = \mu_4 t_3 - \mu_4 - \mu_4\mu_5 - \mu_6 + \mu_4\mu_6 \tag{5}$$

$$0 = \mu_3 + t_3 - \mu_4 t_3 - 1 + \mu_4 - \mu_4 t_5 - t_6 + \mu_4 t_6 \tag{6}$$

The same argument, using the fact that $0 \equiv (ca)b - c(ab) \pmod{J(A)^4}$ gives

$$0 = \lambda_5 + t_5 - \lambda_3\mu_4 - \lambda_6 + \lambda_6\mu_4 \tag{7}$$

$$0 = \mu_4 t_5 - \mu_3\mu_4 - \mu_6 + \mu_4\mu_6 \tag{8}$$

$$0 = \mu_5 + t_5 - \mu_4 t_5 - 1 - \mu_4 t_3 - t_6 + \mu_4 t_6 \tag{9}$$

Adding (5) and (6) and subtracting (8) and (9) from the result, we have

$$0 = (\mu_4 + 1)(\mu_3 - \mu_5 + t_3 - t_5)$$

and hence $\mu_4 - -1$. Therefore, subtracting (7) from (4) we obtain $t_3 - t_5 - 1 = 0$, proving that $t_5 = t_3 - 1$. Next we subtract (9) from (6) to obtain $\mu_3 - \mu_5 = 0$, i.e. $\mu_5 = \mu_3$. Using the fact that $b(ba) \equiv (b^2)a \pmod{J(A)^4}$, we obtain $\lambda_6 = \lambda_2 + t_2, \mu_6 = -t_2$ and $t_6 = \mu_2 + 2t_2$. It follows from (5) that $\mu_3 = t_3 - 1 - 2t_2$. Applying the fact that

$$0 \equiv (c^2)a - c(ca) \pmod{J(A)^4}$$

we derive

$$0 = \mu_2 - t_2 - 3 - 4\lambda_2 + 2\lambda_3 + 2\lambda_5 \tag{10}$$

$$0 = 4t_3 - 2\mu_2 - 1 - 6t_2 \tag{11}$$

and, in particular, $char F \neq 2$.

Applying (6), we then have $0 = 4 + 2\mu_2 + 6t_2 - 4t_3$, so $\mu_2 = 2t_3 - 3t_2 - 2$. Next we multiply (4) by 2 and subtract (10) to obtain $0 = 3$, so $char F = 3$. But $A = F\{1, a, b, c, a^2, ab, ac\} + J(A)^3$, so Step 1 implies that

$$
\begin{aligned}
J(A)^2 &= [A, A] + J(A)^4 \\
&= F\{[a, b], [a, c], [b, c], [a, ab], [a, ac], [b, a^2], \\
&\quad\; [b, ab], [b, ac], [c, a^2], [c, ab], [c, ac]\} + J(A)^4 \\
&\subseteq F\{[a, b], [a, c], [b, c], a^3, a^2b - a^2c\} + J(A)^4
\end{aligned}
$$

as is easily verified. This is, however, contrary to the fact that

$$
dim_F J(A)^2/J(A)^4 = 6
$$

Assume that $dim_F J(A)^3/J(A)^4 = 2$. We consider two subcases, namely $a^2b \in Fa^3 + J(A)^4$ and $a^2b \notin Fa^3 + J(A)^4$.

So assume that $a^2b \in Fa^3 + J(A)^4$. Then

$$
\begin{aligned}
J(A)^3 &= F\{a^3, a^2b, a^2c\} + J(A)^4 \\
&= F\{a^3, a^2c\} + J(A)^4
\end{aligned}
$$

and

$$
a^2b \equiv \lambda a^3 \pmod{J(A)^4}
$$

for some $\lambda \in F$. Because $a^3c \equiv a^2ba \equiv \lambda a^4 \pmod{J(A)^5}$, it follows that

$$
J(A)^4 = aJ(A)^3 = Fa^4 + Fa^3c + J(A)^5 = Fa^4 + J(A)^5
$$

and, since $J(A)^4 \neq J(A)^5$, the latter implies that $dim_F J(A)^4/J(A)^5 = 1$. Applying Step 4 in the proof of Theorem 1.2, we see that $J(A)^3 \subseteq Z(A)$ and, in particular, $a^2c \in Z(A)$. This, however, leads to the contradiction

$$
0 \equiv (a^2c)a - a(a^2c) \equiv a^2(ca) - a^3c \equiv a^4 \pmod{J(A)^5}
$$

Next assume that $a^2b \notin Fa^3 + J(A)^4$. Because $a^3 \notin J(A)^4$ and $dim_F J(A)^3/J(A)^4 = 2$, the elements $a^3 + J(A)^4$ and $a^2b + J(A)^4$ then form a basis of $J(A)^3/J(A)^4$. Now write $a^2c \equiv \lambda a^3 + \mu a^2 b \pmod{J(A)^4}$ for some $\lambda, \mu \in F$. Taking into account that

$$J(A)^4 = aJ(A)^3 = Fa^4 + Fa^3b + J(A)^5$$

and $J(A)^4 \neq J(A)^5$, we obtain $dim_F J(A)^4/J(A)^5 \in \{1, 2\}$.

Assume that $dim_F J(A)^4/J(A)^5 = 2$. Then the elements $a^4 + J(A)^5$ and $a^3b + J(A)^5$ form a basis of $J(A)^4/J(A)^5$. Because

$$\begin{aligned}
0 \equiv\ & (a^2c)a - a^2(ca) \equiv (\lambda\mu + \mu_4\lambda - 1)a^4 \\
& + (\mu - 1)(\mu + \mu_4)a^3b \pmod{J(A)^5}
\end{aligned}$$

this implies that $\lambda\mu + \mu_4\lambda - 1 = 0$ and $(\mu - 1)(\mu + \mu_4) = 0$. The first equation implies that $\mu \neq -\mu_4$, so $\mu = 1$ by the second equation. Hence, applying

$$0 \equiv (bc)a - b(ca) + c(ba) - (cb)a \pmod{J(A)^4}$$

we derive the contradiction $0 = (1 + \mu_4)(\mu_3 - \mu_5 + t_3 - t_5)$.

Finally, assume that $dim_F J(A)^4/J(A)^5 = 1$. Then, by Step 4 in the proof of Theorem 1.2, we have $J(A)^3 \subseteq Z(A)$ which implies that $a^2b, a^2c \in Z(A)$. Thus

$$a^3b \equiv a^2ba \equiv a^3c \pmod{J(A)^5}$$

and

$$a^3b \equiv a^3c \equiv a^2ca \equiv a^4 + a^3b \pmod{J(A)^5}$$

which implies $a^4 \in J(A)^5$ and $J(A)^4 = Fa^3b + J(A)^5$. Moreover, since

$$a^3b \equiv a^3c \equiv \mu a^3b \pmod{J(A)^5}$$

we must have $\mu = 1$.

Using the fact that

$$0 \equiv (bc)a - b(ca) + c(ba) - (cb)a \pmod{J(A)^4},$$

we derive

$$0 = (1 + \mu_4)(\mu_3 - \mu_5 + t_3 - t_5),$$

proving that $\mu_4 = -1$. Similarly, applying the fact that

$$0 \equiv (b^2)a - b(ba) \pmod{J(A)^4}$$

we obtain $t_6 = \mu_2 + t_2 - \mu_6$. Then

$$0 \equiv (a^2 c)b - b(a^2 c) \equiv (a^2 c)b - (ba)ac \pmod{J(A)^5}$$

gives $\lambda = 1 + \mu_2 + t_2 - \mu_3 - t_3$. Since $0 \equiv a^2(c^2) - (a^2 c)c \pmod{J(A)^5}$, this leads to a contradiction.

Step 5. Here we complete the proof by demonstrating that case (3) does not occur. We argue by contradiction and distinguish two cases, namely $dim_F J(A)^3 / J(A)^4 = 3$ and $dim_F J(A)^3 / J(A)^4 = 2$.

Assume that $dim_F J(A)^3 / J(A)^4 = 3$. Then the elements $a^3 + J(A)^4, a^2 b + J(A)^4, a^2 c + J(A)^4$ form a basis of $J(A)^3 / J(A)^4$. Because $\mu_1 \neq 1$ and

$$
\begin{aligned}
0 &\equiv (bc)a - b(ca) \\
&\equiv (1 - \mu_1^2)a^3 + \mu_3(\mu_1 - \mu_1^2)a^2 b + t_3(\mu_1 - \mu_1^2)a^2 c \pmod{J(A)^4}
\end{aligned}
$$

this implies that $\mu_1 = -1$ and, in particular, $char F \neq 2$. Thus $0 = 2\mu_3 = 2t_3$, so $\mu_3 = t_3 = 0$. Then, applying similarly the fact that $0 \equiv c(ba) - (cb)a \pmod{J(A)^4}$, we obtain $\mu_5 = t_5 = 0$. Since $0 \equiv (bc)b - b(cb) \pmod{J(A)^4}$, this leads to a contradiction.

Now assume that $dim_F J(A)^3 / J(A)^4 = 2$. We distinguish two subcases, namely $a^2 b \in Fa^3 + J(A)^4$ and $a^2 b \notin Fa^3 + J(A)^4$.

Assume that $a^2 b \in Fa^3 + J(A)^4$. Then we have

$$J(A)^3 = F\{a^3, a^2 b, a^2 c\} + J(A)^4 = Fa^3 + Fa^2 c + J(A)^4$$

and

$$J(A)^4 = aJ(A)^3 = Fa^4 + Fa^3 c + J(A)^5$$

Assume by way of contradiction that $a^4 \in J(A)^5$. Then $J(A)^4 = Fa^3 c + J(A)^5$ and, in particular, $dim_F J(A)^4 / J(A)^5 = 1$ since $J(A)^4 \neq J(A)^5$. It follows from Step 4 in the proof of Theorem 1.2 that $J(A)^3 \subseteq Z(A)$ and, in particular, $a^2 c \in Z(A)$. This, however, leads to the contradiction $a^3 c \equiv a^2 ca \equiv \mu_1 a^3 c \pmod{J(A)^5}$.

Now write $a^2 b \equiv \lambda a^3 \pmod{J(A)^4}$ for some $\lambda \in F$. Then

$$\lambda a^4 \equiv a^2 ba \equiv \mu_1 a^3 b \equiv \mu_1 \lambda a^4 \pmod{J(A)^5}$$

and, since $a^4 \notin J(A)^5$ and $\mu_1 \neq 1$, this implies that $\lambda = 0$. As in the previous case, we now apply the fact that $0 \equiv (bc)a - b(ca) \pmod{J(A)^4}$ to obtain $\mu_1 = -1$, $char F \neq 2$ and $t_3 = 0$. Similarly, using the fact that

$$0 \equiv (cb)a - c(ba) \equiv (b^2)a - b(ba) \pmod{J(A)^4},$$

we derive $0 = 2t_5 = 2t_2$, hence $t_5 = t_2 = 0$. This, however, yields a contradiction using the fact that $0 \equiv (bc)c - b(c^2) \pmod{J(A)^4}$.

Assume that $a^2 b \notin F a^3 + J(A)^4$. Because $a^3 \notin J(A)^4$ and

$$dim_F J(A)^3 / J(A)^4 = 2$$

the elements $a^3 + J(A)^4$ and $a^2 b + J(A)^4$ form a basis of $J(A)^3 / J(A)^4$ and

$$J(A)^4 = a J(A)^3 = F a^4 + F a^3 b + J(A)^5$$

Assume by way of contradiction that $a^4 \in J(A)^5$. Then $J(A)^4 = F a^3 b + J(A)^5$ and, in particular, $dim_F J(A)^4 / J(A)^5 = 1$ since $J(A)^4 \neq J(A)^5$. It follows from Step 4 in the proof of Theorem 1.2 that $J(A)^3 \subseteq Z(A)$ and, in particular, $a^2 b \in Z(A)$. This leads to the contradiction

$$a^3 b \equiv a^2 ba \equiv \mu_1 a^3 b \pmod{J(A)^3}$$

Thus $a^4 \notin J(A)^5$ and we write $a^2 c = \lambda a^3 + \mu a^2 b \pmod{J(A)^4}$ for some $\lambda, \mu \in F$. Then

$$0 \equiv (a^2 c)a - a^2(ca) \equiv (1 - \mu_1)\lambda a^4 \pmod{J(A)^5}$$

which implies $\lambda = 0$ since $\mu_1 \neq 1$ and $a^4 \notin J(A)^5$. As in the case $dim_F J(A)^3 / J(A)^4 = 3$, we now use the fact that

$$0 \equiv (bc)a - b(ca) \pmod{J(A)^4}$$

to deduce that $\mu_1 = -1$ and $char F \neq 2$. Finally, we distinguish two cases, namely $dim_F J(A)^4 / J(A)^5 = 2$ and $dim_F J(A)^4 / J(A)^5 = 1$.

Assume that $dim_F J(A)^4/J(A)^5 = 2$. Then the elements $a^4 + J(A)^5$ and $a^3 b + J(A)^5$ form a basis of $J(A)^4/J(A)^5$. Applying the fact that

$$0 \equiv b(a^2 c) - (ba)ac \pmod{J(A)^5}$$

we obtain $\lambda_2 \mu = 1$. This, however, leads to the contradiction using the fact that $0 \equiv (a^2 c)b - a^2(cb) \pmod{J(A)^4}$.

Assume that $dim_F J(A)^4/J(A)^5 = 1$. Then, since $a^4 \notin J(A)^5$, we have

$$J(A)^4 = Fa^4 + J(A)^5$$

Moreover, it follows from Step 4 in the proof of Theorem 1.2 that $J(A)^3 \subseteq Z(A)$ and, in particular, $a^2 b \in Z(A)$. Hence

$$a^3 b \equiv a^2 ba \equiv -a^3 b \pmod{J(A)^5}$$

which implies $a^3 b \in J(A)^5$ by using the fact that $char F \neq 2$. But this leads to a final contradiction using the fact that

$$0 \equiv (a^2 c)b - b(a^2 c) \equiv a^2(cb) - (ba)ac \pmod{J(A)^5}$$

The theorem is therefore established. ∎

4. Applications to modular representations

Throughout this section, G denotes a finite group and F an algebraically closed field.

Let B be a block of FG, where $char F = p > 0$, let $l(B)$ be the number of nonisomorphic irreducible FG-modules in B and let $k(B)$ be the dimension of $Z(B)$ over F. The numbers $l(B)$ and $k(B)$ play a fundamental role in modular representaion theory. In fact it can be easily shown (see Brauer (1962, 2G)), that $k(B)$ is equal to the number of irreducible complex characters in B. The determination of the numbers $l(B)$ and $k(B)$ is one of the major problems in modular representation theory. The following well known conjecture of Brauer is still wide open:

$$k(B) \leq |D|, \quad \text{where} \quad D \quad \text{is a defect group of} \quad B$$

One of Brauer's results asserts that $k(B) = 1$ if and only if $D = 1$. The Brauer conjecture has been verified for D cyclic by Dade (1966) and for $|D| \leq p^2$ by Brauer and Feit (1959). Brandt (1982) showed that $k(B) = 2$ or $l(B) = 1$ and $k(B) = 3$ imply $k(B) = |D|$. In our first application of properties of symmetric local algebras, we restrict our attention to the case $l(B) = 1$ and extend Brandt's result to the case $k(B) \leq 4$. As a second application, we demonstrate that if $l(B) = 1$ and $k(B) = 5$, then either $|D| = 5$ or D is nonabelian of order 8. The following lemma will clear our path.

4.1. Lemma. *Let F be an algebraically closed field of character-istic $p > 0$ and let B be a block of the group algebra FG with defect group D and exactly one irreducible FG-module. If e is a primitive idempotent in B, $P = FGe$ and $A = End_{FG}(P)$, then*
    (i) $B \cong M_n(eFGe) \cong M_n(A^\circ)$ for some $n \geq 1$.
    (ii) $Z(B) \cong Z(A)$.
    (iii) A is a symmetric local algebra of F-dimension $|D|$.
    (iv) $k(B) \leq |D|$ and $k(B) = |D|$ if and only if A is commutative.

Proof. We know, from Lemma 1.6.8, that $A^\circ \cong eFGe$. Also, by Proposition 2.4.3, A is a symmetric F-algebra. Since P is indecomposable, A must be local. By Theorem 2.5.12(i), $dim_F(eFGe) = dim_F A$ is the only Cartan invariant of B, which by a result of Brauer (1956) must be the order $|D|$ of D. We are therefore left to verify that $B \cong M_n(A^\circ)$ for some $n \geq 1$. But, by hypothesis, $B \cong P^n$ for some $n \geq 1$, hence $End_B(B) \cong M_n(End_B(P)) = M_n(A)$ by Proposition 1.6.2. The desired assertion now follows by virtue of Lemma 1.6.8. ∎

4.2. Corollary. *(Külshammer (1984a)). Let F be an algebraically closed field of characteristic $p > 0$ and let B be a block of the group algebra FG with defect group D. If $l(B) = 1$ and $k(B) \leq 4$, then $k(B) = |D|$.*

Proof. Let P be a projective indecomposable module in B and let $A = End_{FG}(P)$. By Lemma 4.1, it suffices to show that the symmetric local algebra A is commutative. But, by Lemma 4.1(ii), $dim_F Z(A) = dim_F Z(B) = k(B) \leq 4$. The desired assertion is therefore a consequence of Theorem 1.2. ∎

As a preliminary to our next application of symmetric local algebras, we record the following result.

4.3. Proposition. *(Chlebowitz and Külshammer (1989)). Let F be an algebraically closed field, let G be a finite group and let P be an indecomposable projective FG-module. If $A = End_{FG}(P)$ and $dim_F Z(A) = 5$, then*

$$dim_F A = 5 \quad or \quad dim_F A = 8$$

Proof. Owing to Proposition 2.4.3, A is a symmetric F-algebra. Since P is indecomposable, A is local, i.e. $dim_F A/J(A) = 1$. Now apply Theorem 3.2. ∎

4.4. Theorem. *(Chlebowitz and Külshammer (1989)). Let F be an algebraically closed field of characteristic $p > 0$, let G be a finite group and let B be a block of FG such that $l(B) = 1$ and $k(B) = 5$. If D is a defect group of B, then either $|D| = 5$ or D is nonabelian of order 8.*

Proof. Let P denote the only indecomposable projective FG-module in B and let $A = End_{FG}(P)$. By Lemma 4.1, $dim_F Z(A) = dim_F Z(B) = k(B) = 5$. Hence, by Proposition 4.3, $dim_F A \in \{5, 8\}$. On the other hand, by Lemma 4.1(iii), $dim_F A = |D|$. Thus $|D| = 5$ or $|D| = 8$.

Assume by way of contradiction that D is abelian of order 8. Then B cannot be nilpotent in the sense of Broué and Puig (1981), since otherwise B would contain 8 irreducible complex characters by the main result of Broué and Puig (1981). Hence D must be elementary abelian. This, however, is contrary to the results in Landrock (1981). Thus the theorem is proved. ∎

Chapter 4

G-algebras and their applications

In this chapter we provide a number of general properties of G-algebras together with some applications to modular representations and block theory. After introducing the trace map, we examine in detail permutation G-algebras. We then introduce defect groups in G-algebras and demonstrate that vertex theory can be subsumed under the theory of G-algebras.

Special attention is drawn to the G-algebra $End_R((1_H)^G)$, where H is a subgroup of G. Among other properties, we obtain some general results linking the structure of $End_{RG}((1_H)^G)$ with group-theoretic properties such as control of fusion. As an application, we establish a remarkable result due to Robinson (1983, Corollary 1) regarding defect groups of blocks. We then introduce the Brauer morphism and present the theory of points and pointed groups contained in a work of Puig (1981).

In most G-algebras A, which occur in the applications, the action of G on A is induced by a group homomorphism from G to $U(A)$. Such G-algebras, called interior G-algebras, play an important role. We devote a separate section to the study of interior G-algebras and provide an application, due to Ikeda (1987). This application gives a criterion for an indecomposable two-sided ideal of RG to be a block of RG.

The chapter culminates in examining some general properties of bilinear forms on G-algebras. The material presented is based on an

important work of Broué and Robinson (1986) which contains a number
of further applications and demonstrates how the techniques developed
unify some diverse results (such as results on height 0 characters, multiplicities of Scott modules, etc.).

1.   The trace map

Throughout this section, G denotes a finite group and R a commutative
ring. For any R-algebra A, $Aut_R(A)$ denotes the group of all R-algebra
automorphisms of A.

Let A be an R-algebra. Following Green (1968), we say that A is
a *G-algebra* if G acts as a group of R-algebra automorphisms of A, i.e.
if there is a homomorphism $G \to Aut_R(A)$. Expressed otherwise, each
$g \in G$ acts on $a \in A$ to give ${}^g a \in A$ such that this G-action makes A
into a left RG-module and

$$ {}^g(ab) = {}^g a\, {}^g b \quad \text{for all} \quad a, b \in A, g \in G $$

By a *homomorphism* of G-algebras, we understand an R-algebra homomorphism which preserves the action of G. Two G-algebras A and
B are said to be *isomorphic* if there exists a bijective homomorphism
$A \to B$ of G-algebras. An ideal I of a G-algebra A is called *G-stable* if

$$ {}^g I = I \qquad \text{for all} \quad g \in G $$

It is clear that if I is a G-stable ideal of A, then A/I is a G-algebra via

$$ {}^g(a + I) = {}^g a + I \qquad \text{for all} \quad g \in G, a \in A $$

It is also clear that if $f : A \to B$ is a surjective homomorphism of
G-algebras, then $Ker f$ is a G-stable ideal of A such that

$$ A/I \cong B \qquad \text{as } G\text{-algebras} $$

Let V be an RG-module. We say that V is a *permutation module* if
V is R-free with an R-basis on which G acts as a permutation group.
By a *permutation G-algebra*, we understand a G-algebra A over R such
that A is a permutation RG-module.

The group algebra RG is a typical example of a permutation G-algebra: indeed, for any $x \in RG$ and $g \in G$, define

$$^g x = gxg^{-1}$$

Then RG becomes a G-algebra with an R-basis G on which G acts as a permutation group by conjugation.

A typical example of a G-algebra is the R-algebra

$$A = End_R(V)$$

for any RG-module V. Here for any $g \in G$, $f \in End_R(V)$, we define $^g f \in End_R(V)$ by

$$^g f(v) = g(f(g^{-1}v)) \quad \text{for all} \quad v \in V \tag{1}$$

A particular case of the above, where $V = RG$ is treated as an $R(G \times G)$-module by the rule

$$(x, y)v = xvy^{-1} \qquad x, y \in G \quad \text{and} \quad v \in RG$$

is also important. Thus the R-algebra $End_R(RG)$ becomes a $G \times G$-algebra: if $f \in End_R(RG)$ and $(x, y) \in G \times G$, then $^{(x,y)} f \in End_R(RG)$ is defined by

$$^{(x,y)} f(v) = x(f(x^{-1}vy))y^{-1} \quad \text{for all} \quad v \in RG$$

Let A be a G-algebra. For any subgroup H of G, we put

$$A^H = \{a \in A |\, ^h a = a \quad \text{for all} \quad h \in H\}$$

For example, if $A = RG$ and G acts on A by conjugation, then

$$A^H = C_{RG}(H)$$

while if $A = End_R(V)$ and G acts by (1), then

$$A^H = End_{RH}(V)$$

It is clear that A^H is a subalgebra of A and that if $H \subseteq K$ are subgroups of G, then

$$A^K \subseteq A^H$$

1.1. Lemma.   *Let $H \subseteq K$ be subgroups of G, let T be a left transversal for H in K and let A be a G-algebra. Then the map*

$$Tr_H^K : A^H \to A^K$$

given by

$$Tr_H^K(a) = \sum_{t \in T} {}^t a \qquad for\ all \quad a \in A^H$$

is R-linear and is independent of the choice of T.

Proof.   Given $g \in G$, say $g = th$ with $t \in T, h \in H$ and given $a \in A^H$, we have

$$^g a = {}^{th} a = {}^t({}^h a) = {}^t a$$

Hence $Tr_H^K(a)$ is independent of the choice of T. Since, for any $g \in K, gT$ is another left tranversal for H in K, we have

$$^g Tr_H^K(a) = Tr_H^K(a) \qquad for\ all \quad a \in A^H, g \in K$$

Thus $Tr_H^K(a) \in A^K$ for all $a \in A$. Since Tr_H^K is obviously R-linear, the result follows.   ∎

In what follows, given subgroups H, K of G with $H \subseteq K$, we put

$$A_H^K = Tr_H^K(A^H)$$

and refer to Tr_H^K as the *trace map*. We next record a number of elementary properties of this map.

1.2. Lemma.   *Let A be a G-algebra, let $D \subseteq H \subseteq K$ be subgroups of G and let $a \in A^H, b \in A^K$ and $g \in G$. Then*
(i) $Tr_H^K(ab) = Tr_H^K(a)b$ and $Tr_H^K(ba) = bTr_H^K(a)$.
In particular, A_H^K is an ideal of A^K.
(ii) $Tr_D^K = Tr_H^K \circ Tr_D^H$.
(iii) $^g(A^H) = A^{gHg^{-1}}$.
(iv) $^g[Tr_H^K(a)] = Tr_{gHg^{-1}}^{gKg^{-1}}({}^g a)$.

Proof.   (i) Let T be a left transversal for H in K. Then

$$Tr_H^K(ab) = \sum_{t \in T} {}^t(ab) = \sum_{t \in T} {}^t a\, {}^t b = \sum_{t \in T} {}^t ab = Tr_H^K(a)b$$

and similarly $Tr_H^K(ba) = b Tr_H^K(a)$.

(b) Let S be a left transversal for D in H and let T be a left transversal for H in K. Then TS is a left transversal for D in K and hence for any $a \in A^D$ we have

$$
\begin{aligned}
(Tr_H^K \circ Tr_D^H)(a) &= Tr_H^K(\sum_{s \in S} {}^s a) = \sum_{t \in T} \sum_{s \in S} {}^{ts} a \\
&= Tr_D^K(a),
\end{aligned}
$$

as required.

(iii) For any $h \in H$, we have

$$
{}^{ghg^{-1}}({}^g a) = {}^{gh} a = {}^g a
$$

and so ${}^g a \in A^{gHg^{-1}}$. Similarly, if $c \in A^{gHg^{-1}}$, then $c \in {}^g(A^H)$.

(iv) Let T be a left transversal for H in K. Then gTg^{-1} is a left transversal for gHg^{-1} in gKg^{-1}. Hence

$$
\begin{aligned}
Tr_{gHg^{-1}}^{gKg^{-1}}({}^g a) &= \sum_{t \in T} {}^{gtg^{-1}}({}^g a) = \sum_{t \in T} {}^{gt} a \\
&= {}^g(\sum_{t \in T} {}^t a) = {}^g\left[Tr_H^K(a)\right],
\end{aligned}
$$

as required. ∎

1.3. Lemma. *Let K be a subgroup of G, let D, H be subgroups of K and let S be a (D, H)-transversal of K, i.e. S is a complete set of representatives of the double cosets DsH in K. Then, for all $a \in A^H$ and $b \in A^D$, we have*

(i) $Tr_H^K(a) = \sum_{s \in S} Tr_{sHs^{-1} \cap D}^D({}^s a)$.

(ii) $Tr_H^K(a) Tr_D^K(b) = \sum_{s \in S} Tr_{sHs^{-1} \cap D}^K({}^s a b)$.

Proof. (i) For each $s \in S$, let $T(s)$ be a left transversal for $sHs^{-1} \cap D$ in D. Then it is easy to see that

$$
T = \cup_{s \in S} T(s) s
$$

is a left tranversal for H in K. Hence

$$Tr_H^K(a) = \sum_{s \in S} \left(\sum_{x \in T(s)} {}^{xs}a \right) = \sum_{s \in S} Tr_{sHs^{-1} \cap D}^D({}^s a),$$

as required.

(ii) We have

$$
\begin{aligned}
Tr_H^K(a)Tr_D^K(b) &= Tr_D^K(Tr_H^K(a)b) \quad \text{(by Lemma 1.2(i))} \\
&= Tr_D^K(\sum_{s \in S} Tr_{sHs^{-1} \cap D}^D({}^s ab)) \quad \text{(by (i) and Lemma 1.2(i))} \\
&= \sum_{s \in S} Tr_{sHs^{-1} \cap D}^K({}^s ab), \quad \text{(by Lemma 1.2(ii))}
\end{aligned}
$$

as required. ∎

1.4. Corollary. *Let A be a G-algebra and let $D \subseteq H \subseteq K$ be subgroups of G.*
(i) $Tr_H^K(A_D^H) = A_D^K$ and $A_D^K \subseteq A_H^K$.
(ii) ${}^g(A_H^K) = A_{gHg^{-1}}^{gKg^{-1}}$ for all $g \in G$.

Proof. (i) This is a direct consequence of Lemma 1.2(ii).
(ii) For any $g \in G$,

$$
\begin{aligned}
{}^g(A_H^K) &= {}^g\left[Tr_H^K(A^H)\right] = Tr_{gHg^{-1}}^{gKg^{-1}}({}^g(A^H)) \quad \text{(by Lemma 1.2(iv))} \\
&= Tr_{gHg^{-1}}^{gKg^{-1}}(A^{gHg^{-1}}) \quad \text{(by Lemma 1.2(iii))} \\
&= A_{gHg^{-1}}^{gKg^{-1}},
\end{aligned}
$$

as asserted. ∎

1.5. Corollary. *Let A be a G-algebra, let K be a subgroup of G and let D, H be subgroups of K. Then*
(i) $A_H^K \subseteq \sum_{k \in K} A_{kHk^{-1} \cap D}^D$.

(ii) $A_H^K A_D^K \subseteq \sum_{k \in K} A_{kHk^{-1} \cap D}^K$.

Proof. (i) In the notation of Lemma 1.3, we have

$$Tr_{sHs^{-1} \cap D}^D({}^s a) \in A_{sHs^{-1} \cap D}^D$$

for all $a \in A^H$, $s \in S$. Since $S \subseteq K$, it follows from Lemma 1.3(i) that

$$Tr_H^K(a) \in \sum_{k \in K} A_{kHk^{-1} \cap D}^D \quad \text{for all} \quad a \in A^H$$

proving (i).

(ii) This follows by applying Lemma 1.3(ii). ∎

1.6. Corollary. *Let A be a G-algebra and let H be a subgroup of G.*

(i) $A_H^G = A_{gHg^{-1}}^G$ for all $g \in G$.

(ii) If D is a subgroup of G with $D \subseteq gHg^{-1}$ for some $g \in G$, then

$$A_D^G \subseteq A_H^G$$

Proof. (i) Since $A_H^G \subseteq A^G$, we have $^g(A_H^G) = A_H^G$ for all $g \in G$. Now apply Corollary 1.4(ii) for $K = G$.

(ii) Apply (i) and Corollary 1.4(i). ∎

1.7. Lemma. *Let A be a G-algebra over R and let $H \subseteq K$ be subgroups of G. If $(K : H)$ is invertible in R, then $A_H^K = A^K$.*

Proof. We must show that the trace map $Tr_H^K : A^H \to A^K$ is surjective. Let $a \in A^K$ and let T be a left transversal for H in K. Since $^ta = a$ for all $t \in T$, we have

$$Tr_H^K(a) = (K : H)a$$

and so

$$Tr_H^K((K : H)^{-1}a) = a,$$

as required. ∎

1.8. Lemma. *Let A be a G-algebra over R and let e be a nonzero idempotent of A^G. Then, with respect to the induced action of G on eAe, eAe is a G-algebra with $(eAe)^G = eA^Ge$. Furthermore, if e is primitive in A^G, then e is primitive in $(eAe)^G$.*

Proof. For any $a \in A$ and $g \in G$, we have

$$^g(eae) = {}^ge \, {}^ga \, {}^ge = e^g ae \tag{2}$$

since $e \in A^G$. Hence, for each $g \in G$, $eae \mapsto e^g ae$ is an automorphism of eAe. This shows that eAe is a G-algebra where G acts on eAe by restriction from A to eAe. By (2), we obviously have $eA^Ge \subseteq (eAe)^G$. Conversely, assume that $x = eae \in (eAe)^G$. Then $x = ex \in eA^G$ and $x = xe \in A^Ge$, so

$$x \in eA^G \cap A^Ge = eA^Ge,$$

proving that $(eAe)^G = eA^Ge$. The final assertion being obvious, the result is established. ∎

1.9. Lemma. *Let A be a G-algebra over R and let e be a nonzero idempotent of A^G. Then, for any subgroup H of G,*

$$e \in A_H^G \quad \text{if and only if} \quad e \in (eAe)_H^G$$

Proof. By Lemma 1.8, eAe is a G-algebra with $(eAe)^G = eA^Ge$. Let H be a subgroup of G and let

$$Tr_H^G : A^H \to A^G$$

and

$$tr_H^G : (eAe)^H \to (eAe)^G$$

be the corresponding trace maps. Then

$$tr_H^G(x) = Tr_H^G(x) \quad \text{for all} \quad x \in (eAe)^H$$

since the action of G on eAe is induced by the action of G on A. Furthermore, if $e = Tr_H^G(x)$ for some $x \in A^H$, then by Lemma 1.2(i),

$$tr_H^G(exe) = Tr_H^G(exe) = eTr_H^G(x)e = e$$

Hence $e \in A_H^G$ if and only if $e \in (eAe)_H^G$, as required. ∎

1.10. Lemma. *Let V be an RG-module and let e be an idempotent of $End_{RG}(V)$. Then the G-algebras $End_R(eV)$ and $eEnd_R(V)e$ are isomorphic.*

Proof. We obviously have $V = eV \oplus (1 - e)V$. For each $f \in End_R(eV)$, let $\lambda(f) \in End_R(V)$ be defined by $f = \lambda(f)$ on eV and

$\lambda(f)((1-e)V) = 0$. Then, by the proof of Lemma 2.4.1 (with $S = R$), we have

$$eEnd_R(V)e = \{\lambda(f)|f \in End_R(eV)\}$$

Hence the map

$$\begin{cases} End_R(eV) & \to & eEnd_R(V)e \\ f & \mapsto & \lambda(f) \end{cases}$$

is an isomorphism of R-algebras. Furthermore, for any $v \in V$, $f \in End_R(eV)$ and $g \in G$, we have

$${}^g\lambda(f)(ev) = g\lambda(f)g^{-1}(ev) = gfg^{-1}(ev) = {}^gf(ev) = \lambda({}^gf)(ev),$$

proving that ${}^g\lambda(f) = \lambda({}^gf)$, as required. ∎

2. Permutation G-algebras

Our aim here is to examine in detail permutation G-algebras introduced in the previous section. Throughout, G denotes a finite group and R a commutative ring.

Let A be a permutation G-algebra over R. Then, by definition, there exists an R-basis X of A on which G acts as a permutation group. For any given $x \in X$ and any subgroup H of G, we write ${}^Hx$ and $H(x)$ for the H-orbit of x and the stabilizer of x in H, respectively. Thus

$${}^Hx = \{{}^hx|h \in H\}$$

and

$$H(x) = \{h \in H|{}^hx = x\}$$

It is clear that

$$|{}^Hx| = (H : H(x))$$

For any subset S of A, we put

$$S^+ = \sum_{s \in S} s$$

2.1. Lemma.   *Let A be a permutation G-algebra, let X be an R-basis of A which is permuted by the action of G and let $H \subseteq K$ be subgroups* *of G.*

(i) *If $x_1, \ldots, x_n$ are all representatives for the K-orbits of X, then*

$$\{ (^K x_i)^+ | 1 \le i \le n \}$$

is an R-basis of A^K.

(ii) *For any $x \in X$,*

$$Tr_H^K((^H x)^+) = (K(x) : H(x))(^K x)^+$$

Proof. (i) It is clear that the $(^K x_i)^+$ are R-linearly independent elements of A^K. Let $a = \sum a_x x$, $a_x \in R, x \in X$, be an arbitrary element of A^K. If $x, y \in X$ are such that $a_x \ne 0, a_y \ne 0$ and $y = {}^k x$ for some $k \in K$, then ${}^k a = a$ implies that $a_x = a_y$. Hence a is an R-linear combination of the $(^K x_i)^+$, $1 \le i \le n$, as required.

(ii) Let S be a left transversal for $H(x)$ in H and let T be a left transversal for H in K. Then TS is a left transversal for $H(x)$ in K and so

$$
\begin{aligned}
Tr_H^K((^H x)^+) &= Tr_H^K(\sum_{s \in S} {}^s x) = \sum_{s \in S}(\sum_{t \in T} {}^{ts} x) \\
&= Tr_{H(x)}^K(x) = Tr_{K(x)}^K Tr_{H(x)}^{K(x)}(x) \\
&= (K(x) : H(x)(^K x)^+,
\end{aligned}
$$

as required. ∎

In what follows, the zero module is regarded as a free R-module on an empty basis.

2.2. Proposition. *(Green (1968)). Let A be a permutation G-algebra, let X be an R-basis of A which is permuted by the action of G and let $H \subseteq K$ be subgroups of G. If $x_1, \ldots, x_n$ are all representatives for the K-orbits of X and t_i is the greatest common divisor of $(K(x_i) : (kHk^{-1} \cap K(x_i)))$ with k running over K, then the nonzero elements of the set*

$$\{ t_i(^K x_i)^+ | 1 \le i \le n \}$$

form an R-basis of A_H^K.

Proof. Owing to Lemma 2.1, A_H^K is the R-linear span of the set

$$\{(K(x) : H(x))(^K x)^+ | x \in X\}$$

Each $(^K x)^+$ is equal to $(^K x_i)^+$ for exactly one $i \in \{1, \ldots, n\}$; in fact $(^K x)^+ = (^K x_i)^+$ if and only if $x \in {}^K x_i$, i.e. if and only if $x = {}^k x_i$ for some $k \in K$. Hence A_H^K has an R-basis $\{t_i(^K x_i)^+ | 1 \leq i \leq n\}$, where t_i is the greatest common divisor of the set of integers

$$\{(K(^k x_i) : H(^k x_i)) | k \in K\}$$

Note that $G(^g x) = gG(x)g^{-1}$ for all $x \in X, g \in G$ and therefore

$$K(^g x) = K \cap G(^g x) = g(g^{-1} K g \cap G(x))g^{-1}$$

Applying this formula, we find that for each $k \in K$

$$
\begin{aligned}
(K(^k x_i) : H(^k x_i)) &= ((k^{-1} K k \cap G(x_i)) : (k^{-1} H k \cap G(x_i))) \\
&= (K(x_i) : (k^{-1} H k \cap K(x_i))),
\end{aligned}
$$

which gives the result. ∎

2.3. Corollary. *Further to the assumptions and notation of Proposition 2.2, assume that p is a prime number such that every p'-number is a unit of R. For any positive integer m, write $m = p^{\nu(m)} \cdot m'$ where $\mu(m) \geq 0$ and $(p, m') = 1$, and let*

$$\lambda_i = min\nu_{k \in K}(K(x_i) : (kHk^{-1} \cap K(x_i))) \qquad (1 \leq i \leq n)$$

Then the nonzero elements of the set

$$\{p^{\lambda_i}(^K x_i)^+ | 1 \leq i \leq n\}$$

form an R-basis of A_H^K. In particular, if $char R = p$ and K is a p-group, then the elements $(^K x_i)^+$ with $K(x_i) \subseteq kHk^{-1}$ for some $k \in K$, form an R-basis of A_H^K.

Proof. For any positive integer m, we have $mR = p^{\nu(m)} m' R = p^{\nu(m)} R$. The desired conclusion is therefore a consequence of Proposition 2.2. ∎

The special case of the above results in which $A = RG$ and G acts on A by conjugation will be recorded below. Recall that in this case for any subgroup H of G, we have

$$A^H = C_{RG}(H)$$

and the trace map Tr_H^K, where $H \subseteq K$ are subgroups of G, is given by

$$Tr_H^K(a) = \sum_{t \in T} tat^{-1} \qquad (a \in C_{RG}(H))$$

where T is a left transversal for H in K. Note also that, by definition,

$$(RG)_H^K = Tr_H^K(C_{RG}(H))$$

2.4. Corollary. *Let $H \subseteq K$ be subgroups of G, let $C_1, \ldots, C_n$ be all K-conjugacy classes of G and let $g_i \in C_i, 1 \leq i \leq n$.*

(i) If t_i is the greatest common divisor of $(C_K(g_i) : (kHk^{-1} \cap C_K(g_i)))$ with k running over K, then the nonzero elements of the set $\{t_1 C_1^+, \ldots, t_n C_n^+\}$ form an R-basis of $(RG)_H^K$.

(ii) If p is a prime and every p'-number is a unit of R, then for

$$\lambda_i = min_{k \in K} \nu(C_K(g_i) : (kHk^{-1} \cap C_K(g_i))) \qquad (1 \leq i \leq n)$$

the nonzero elements of the set

$$\{p^{\lambda_i} C_i^+ | 1 \leq i \leq n\}$$

form an R-basis of $(RG)_H^K$.

(iii) If $char R = p$, p prime, and K is a p-group, then the elements C_i^+ with $C_K(g_i) \subseteq kHk^{-1}$ for some $k \in K$, form an R-basis of $(RG)_H^K$.

Proof. This is a special case of Proposition 2.2 and Corollary 2.3, in which $A = RG$, $X = G$ and G acts on A by conjugation. ∎

Given subgroups H, K of G, we write $H \subseteq_G K$ to mean that $H \subseteq gKg^{-1}$ for some $g \in G$. We now fix a prime p and, for any conjugacy class C of G, denote by $\delta(C)$ a defect group of C (with respect to p). If $C_1, \ldots, C_n$ are all conjugacy classes of G with $\delta(C_i) \subseteq_G D$, where D

is a given p-subgroup of G, then we write $I_D(G)$ for the R-linear span of $C_1^+, \ldots, C_n^+$.

2.5. Corollary. *Let p be a prime, let D be a p-subgroup of G and let $\operatorname{char} R = p$. Then*
$$(RG)_D^G = I_D(G)$$
Proof. We apply Corollary 2.4(ii) for $K = G$ and $H = D$. Then $\lambda_i = 0$ if and only if $gDg^{-1} \cap C_G(g_i) = \delta(C_i)$ for some $g \in G$. Since the latter is equivalent to $\delta(C_i) \subseteq_G D$, the result follows by virtue of Corollary 2.4(ii). ∎

The rest of this section, will be devoted to the investigation of another type of permutation G-algebras.

Given a subgroup H of G, we write 1_H for the trivial RH-module, i.e. $1_H = R$ as an R-module and
$$hr = r \quad \text{for all} \quad r \in R, h \in H$$
Given an RH-module V, we write V^G for the induced RG-module given by $V^G = RG \otimes_{RH} V$. For any subset X of G, we define $X^+ \in RG$ by
$$X^+ = \sum_{x \in X} x$$

2.6. Lemma. *Let H be a subgroup of G.*
(i) $1_H \cong RH^+$.
(ii) $(1_H)^G \cong RG \cdot H^+$.

Proof. (i) This is obvious.

(ii) Let T be a left transversal for H in G. Then each element of $(1_H)^G$ can be uniquely written in the form $\sum_{t \in T} t \otimes r_t$, $r_t \in R$. Hence the map
$$\begin{cases} (1_H)^G & \to & RG \cdot H^+ \\ \sum_{t \in T} t \otimes r_t & \mapsto & \sum_{t \in T} r_t t H^+ \end{cases}$$
is an isomorphism of RG-modules. ∎

Applying Lemma 2.6, we shall identify $(1_H)^G$ with the left ideal $RG \cdot H^+$ of RG. If T is a left transversal for H in G, then
$$\{t H^+ | t \in T\}$$

is an R-basis of $RG \cdot H^+$ which is permuted transitively by G via left multiplication. In particular, the R-module $End_R((1_H)^G)$ has an R-basis

$$\{f_{tu}|t, u \in T\}$$

given by

$$f_{tu}(xH^+) = \begin{cases} uH^+ & if \ \ xH = tH \\ 0 & if \ \ xH \neq tH \end{cases} \tag{1}$$

for all $x \in G$.

Recall that $End_R((1_H)^G)$ is a G-algebra with the G-action given by

$$(^g f)(v) = gf(g^{-1}v) \tag{2}$$

for all $f \in End_R((1_H)^G)$ and all $v \in (1_H)^G$.

2.7. Theorem. *Let T be a left transversal for H in G, let $S \subseteq T$ be a complete set of double coset representatives for (H, H) in G and let $h \in T \cap H$.*

(i) $End_R((1_H)^G)$ is a permutation G-algebra with permutation basis $\{f_{tu}|t, u \in T\}$.

(ii) The elements $a_H(s) = Tr^G_{H \cap sHs^{-1}}(f_{hs})$, with s running over S, form an R-basis of $End_{RG}((1_H)^G)$.

(iii) For each $t \in T$ and $s \in S$, $a_H(s)(tH^+) = t(HsH)^+$.

Proof. (i) Given $t, u \in T$ and $g \in G$, let $\lambda, \mu \in T$ be defined by

$$uH = guH \quad and \quad \lambda H = gtH$$

It will be shown that

$$^g f_{tu} = f_{\lambda \mu} \tag{3}$$

and

$$G(f_{tu}) = tHt^{-1} \cap uHu^{-1} \tag{4}$$

In particular, (i) will follow from (3). Given $x \in G$, it follows from (1) and (2) that

$$^g f_{ts}(xH^+) = gf_{ts}(g^{-1}xH^+) = \begin{cases} guH^+ & ,if \ \ g^{-1}xH = tH \\ 0 & ,if \ \ g^{-1}xH \neq tH \end{cases}$$

$$= f_{\lambda \mu}(xH^+),$$

proving (3).

To prove (4), note that by (3) $^g f_{tu} = f_{tu}$ if and only if $tH = gtH$ and $uH = guH$, i.e. if and only if $g \in tHt^{-1} \cap uHu^{-1}$, as required.

(ii) We first claim that $\{f_{hs} | s \in S\}$ are all distinct representatives for the G-orbits of $\{f_{tu} | t, u \in T\}$. Indeed, assume that $x, y \in S$ and $g \in G$ are such that $^g f_{hx} = f_{hy}$. Then, by (3), $H = gH$ and $yH = gxH$, so $HyH = HgxH = HxH$ and thus $x = y$. Now fix $t, u \in T$ and let $h_1 \in H$, $s \in S$ be chosen so that $t^{-1}uH = h_1 sH$. Then for $g = th_1$ we have $tH = gH$ and $uH = gsH$ and thus $^g f_{hs} = f_{tu}$, proving our claim.

By (4), we have

$$G(f_{hs}) = H \cap sHs^{-1} \qquad \text{for all} \quad s \in S$$

and hence

$$a_H(s) = Tr^G_{G(f_{hs})}(f_{hs}),$$

proving that $a_H(s)$ is the sum of all elements in the G-orbit of f_{hs}. Since, by the claim proved above, $\{f_{hs} | s \in S\}$ are all distinct representatives for the G-orbits of $\{f_{tu} | t, u \in T\}$, the desired conclusion follows by virtue of Lemma 2.1(i).

(iii) Since $a_H(s) \in End_{RG}((1_H)^G)$, it suffices to show that for any given $s \in S$,

$$a_H(s)(H^+) = (HsH)^+$$

To this end, let X denote a left transversal for $H \cap sHs^{-1}$ in G. Then

$$HsH = \cup_{x \in X \cap H} xsH \qquad \text{(disjoint union)}$$

and $^x f_{hs}(H^+) = xsH^+$ for $x \in X \cap H$, while $^x f_{hs}(H^+) = 0$ for $x \in X - H$. Hence

$$a_H(s)(H^+) = \sum_{x \in X} {}^x f_{hs}(H^+) = \sum_{x \in X \cap H} xsH^+ = (HsH)^+$$

thus completing the proof. ∎

We close by providing circumstances under which $(1_H)^G$ is indecomposable.

2.8. Proposition. *Let p be a prime, let G be a p-group and let R be any commutative noetherian ring such that $R/J(R)$ is a field of characteristic p. Then $(1_H)^G$ is an indecomposable RG-module.*

Proof. Put $F = R/J(R)$, $V = (1_H)^G$ and for any nonzero submodule W of V, let $\bar{W} = W/J(R)W$. By Nakayama's lemma, $\bar{W}$ is a nonzero FG-module. If $V = V_1 \oplus V_2$ for some nonzero RG-modules V_1 and V_2, then obviously $\bar{V} \cong \bar{V}_1 \oplus \bar{V}_2$ with $\bar{V} \neq 0$ and $\bar{V}_2 \neq 0$. Hence it suffices to show that the FG-module $\bar{V}$ is indecomposable.

To this end, let T denote a left transversal for H in G. By Lemma 2.6(ii), we may identify V with the left ideal $RG \cdot H^+$ of RG, in which case $\{tH^+ | t \in T\}$ is an R-basis of V which is permuted transitively by G via left multiplication. Hence $\bar{V}$ is an FG-module with an F-basis which is permuted transitively by G. Therefore the fixed-point space of $\bar{V}$ has F-dimension 1. Since F is a field of characteristic p and G is a p-group, we conclude that $\bar{V}$ is an indecomposable FG-module, as required. ∎

3.   Algebras over complete noetherian local rings

Throughout this section, R denotes a commutative ring. We begin by recording the following piece of information.

A *metric space* (M, ρ) is a set M together with a map

$$\rho : M \times M \to \mathbb{R}$$

such that for all $x, y, z \in M$
  (i) $\rho(x, y) = \rho(y, x) \geq 0$.
  (ii) $\rho(x, y) = 0$ if and only if $x = y$.
  (iii) $\rho(x, z) \leq \rho(x, y) + \rho(y, z)$.
A *Cauchy sequence* in M is a sequence $\{x_i\}$ of elements of M such that for any $\varepsilon > 0$, there exists a positive integer k such that

$$\rho(x_m, x_n) < \varepsilon \quad \text{for all} \quad m, n > k$$

A sequence $\{x_i\}$ is said to be *converge* to $x \in M$, written $lim x_i = x$, provided

$$lim_{i \to \infty} \rho(x, x_i) = 0$$

Assume that I is an ideal of a ring S such that

$$\cap_{n=1}^{\infty} I^n = 0$$

Then, for any distinct $x, y \in S$, we may find a unique $m \geq 0$ such that

$$x - y \in I^m - I^{m+1}$$

where, by convention, $I^0 = S$. We define

$$\rho_I : S \times S \to \mathbb{R}$$

by

$$\rho_I(x, y) = \begin{cases} 0 & if \quad x = y \\ 2^{-m} & if \quad x \neq y \quad \text{and} \quad x - y \in I^m - I^{m+1} \end{cases}$$

It is then easily verified that (S, ρ_I) is a metric space. We refer to S as being *complete in the I-adic topology* if each Cauchy sequence in (S, ρ_I) converges.

Let $\{s_i\}$ be a sequence in S. The following properties are easy consequences of the definitions:

(i) $\{s_i\}$ is Cauchy if and only if for any $n > 0$, there exists $m > 0$, such that $i, j > m$ implies $s_i - s_j \in I^n$.

(ii) $lim s_i = s$ if and only if for any $n > 0$, there exists $m > 0$ such that $i > m$ implies $s_i - s \in I^n$.

(iii) If $lim s_i = s$ and $lim s_i' = s'$, then

$$lim(s_i + s_i') = s + s' \quad \text{and} \quad lim(s_i s_i') = ss'$$

3.1. Lemma. *Suppose that R is a noetherian ring and that A is an R-algebra which is finitely generated as R-module. If I is an ideal of R such that $I \subseteq J(A)$, then $\cap_{n=1}^{\infty} I^n A = 0$. In particular, if R is a noetherian local ring with maximal ideal $P = J(R)$, then $\cap_{n=1}^{\infty} P^n = 0$.*

Proof. Let L be an arbitrary ideal of R and let V be a finitely generated R-module. It is well known (see Bourbaki (1961)) that $LW = W$ for $W = \cap_{n=1}^{\infty} L^n V$. Applying this fact for $V = A$, $L = I$ and $W = \cap_{n=1}^{\infty} I^n A$, we conclude that $IW = W$ and hence $J(A)W = W$.

Since W is a finitely generated A-module, it follows from Nakayama's lemma that $W = 0$. The proof is therefore complete. ∎

3.2. Lemma. *Let A be an R-algebra which is finitely generated as an R-module. Then $J(R)A \subseteq J(A)$.*

Proof. Let V be a simple A-module. Then $V = Av$ for each nonzero $v \in V$, so V is a finitely generated R-module. Now $J(R)V$ is an A-submodule of V, hence is 0 or V. But $J(R)V \neq V$, otherwise by Nakayama's lemma we may find that $V = 0$. Hence $J(R)V = 0$ and so $J(R)$ annihilates every such V. Thus the same is true for $J(R)A$ and so $J(R)A \subseteq J(A)$. ∎

3.3. Lemma. *Let R be a local ring with maximal ideal P and residue class field $F = R/P$. Assume that A is an R-algebra which is finitely generated as R-module and put $\bar{A} = A/PA$. Then*
(i) $J(\bar{A}) = J(A)/PA$ and $A/J(A) \cong \bar{A}/J(\bar{A})$ as F-algebras.
(ii) For any finite group G, $RG/J(RG) \cong FG/J(FG)$.

Proof. (i) Owing to Lemma 3.2, $PA \subseteq J(A)$ and so, by Corollary 1.4.7(ii), $J(\bar{A}) = J(A)/PA$. The latter obviously implies the second assertion.
(ii) Put $A = RG$ and observe that

$$\bar{A} = RG/P(RG) \cong (R/P)G = FG$$

Now apply (i). ∎

Suppose that R is a noetherian local ring with maximal ideal P. By Lemma 3.1, $\cap_{n=1}^{\infty} P^n = 0$. We say that R is *complete* if R is complete in the P-adic topology.

3.4. Lemma. *Let R be a complete noetherian local ring with maximal ideal P, and let A be an R-algebra which is finitely generated as R-module. Then, for any ideal $J \subseteq J(A)$ of A, $\cap_{n=1}^{\infty} J^n = 0$ and A is complete in the J-adic topology.*

Proof. Owing to Lemma 3.2, $PA \subseteq J(A)$, while by Lemma 3.3(i),

$$J^n \subseteq J(A)^n \subseteq PA$$

for some $n \geq 1$. Thus we may harmlessly assume that $J = PA$. Note that, by Lemma 3.1, the intersection of all $(PA)^n = P^n A$ is zero. Now write

$$A = Rv_1 + \cdots + Rv_s$$

with $v_i \in A, 1 \leq i \leq s$, and let $\{w_i\}$ be a Cauchy sequence in A. Given an integer $n > 0$, let $m(n)$ denote the largest integer such that for all $i, j \geq n$

$$w_i - w_j \in P^{m(n)} A$$

(if no such $m(n)$ exists, put $m(n) = n$). Taking into account that $\{w_i\}$ is Cauchy, we have $\lim_{n \to \infty} m(n) = \infty$. Let

$$w_1 = \sum_{t=1}^{s} b_{1t} v_t \qquad (b_{1t} \in R)$$

and write

$$w_{n+1} - w_n = \sum_{t=1}^{s} a_{nt} v_t \qquad (a_{nt} \in P^{m(n)} A \quad \text{for all} \quad n)$$

Then we have

$$w_n = w_1 + \sum_{i=1}^{n-1} (w_{i+1} - w_i) = \sum_{t=1}^{s} (b_{1t} + \sum_{t=1}^{n-1} a_{it}) v_t = \sum_{t=1}^{s} b_{nt} v_t,$$

where

$$b_{nt} = b_{1t} + \sum_{i=1}^{n-1} a_{it},$$

and

$$b_{n+k,t} - b_{nt} = \sum_{i=n}^{n+k-1} a_{it} \in P^{m(n)},$$

so for any t, $\{b_{nt}\}$ is a Cauchy sequence in R. Because R is complete, $\lim b_{nt} = b_t$ for some $b_t \in R$. Setting $w = \sum_{t=1}^{s} b_t v_t$, it follows that

$$w - w_n = \sum_{t=1}^{s} (b_t - b_{nt}) v_t \to 0 \quad \text{as} \quad n \to \infty$$

thus completing the proof. ∎

We are now ready to prove the following important result.

3.5. Theorem. *Let R be a complete noetherian local ring, let A be an R-algebra which is finitely generated as R-module and let $J = PA$ where P is the maximal ideal of R. For each $x \in A$, let $\bar{x}$ be the image of x in $\bar{A} = A/J$.*

(i) Each idempotent $\varepsilon \in \bar{A}$ can be lifted to an idempotent $e \in A$, that is $\bar{e} = \varepsilon$.
Moreover,

(a) $e \in Z(A)$ if and only if $\bar{e} \in Z(\bar{A})$.

(b) e is primitive (respectively, centrally primitive) if and only if $\bar{e}$ is primitive (respectively, centrally primitive).

(ii) If e_1, e_2 are idempotents of A, then $Ae_1 \cong Ae_2$ as left A-modules if and only if $\bar{A}\bar{e}_1 \cong \bar{A}\bar{e}_2$ as left $\bar{A}$-modules (equivalently, $e_1 A \cong e_2 A$ as right A-modules if and only if $\bar{e}_1\bar{A} \cong \bar{e}_2\bar{A}$ as right $\bar{A}$-modules).

(iii) If $1 = \varepsilon_1 + \cdots + \varepsilon_n$ is a decomposition of 1 into orthogonal idempotents in $\bar{A}$, then there exist orthogonal idempotents $e_1, \ldots, e_n \in A$ such that

$$1 = e_1 + \cdots + e_n \quad and \quad \bar{e}_i = \varepsilon_i$$

(iv) The algebra A is expressible as a finite direct sum

$$A = Ae_1 \oplus \cdots \oplus Ae_m$$

of indecomposable left ideals, where $\{e_1, \ldots, e_n\}$ is a complete system of orthogonal primitive idempotents of A. Moreover,

$$\bar{A} = \bar{A}\bar{e}_1 \oplus \cdots \oplus \bar{A}\bar{e}_m$$

is a direct decomposition of $\bar{A}$ into indecomposable left ideals of $\bar{A}$.

(v) The algebra A admits a finite block decomposition

$$A = Av_1 \oplus \cdots \oplus Av_s$$

where $\{v_1, \ldots, v_s\}$ is a complete system of block idempotents of A. Moreover,

$$\bar{A} = \bar{A}\bar{v}_1 \oplus \cdots \oplus \bar{A}\bar{v}_s$$

is a block decomposition of $\bar{A}$ (i.e. $\bar{v}_1, \ldots, \bar{v}_s$ is a complete system of block idempotents of $\bar{A}$).

Proof. (i) Given an idempotent $\varepsilon \in \bar{A}$, choose $x_1 \in A$ such that $\bar{x}_1 = \varepsilon$. Then $n_1 = x_1^2 - x_1 \in J$. Once x_i and n_i are chosen, put

$$x_{i+1} = x_i + n_i - 2x_i n_i, \quad n_{i+1} = x_{i+1}^2 - x_{i+1} \tag{1}$$

Then $n_i \in J^{2^i}$ and thus $\{x_i\}$ is a Cauchy sequence. By Lemma 3.4, A is complete in the J-adic topology, so $\lim x_i = e$ for some $e \in A$. Then

$$e^2 - e = \lim n_i = 0$$

and, moreover, $\bar{e} = \bar{x}_1 = \varepsilon$, so e is a required idempotent of A.

If $e \in Z(A)$, then obviously $\bar{e} \in Z(\bar{A})$. Conversely, suppose that $\bar{e} \in Z(\bar{A})$. Then $eA(1 - e) \subseteq J$ and therefore

$$eA(1 - e) \subseteq eJ(1 - e) = eA(1 - e)P \subseteq J^2,$$

proving, by induction, that $eA(1 - e) \in J^n$ for all n. Applying Lemma 3.4, we conclude that $eA(1 - e) = 0$. Since $\overline{1 - e} \in Z(\bar{A})$, the same argument shows that $(1 - e)Ae = 0$. Hence, for any $a \in A$, $ae = eae = ea$ and thus $e \in Z(A)$, proving (a).

Let e be an idempotent in A and let $\bar{e} = \varepsilon_1 + \varepsilon_2$, where ε_1 and ε_2 are orthogonal idempotents of A. We claim that there exist orthogonal idempotents e_1, e_2 in A such that

$$e = e_1 + e_2 \quad \text{and} \quad \bar{e}_1 = \varepsilon_1, \bar{e}_2 = \varepsilon_2 \tag{2}$$

Indeed, choose $a \in A$ such that $\bar{a} = \varepsilon_1$ and put $x_1 = eae$. Then $\bar{x}_1 = \bar{e}\bar{a}\bar{e} = \varepsilon_1$, so we may consider the Cauchy sequence $\{x_i\}$ given by (1). Because $x_1 \in eAe$, each $x_i \in eAe$ and so $ex_i = x_i = x_i e$ for all i. Accordingly, setting $e_1 = \lim x_i$, we deduce that e_1 is an idempotent in A such that $\bar{e}_1 = \varepsilon_1$ and (by taking limits) $ee_1 = e_1 = e_1 e$. Letting $e_2 = e - e_1$, we now have $e_2^2 = e_2 \neq 0, \bar{e}_2 = \varepsilon_2$ and $e_1 e_2 = e_2 e_1 = 0$, proving (2).

If $\bar{e}$ is not primitive, then so is e, by virtue of (2). The converse being obvious, it follows that e is primitive if and only if so is $\bar{e}$. Now assume that e is a central idempotent of A. If e is not centrally primitive, then obviously $\bar{e}$ is not centrally primitive. Conversely, assume that $\bar{e} = \varepsilon_1 + \varepsilon_2$, where ε_1 and ε_2 are orthogonal central idempotents of $\bar{A}$. Then, by (a) the idempotents e_1 and e_2 in (2) are central and hence e is not centrally primitive.

(ii) Assume that e_1, e_2 are idempotents in A. Any isomorphism $Ae_1 \cong Ae_2$ of left A-modules carries Je_1 onto Je_2 and thus induces an isomorphism $\bar{A}\bar{e}_1 \cong \bar{A}\bar{e}_2$ of left $\bar{A}$-modules. Conversely, let $\bar{f} : \bar{A}\bar{e}_1 \to \bar{A}\bar{e}_2$ be an isomorphism of left $\bar{A}$-modules and let $\bar{g} = \bar{f}^{-1}$. Since $A = Ae_i \oplus A(1 - e_i)$, each Ae_i is a projective A-module. Thus we can find A-homomorphisms $f : Ae_1 \to Ae_2$ and $g : Ae_2 \to Ae_1$ lifting $\bar{f}$ and $\bar{g}$, repectively. We claim that $\theta = g \circ f$ is an automorphism of Ae_1; if sustained, it will follow, by symmetry, that $f \circ g$ is an automorphism of Ae_2 and hence that f is an isomorphism.

It is clear that θ lifts $\bar{g} \circ \bar{f}$ and therefore $(\theta - 1)Ae_1 \subseteq Je_1$. Setting $\beta = 1 - \theta$, we have $\beta^k Ae_1 \subseteq J^k e_1$ for all $k \geq 1$. Thus $1 + \beta + \beta^2 + \cdots$ is a well defined endomorphism of Ae_1 and is a two-sided inverse of $1 - \beta$. Because $1 - \beta = \theta$, this shows that θ is an A-automorphism of Ae_1, as desired.

(iii) The case $n = 1$ being obvious, we argue by induction on n. So assume that $n > 1$ and that the result is true for $n-1$. Put $\delta = \varepsilon_{n-1} + \varepsilon_n$, so that $1 = \varepsilon_1 + \cdots + \varepsilon_{n-2} + \delta$ is an orthogonal decomposition. By the induction hypothesis, we may find an orthogonal decomposition:

$$1 = e_1 + \cdots + e_{n-2} + e, \bar{e}_1 = \varepsilon_1, \ldots, \bar{e}_{n-2} = \varepsilon_{n-2}, \bar{e} = \delta$$

Applying (2) to the decomposition $\bar{e} = \delta = \varepsilon_{n-1} + \varepsilon_n$, it follows that $e = e_{n-1} + e_n$ for some orthogonal idempotents e_{n-1}, e_n in A such that $\bar{e}_{n-1} = \varepsilon_{n-1}$ and $\bar{e}_n = \varepsilon_n$. Thus $1 = e_1 + \cdots + e_n$ is the desired orthogonal decomposition.

(iv) Since A is a left noetherian, A is expressible as a finite direct sum $\oplus_{i=1}^n L_i$ of indecomposable left ideals. Writing $1 = \sum e_i, e_i \in L_i$, we obtain a decomposition of 1 into orthogonal primitive idempotents. Then, by (i) (b), $\bar{1} = \sum \bar{e}_i$ is such a decomposition in $\bar{A}$, and so each $\bar{A}\bar{e}_i$ is an indecomposable left ideal.

(v) Note that $Z(A)$ is a finitely generated R-submodule of A and therefore $Z(A)$ is noetherian. Hence there exist orthogonal primitive idempotents $v_1, \ldots, v_s$ in $Z(A)$ such that $1 = v_1 + \cdots + v_s$. Therefore

$A = \oplus_{i=1}^{s} Av_i$ is a block decomposition of A. The final assertion being a consequence of (i)(b), the result follows. ∎

3.6. Corollary. *Let R be a complete noetherian local ring and let A be an R-algebra which is finitely generated as R-module. Then A is local if and only if 1 is the only nonzero idempotent of A.*

Proof. If A is local, then 0 and 1 are the only idempotents of A, by Corollary 1.5.2(i). Conversely, assume that 1 is the only nonzero idempotent of A. Let P be the maximal ideal of R and let $\bar{A} = A/PA$. By Lemma 3.3(i), $A/J(A) \cong \bar{A}/J(\bar{A})$ and so it suffices to show that $\bar{A}$ is local. By Theorem 3.5, $\bar{1}$ is the only nonzero idempotent of $\bar{A}$. Since $\bar{A}$ is a finite-dimensional algebra over a field, the result follows. ∎

The next lemma will enable us to take full advantage of the results so far obtained.

3.7. Lemma. *Let R be a noetherian ring and let V be a finitely generated R-module. Then the R-algebra $End_R(V)$ is a finitely generated R-module*

Proof. We may write $V = Rv_1 + \cdots + Rv_n$ for some $v_i \in V$. Let W be a free R-module with basis $w_1, \ldots, w_n$, let $\varphi : W \to V$ be the R-homomorphism such that $\varphi(w_i) = v_i$, and put $K = Ker\varphi$. Because $End_R(W) \cong M_n(R)$, $End_R(W)$ is a finitely generated R-module.

Put $E_0 = \{f \in End_R(W) | f(K) \subseteq K\}$. If $f \in E_0$, then f induces an $\bar{f} \in End_R(V)$ by

$$\bar{f}(x + K) = f(x) + K \qquad (x \in W)$$

For any given $\bar{g} \in End_R(V)$, write

$$\bar{g}(v_i) = \sum_{j=1}^{n} a_{ij}v_j \qquad (a_{ij} \in R)$$

and define $g \in End_R(W)$ by

$$g(w_i) = \sum_{j=1}^{n} a_{ij}w_j$$

If $\sum_j b_j w_j \in K$, then $\sum_j b_j v_j = 0$ and

$$0 = \bar{g}(\sum_j b_j v_j) = \sum_{j,k} b_j a_{jk} v_k$$

so

$$g(\sum_j b_j w_j) = \sum_{j,k} b_j a_{jk} w_k \in K,$$

proving that $g \in E_0$. It follows that $g \mapsto \bar{g}$ is a surjective R-homomorphism $E_0 \to End_R(V)$. Now $End_R(W)$ is a finitely generated R-module and R is noetherian. Hence E_0 and $End_R(V)$ are finitely generated R-modules, as required. ■

3.8. Corollary. *Let R be a complete noetherian local ring and let A be an R-algebra which is finitely generated as R-module. Then every finitely generated indecomposable R-module is strongly indecomposable.*

Proof. Apply Lemma 3.7 and Corollary 3.6. ■

3.9. Corollary. *Let R be a complete noetherian local ring and let A be an R-algebra which is finitely generated as R-module. Then every nonzero finitely generated A-module has the unique decomposition property.*

Proof. By Corollary 1.2.8, A is noetherian. Now apply Corollaries 1.5.5 and 3.8. ■

4. Defect groups in G-algebras

Throughout this section, G denotes a finite group and R a complete noetherian local ring with residue class field $\bar{R} = R/J(R)$ of characteristic $p > 0$. In particular, if $J(R) = 0$ then $R = \bar{R}$ is a field of characteristic $p > 0$. All algebras over R are assumed to be finitely generated as R-modules.

Let A be a G-algebra over R and let e be a primitive idempotent of A^G. Following Green (1968), we say that a subgroup D of G is a *defect group* of e in the G-algebra A if D is a minimal element in the set of all subgroups H of G such that $e \in A_H^G$. Such a subgroup D

exists because the set of all subgroups H of G with $e \in A_H^G$ contains G and hence is nonempty. The following lemma holds for an arbitrary complete noetherian local ring R.

4.1. Lemma. *(Rosenberg's lemma). Let A be an algebra over R and let e be a primitive idempotent of A. If $I_1, \ldots, I_n$ are ideals of A such that $e \in I_1 + \cdots + I_n$, then $e \in I_i$ for some $i \in \{1, \ldots, n\}$.*

Proof. It suffices to treat the case $n = 2$. Since e is primitive, Ae is a finitely generated indecomposable A-module. Hence, by Corollary 3.8, Ae is strongly indecomposable. Thus, by Lemma 1.6.8, eAe is a local ring. Since $e \in I_1 + I_2$, we have $e \in eI_1e + eI_2e$. Since e is the identity element of the local ring eAe, we may assume that A is local and that $e = 1$ is the identity element of A. If $1 = a + b$ with $a \in I_1, b \in I_2$, then not both a and b lie in $J(A)$, say $a \notin J(A)$. Since A is local, a has inverse a^{-1}, which gives $1 = aa^{-1} \in I_1$, as required. ∎

4.2. Theorem. *(Green (1968)). Let A be a G-algebra over R, let e be a primitive idempotent of A^G and let D be a defect group of e.*

(i) If H is a subgroup of G such that $e \in A_H^G$, then $D \subseteq gHg^{-1}$ for some $g \in G$.

(ii) D is a p-subgroup of G and any other defect group of e is G-conjugate to D.

Proof. (i) Let H be a subgroup of G such that $e \in A_H^G$. Then, by Corollary 1.5(ii) (with $K = G$), we have

$$e = ee \in A_H^G A_D^G \subseteq \sum_{g \in G} A_{gHg^{-1} \cap D}^G \tag{1}$$

By Lemma 1.2(i), A_H^G and A_D^G are ideals of A^G. Hence, by (1) and Lemma 4.1,

$$e \in A_{gHg^{-1} \cap D}^G \qquad \text{for some} \quad g \in G$$

Then the minimality of D shows that $D \subseteq gHg^{-1}$, as required.

(ii) Let S be a Sylow p-subgroup of G. Since p is prime to $(G : S)$, $(G : S)$ is invertible in R. Hence, by Lemma 1.7 applied to $K = G$ and $H = S$, we see that $A_S^G = A^G$. Thus $e \in A_S^G$ and so, by (i),

$$D \subseteq gSg^{-1}$$

proving that D is a p-subgroup of G.

Finally, let L be any other defect group of e. Then

$$e \in A_D^G \cap A_L^G$$

and hence, by (1), $D \subseteq xLx^{-1}$ and $L \subseteq yDy^{-1}$ for some $x, y \in G$. Thus D and L are G-conjugate, as required. ∎

For future use, we now record the following useful observation.

4.3. Lemma. *Let A be a G-algebra over R and let e be a primitive idempotent of A^G. Then e is a primitive idempotent of the G-algebra eAe and the defect groups of e in G-algebras A and eAe coincide.*

Proof. By Lemma 1.8, eAe is a G-algebra and e is a primitive idempotent of $(eAe)^G$. Now apply Lemma 1.9. ∎

We now examine the special case, where $A = RG$. Then $A^G = Z(RG)$ and hence primitive idempotents of A^G are the same as the block idempotents of RG. By a *defect group* of a block idempotent e of RG, we understand a defect group of the primitive idempotent e of $(RG)^G$ in the G-algebra RG.

If R is a field of characteristic $p > 0$, then our earlier definition of the defect group of a block idempotent e of RG coincides with the definition given above. This follows immediately from Corollary 2.5 and Proposition 2.12.9.

Again assume that R is a complete noetherian local ring with residue class field $\bar{R} = R/J(R)$ of characteristic $p > 0$. For each

$$x = \sum x_g g \in RG \qquad (x_g \in R, g \in G)$$

put

$$\bar{x} = \sum \bar{x}_g g$$

where $\bar{x}_g = x_g + J(R)$. Then the map $RG \to \bar{R}G, x \mapsto \bar{x}$ is a surjective homomorphism with kernel $J(R)G$.

4.4. Proposition. *With the notation above, the following properties hold:*

(i) *The algebra RG admits a finite block decomposition*

$$RG = RGe_1 \oplus \cdots \oplus RGe_n$$

where $\{e_1, \ldots, e_n\}$ is a complete system of block idempotents of A.

(ii) $\bar{R}G = \bar{R}G\bar{e}_1 \oplus \cdots \oplus \bar{R}G\bar{e}_n$
is a block decomposition of $\bar{R}G$.

(iii) *For any subgroup H of G, the map $RG \to R\bar{G}$ carries $(RG)^G_H$ onto $(\bar{R}G)^G_H$.*

(iv) *For any $e \in \{e_1, \ldots, e_n\}$, the defect groups of e and $\bar{e}$ are the same.*

Proof. (i) and (ii): This is a special case of Theorem 3.5(v).

(iii) We use the notation of Corollary 2.4 with $K = G$. Let $C_1, \ldots, C_r$ be all conjugacy classes of G with $\lambda_i = 0$, $1 \leq i \leq r$. Then, by Corollary 2.4(ii), $C_1^+, \ldots, C_r^+$ are in $(RG)^G_H$ and $\bar{C}_1^+, \ldots, \bar{C}_r^+$ is an $\bar{R}$-basis of $(\bar{R}G)^G_H$. This obviously implies the required assertion.

(iv) Let ε be a block idempotent of $\bar{R}G$ contained in $(\bar{R}G)^G_H$. Since, by (iii), $(RG)^G_H$ is mapped onto $(\bar{R}G)^G_H$, it follows from Theorem 3.5 that ε lifts uniquely to a block idempotent of RG contained in $(RG)^G_H$. Thus, for any subgroup H of G,

$$e \in (RG)^G_H \quad \text{if and only if} \quad \bar{e} \in (\bar{R}G)^G_H,$$

proving the required assertion. ∎

5. Relative projective and injective modules

Our aim here is to record a number of standard facts concerning relative projective and injective modules over group algebras. All the information obtained will be needed in the next section. Throughout, G denotes a finite group and R an arbitrary commutative ring.

Let H be a subgroup of G and let V be an RG-module. We say that V is *H-projective* if every exact sequence of RG-modules

$$0 \to U \to W \to V \to 0$$

for which the associated sequence of RH-modules

$$0 \to U_H \to W_H \to V_H \to 0$$

splits is also a split sequence of RG-modules.

The RG-module V is said to be *H-injective* if every exact sequence of RG-modules

$$0 \to V \to W \to U \to 0$$

for which

$$0 \to V_H \to W_H \to U_H \to 0$$

is a split sequence of RH-modules, is also a split sequence of RG-modules. Hence, if R is a field, then an RG-module is projective (respectively, injective) if and only if it is 1-projective (respectively, 1-injective).

5.1. Lemma. *Let H be a subgroup of G, let T be a left transversal for H in G containing 1 and let V be an RG-module. Define*

$$f : (V_H)^G \to V$$

by

$$f(\sum_{t \in T} t \otimes v_t) = \sum_{t \in T} t v_t \qquad (v_t \in V)$$

Then f is a surjective RG-homomorphism such that $Ker f$ is a direct summand of $((V_H)^G)_H$.

Proof. It is obvious that f is a surjective R-homomorphism. To prove that f is an RG-homomorphism, fix $g \in G$. Then, for any $t \in T$, $gt = t'h_t$ for some $t' \in T, h_t \in H$. Hence

$$\begin{aligned} g(\sum_{t \in T} t \otimes v_t) &= \sum_{t \in T} gt \otimes v_t = \sum_{t \in T} t'h_t \otimes v_t \\ &= \sum_{t \in T} t' \otimes h_t v_t \end{aligned}$$

and therefore

$$\begin{aligned} f(g(\sum_{t \in T} t \otimes v_t)) &= \sum_{t \in T} t'h_t v_t = \sum_{t \in T} gt v_t \\ &= gf(\sum_{t \in T} t \otimes v_t), \end{aligned}$$

proving that f is a surjective RG-homomorphism.

Let $i : V_H \to (V_H)^G$ be the canonical injection. It is plain that $i(V_H) \cap Kerf = 0$. Since

$$\sum_{t\in T} t \otimes v_t - 1 \otimes \left(\sum_{t\in T} tv_t\right) \in Kerf,$$

we deduce that

$$((V_H)^G)_H = Kerf \oplus i(V_H),$$

as desired. ∎

5.2. Lemma. *Let H be a subgroup of G, let T be a left transversal for H in G and let V be an RG-module. Then the map*

$$f : V \to (V_H)^G$$

defined by

$$f(v) = \sum_{t\in T} t \otimes t^{-1}v$$

is an injective RG-homomorphism which is independent of the choice of T and $f(V)_H$ is a direct summand of $((V_H)^G)_H$.

Proof. Suppose that T' is another left transversal for H in G. Then, for any given $t \in T$, there exists $h_t \in H$ and $t' \in T'$ such that $t = t'h_t$. Hence

$$\sum_{t\in T} t \otimes t^{-1}v = \sum_{t\in T} t'h_t \otimes h_t^{-1}(t')^{-1}v = \sum_{t'\in T'} t' \otimes (t')^{-1}v,$$

proving that f is independent of the choice of T.

Given $g \in G$, we have

$$
\begin{aligned}
f(gv) &= \sum_{t\in T} t \otimes t^{-1}(gv) = g\sum_{t\in T}(g^{-1}t) \otimes (t^{-1}g)v \\
&= gf(v),
\end{aligned}
$$

which shows that f is an RG-homomorphism (since f is clearly an R-homomorphism). Taking into account that

$$(V_H)^G = \oplus_{t\in T} t \otimes V_H,$$

we see that f is an injection. Furthermore, one immediately verifies that $W' = \oplus_{t \in T, t \notin H} t \otimes V_H$ is an RH-submodule of $((V_H)^G)_H$.

We claim that

$$((V_H)^G)_H = W' \oplus f(V)_H,$$

which will complete the proof. Indeed, if

$$f(v) = \sum_{t \in T} t \otimes t^{-1}v \in f(V) \cap W',$$

then $v = 0$. On the other hand, if $h \in T \cap H$, then

$$\sum_{t \in T} t \otimes v_t - \sum_{t \in T} t \otimes t^{-1}hv_h \in W',$$

thus completing the proof. ■

We are now ready to prove the following classical result due to Higman (1954).

5.3. Theorem. *Let H be a subgroup of G, let R be a commutative ring and let V be an RG-module. Then the following conditions are equivalent:*

(i) V is H-projective.

(ii) V is isomorphic to a direct summand of $(V_H)^G$.

(iii) V is isomorphic to a direct summand of W^G, where W is an RH-module.

(iv) $Tr_H^G : End_{RH}(V) \to End_{RG}(V)$ is surjective.

(v) There exists $\psi \in End_{RH}(V)$ such that $Tr_H^G(\psi) = 1$.

(vi) V is H-injective.

Proof. (i) $\Rightarrow$ (ii): By Lemma 5.1, there is an exact sequence of RG-modules

$$0 \to U \to (V_H)^G \xrightarrow{f} V \to 0$$

such that the associated sequence of RH-modules splits. By assumption, V is H-projective, so $Ker f$ is a direct summand of $(V_H)^G$. Thus

$$(V_H)^G = Ker f \oplus V'$$

where $V' \cong V$.

(ii) $\Rightarrow$ (iii): Obvious.

(iv) $\Leftrightarrow$ (v): This follows from the fact that $Im Tr_H^G$ is an ideal of $End_{RG}(V)$.

(iii) $\Rightarrow$ (v): We first show that any RG-module V of the form $V = W^G$, where W is an RH-module, satisfies (v). To this end, denote by T a left transversal for H in G containing 1 and define $\psi : V \to V$ by

$$\psi(\sum_{t \in T} t \otimes w_t) = 1 \otimes w_1 \qquad (w_t \in W)$$

It is easily verified that ψ is an RH-homomorphism. Furthermore, for all $w \in W$, we have for any $t' \in T$,

$$(\sum_{t \in T} t\psi t^{-1})(t' \otimes w) = \sum_{t \in T} t\psi(t^{-1}t' \otimes w) = t' \otimes w$$

which shows that $Tr_H^G(\psi) = 1$. Turning to the general case, we may harmlessly assume that

$$W^G = V \oplus V' \qquad \text{(direct sum of } RG\text{-modules)}$$

Let $\pi : W^G \to V$ be the projection map and let ψ satisfy (v) with respect to W^G. Then $\pi \circ \psi$ induces an RH-homomorphism $V \to V$ and, for all $v \in V$, we have

$$v = \pi(\sum_{t \in T} t\psi t^{-1}v) = (\sum_{t \in T} t(\pi \circ \psi)t^{-1})v$$

since π is an RG-homomorphism. The desired implication follows.

(v) $\Rightarrow$ (vi): Assume that V is an RG-submodule of an RG-module U such that $U_H = V_H \oplus W$ for some RH-submodule W. Let $\pi : U_H \to V_H$ be the projection map and let $\varphi \in Hom_{RG}(U, V)$ be defined by $\varphi = Tr_H^G(\psi \circ \pi)$, where $\psi \in End_{RH}(V_H)$ satisfies (v). If $v \in V$, then $\varphi(v) = v$ and hence $U = V \oplus Ker\varphi$.

(vi) $\Rightarrow$ (i): Assume that V is H-injective. By Lemma 5.2, V is isomorphic to a direct summand of $(V_H)^G$. Applying implication (iii) $\Rightarrow$ (v), there exists $\psi \in End_{RH}(V_H)$ such that $Tr_H^G(\psi) = 1$.

Now suppose that $f : U \to V$ is a surjective RG-homomorphism for which

$$U = Ker f \oplus W \qquad \text{(direct sum of } RH\text{-modules)}$$

The restriction $f_1 = f|W$ is an RH-isomorphism of W onto V_H. Set $\varphi = f_1^{-1}$ and put $\theta = Tr_H^G(\varphi\psi f)$. Then $\theta \in End_{RG}(U)$ and

$$
\begin{aligned}
(f \circ \theta)(u) &= \sum_{t \in T} ft\varphi\psi ft^{-1}u \\
&= \sum_{t \in T} tf\varphi\psi t^{-1}u \\
&= \sum_{t \in T} t\psi t^{-1}fu = f(u),
\end{aligned}
$$

proving that $U = \theta(U) + Ker f$. If $\theta(u) \in \theta(U) \cap Ker f$, then

$$
\theta(u) = \sum_{t \in T} t\varphi\psi ft^{-1}u = \sum_{t \in T} t\varphi\psi t^{-1}fu = 0
$$

Hence $U = \theta(U) \oplus Ker f$, as required. ∎

5.4. Corollary. *Let H be a subgroup of G of index n such that n is a unit of R. Then any RG-module is H-projective.*

Proof. Apply Lemma 1.7 and Theorem 5.3. ∎

6. Vertices as defect groups

In this section, we demonstrate that vertex theory can be subsumed under the theory of G-algebras. In what follows, R denotes a complete noetherian local ring with $char R/J(R) = p > 0$ and G a finite group. All RG-modules are assumed to be finitely generated.

Let V be an indecomposable RG-module. Then a subgroup Q of G is said to be a *vertex* of V if the following two properties hold:
(a) V is Q-projective.
(b) V is not H-projective for any proper subgroup H of Q.
The following result shows that vertex theory can be subsumed under the theory of G-algebras.

6.1. Proposition. *(Green (1968)). Let V be an RG-module.*
(i) If V is indecomposable and 1_V is the identity endomorphism of V, then the defect groups of 1_V in the G-algebra $End_R(V)$ are the same as the vertices of V.

(ii) If e is a primitive idempotent of $End_{RG}(V)$, then eV is an indecomposable RG-module such that the defect groups of e in $End_R(V)$ are the same as the vertices of eV.

Proof. (i) Consider the G-algebra $A = End_R(V)$. Then, for any subgroup H of G, we have $A^H = End_{RH}(V)$. Hence, by Theorem 5.3, V is H-projective if and only if $1_V \in A_H^G$. But V is indecomposable, hence 1_V is a primitive idempotent of $End_{RG}(V) = A^G$. Comparing the definitions of vertices and defect groups, the required assertion follows.

(ii) It is clear that eV is an indecomposable RG-module and that e restricts to the identity endomorphism of eV. Hence, by (i), the defect groups of e in the G-algebra $End_R(eV)$ are the same as the vertices of eV. By Lemma 1.10, $End_R(eV)$ and $eEnd_R(V)e$ are isomorphic G-algebras. On the other hand, by Lemma 4.3, the defect groups of e in G-algebras $End_R(V)$ and $eEnd_R(V)e$ are the same. The required assertion is therefore established. ∎

6.2. Corollary. *Let V be an indecomposable RG-module and let Q be a vertex of V. Then Q is a p-subgroup of G and any other vertex of V is G-conjugate to Q.*

Proof. Apply Theorem 4.2 and Proposition 6.1(i). ∎

Recall that the R-algebra $End_R(RG)$ is a $G \times G$-algebra: if $f \in End_R(RG)$ and $(x, y) \in G \times G$, then $^{x,y}f \in End_R(RG)$ is defined by

$$[^{x,y}f](a) = xf(x^{-1}ay)y^{-1} \quad \text{for all} \quad a \in RG \tag{1}$$

6.3. Lemma. *Let H be a subgroup of G and, for any $a \in RG$, let $\lambda(a) \in End_R(RG)$ be given by $\lambda(a)(v) = av$ for all $v \in RG$. Then the map $\lambda : RG \to End_R(RG)$ is an injective homomorphism of R-algebras such that*
   (i) $\lambda(RG) = [End_R(RG)]^{1 \times G}$.
   (ii) $^{x,y}\lambda(a) = \lambda(^x a)$ for all $a \in RG$, $(x, y) \in G \times G$.
   (iii) $\lambda((RG)^H) = [End_R(RG)]^{H \times G}$.
   (iv) $\lambda((RG)_H^G) = [End_R(RG)]_{H \times G}^{G \times G}$.

Proof. (i) It is clear that λ is an injective homomorphism of R-

algebras. If $f \in End_R(RG)$, then by (1), $f \in [End_R(RG)]^{1 \times G}$ if and only if $f(ag) = f(a)g$ for all $a \in RG, g \in G$. Since the latter is equivalent to the requirement that f is an endomorphism of the right RG-module RG, the required assertion follows.

(ii) Fix $a \in RG$ and $(x, y) \in G \times G$. Then, for all $v \in RG$, we have by (1) that

$$[^{x,y}\lambda(a)](v) = x\lambda(a)(x^{-1}vy)y^{-1} = xax^{-1}v = \lambda(^xa)(v),$$

as required.

(iii) Given $a \in RG$, it follows from (ii) that $\lambda(a) \in [End_R(RG)]^{H \times G}$ if and only if $a \in (RG)^H$. In particular,

$$\lambda((RG)^H) \subseteq [End_R(RG)]^{H \times G}$$

Conversely, assume that $f \in [End_R(RG)]^{H \times G}$. Then $f \in [End_R(RG)]^{1 \times G}$ and hence, by (i), $f = \lambda(a)$ for some $a \in RG$. Since

$$\lambda(a) \in [End_R(RG)]^{H \times G},$$

it follows that $a \in (RG)^H$ and so $f \in \lambda((RG)^H)$, as required.

(iv) Let T be a left transversal for H in G. Then $T \times 1$ is a left transversal for $H \times G$ in $G \times G$. Hence, for any $a \in (RG)^H$,

$$
\begin{aligned}
Tr_H^G(\lambda(a)) &= \sum_{t \in T} {}^{(t,1)}\lambda(a) \\
&= \sum_{t \in T} \lambda(^ta) \qquad \text{(by (ii))} \\
&= \lambda Tr_H^G(a)
\end{aligned}
$$

and the result follows by applying (iii). ∎

In what follows, for any subgroup H of G, we put

$$\Delta(H) = \{(h, h)|h \in H\}$$

and define $\lambda : RG \to End_R(RG)$ as in Lemma 6.3. We are now ready to prove the following result.

6.4. Theorem. *(Green (1968)). Let e be a block idempotent of RG and let D be a defect group of e. Then $\Delta(D)$ is a defect group of $\lambda(e)$ in the $G \times G$-algebra $End_R(RG)$.*

Proof. By Lemma 6.3(iii), $\lambda(e)$ is a primitive idempotent in

$$[End_R(RG)]^{G \times G}$$

Hence, by Proposition 6.1(ii), a vertex Q of the indecomposable $R(G \times G)$-module $\lambda(e)RG = RGe$ is the same as a defect group of $\lambda(e)$ in the $G \times G$-algebra $End_R(RG)$.

Now we obviously have

$$RG \cong (1_{D(G)})^{G \times G}$$

Since RGe is a direct summand of RG, it follows from Theorem 5.3 that RGe is $D(G)$-projective. Hence we may take $Q \subseteq D(G)$, which implies that $Q = \Delta(P)$ for some p-subgroup P of G.

If H is any subgroup of G, then obviously

$$\Delta(P) \subseteq_{G \times G} H \times G \iff P \subseteq_G H$$

Hence, by Lemma 6.3(iv),

$$\begin{aligned}
e \in (RG)^G_H &\iff \lambda(e) \in [End_R(RG)]^{G \times G}_{H \times G} \\
&\iff \Delta(P) \subseteq_{G \times G} H \times G \\
&\iff P \subseteq_G H
\end{aligned}$$

which shows that P is a defect group of e. This completes the proof of the theorem. ∎

6.5. Corollary. *(Green (1962)). Let e be a block idempotent of RG and let D be a defect group of e. Then $\Delta(D)$ is a vertex of the indecomposable $R(G \times G)$-module RGe.*

Proof. By Proposition 6.1(ii), a vertex Q of the indecomposable $R(G \times G)$-module RGe is the same as a defect group of $\lambda(e)$ in the $G \times G$-algebra $End_R(RG)$. Now apply Theorem 6.4. ∎

7.   The G-algebra $End_R((1_H)^G)$

Throughout, G denotes a finite group and R a commutative ring. Our aim here is to investigate in detail a distinguished G-algebra, namely the G-algebra $End_R((1_H)^G)$, where H is a subgroup of G. We also obtain some general results linking the structure of $End_{RG}((1_H)^G)$ with group-theoretic properties such as control of fusion. An application of some of the results to the block theory will be provided in the next section.

We begin by fixing the following notation. Given a subgroup H of G, we write 1_H for the trivial RH-module. For any subset X of G, we define $X^+ \in RG$ by

$$X^+ = \sum_{x \in X} x$$

Applying Lemma 2.6, we shall identify $(1_H)^G$ with the left ideal $RG \cdot H^+$ of RG. If T is a left transversal for H in G, then

$$\{tH^+ | t \in T\}$$

is an R-basis of $(1_H)^G$ which is permuted transitively by G via left multiplication. Recall, from Theorem 2.7, that $End_R((1_H)^G)$ is a permutation G-algebra. The R-algebra $End_{RG}((1_H)^G)$ is called the *Hecke algebra* associated with the triple (G, R, H).

Let T be a left transversal for H in G and let $S \subseteq T$ be a complete set of double coset representatives for (H, H) in G. By Theorem 2.7(ii), the elements $a_H(s), s \in S$ of $End_{RG}((1_H)^G)$ given by

$$a_H(s)(tH^+) = t(HsH)^+$$

form an R-basis of $End_{RG}((1_H)^G)$. If g is an arbitrary element of G and $s \in S$ is such that $HgH = HsH$, then we put

$$a_H(g) = a_H(s)$$

Thus, by definition,

$$a_H(g)(H^+) = (HgH)^+ \tag{1}$$

and, for any $x, y \in G$, $a_H(x) = a_H(y)$ if and only if $HxH = HyH$.

For future use, it will be convenient to present an alternative description of the maps $a_H(g), g \in G$. Let T be a left transversal for H in G. For any given $x \in RG$, define $\alpha_H(x) \in End_R((1_H)^G)$ by

$$\begin{cases} \alpha_H(x)(tH^+) = xH^+ & if \quad t \in H \\ \alpha_H(x)(tH^+) = 0 & if \quad t \notin H \end{cases}$$

Since $\{tH^+ | t \in T\}$ is an R-basis of $(1_H)^G$, $\alpha_H(x)$ is indeed a uniquely determined element of $End_R((1_H)^G)$. It is clear that α_H is an R-linear map.

7.1. Lemma. *The R-linear map $\alpha_H : RG \to End_R((1_H)^G)$ satisfies the following properties:*

(i) $\alpha_H(x)(gH^+) = xH^+$ if $g \in H$ and $\alpha_H(x)(gH^+) = 0$ if $g \notin H$, for all $x \in RG$.

(ii) $\alpha_H(xh) = \alpha_H(x)$, $h\alpha_H(x)h_1 = \alpha_H(hx)$ for all $h, h_1 \in H$, $x \in RG$.

(iii) α_H is a homomorphism of $R(H \times H)$-modules, where the $R(H \times H)$-module structure of $End_R((1_H)^G)$ is given by

$$(h_1, h_2)f = h_1 f h_2^{-1} \qquad (h_1, h_2 \in H, f \in End_R((1_H)^G))$$

and $(h_1 f h_2^{-1})(v) = h_1(f(h_2^{-1}v))$ for all $v \in (1_H)^G$.

(iv) If $g \in G$ and $h \in gHg^{-1}$, then $\alpha_H(g) = \alpha_H(hg)$ and so

$$\alpha_H(g) \in End_{R(H \cap gHg^{-1})}((1_H)^G)$$

(v) For each $g \in G$,

$$a_H(g) = Tr^G_{H \cap gHg^{-1}}(\alpha_H(g))$$

Proof. (i) This is a direct consequence of the definition of α_H.

(ii) The first equality follows from the definition and the fact that $xhH^+ = xH^+$ for all $h \in H, x \in RG$. To prove the second, we need only show that for all $g \in G, x \in RG, h, h_1 \in H$

$$\alpha_H(hx)(gH^+) = h\alpha_H(x)h_1(gH^+)$$

If $g \notin H$, then both sides of the above are zero. If $g \in H$, then $H^+ = gH^+ = h_1(gH^+)$ and so

$$\alpha_H(hx)(gH^+) = hxH^+ = h\alpha_H(x)h_1(gH^+),$$

as required.

(iii) Given $x \in RG$, $h_1, h_2 \in H$, it follows from (ii) that

$$
\begin{aligned}
\alpha_H((h_1, h_2)x) &= \alpha_H(h_1 x h_2^{-1}) = \alpha_H(h_1 x) \\
&= h_1 \alpha_H(x) h_2^{-1} = (h_1, h_2)\alpha_H(x),
\end{aligned}
$$

as required.

(iv) Fix $g \in G$ and $h = gh_1 g^{-1}$ for some $h_1 \in H$. Then, by (ii),

$$\alpha_H(hg) = \alpha_H(gh_1) = \alpha_H(g)$$

Hence, if $h \in H \cap gHg^{-1}$, then by the above equality and (ii),

$$\alpha_H(g)h = \alpha_H(g) = \alpha_H(hg) = h\alpha_H(g),$$

as required.

(v) Let T be a left transversal for $H \cap gHg^{-1}$ in H. Then

$$HgH = \cup_{t \in T} tgH \qquad \text{(disjoint union)}$$

and so

$$
\begin{aligned}
(HgH)^+ &= \sum_{t \in T} tgH^+ = \sum_{t \in T} \alpha_H(tg)(H^+) \\
&= \sum_{t \in T} t\alpha_H(g)t^{-1}(H^+) \quad \text{(by (ii))} \\
&= Tr^H_{H \cap gHg^{-1}}(\alpha_H(g))(H^+) \\
&= Tr^G_{H \cap gHg^{-1}}(\alpha_H(g))(H^+)
\end{aligned}
$$

where the last equality is a consequence of the fact that $\alpha_H(g)(g_1 H^+) = 0$ if $g_1 \notin H$. This completes the proof of the lemma. ∎

The ordinary trace map

$$tr : End_R((1_H)^G) \to R$$

restricts to the R-linear map

$$tr : End_{RG}((1_H)^G) \to R$$

7.2. Lemma. *With the notation above,*

$$tr a_H(g) = \begin{cases} (G : H) & if \quad g \in H \\ 0 & if \quad g \notin H \end{cases}$$

Proof. If $g \in H$ then by (1), $a_H(g)$ is the identity map and so $tr a_H(g) = (G : H)$ ($= R$-rank of $(1_H)^G$). On the other hand, if $g \notin H$ then for all $t \in G$,

$$a_H(g)(tH^+) = t(HgH)^+$$

and the coefficient of tH^+ in $a_H(g)(tH^+)$ is zero. Thus in this case we have indeed $tr a_H(g) = 0$. ∎

7.3. Lemma. *For all $x, y \in G$,*

$$tr(a_H(x)a_H(y)) = \begin{cases} (G : (H \cap xHx^{-1})) & if \quad HxH = Hy^{-1}H \\ 0 & if \quad HxH \neq Hy^{-1}H \end{cases}$$

Proof. Denote by T a left transversal for $H \cap xHx^{-1}$ in H. Then

$$HxH = \cup_{t \in T} txH \qquad \text{(disjoint union)}$$

and hence

$$\begin{aligned} a_H(y)a_H(x)(H^+) &= a_H(y)(HxH)^+ \\ &= a_H(y)(\sum_{t \in T} txH^+) \\ &= \sum_{t \in T} tx(HyH)^+ \end{aligned}$$

Thus, if μ is the coefficient of H^+ in $a_H(y)a_H(x)(H^+)$, then

$$\mu = \begin{cases} (H : (H \cap xHx^{-1})) & if \quad HxH = Hy^{-1}H \\ 0 & if \quad HxH \neq Hy^{-1}H \end{cases}$$

Because μ is also the coefficient of $a_H(1)$ in $a_H(y)a_H(x)$ and

$$tr(a_H(x)a_H(y)) = tr(a_H(y)a_H(x)),$$

the rsult follows by virtue of Lemma 7.2. ∎

In what follows, we write $I(H)$ for the augmentation ideal of RH, i.e.$I(H)$ is the kernel of the augmentation map $aug : RH \to R$ given by

$$aug(\sum x_h h) = \sum x_h \qquad (x_h \in R, h \in H)$$

7.4. Lemma.   *The left annihilator of H^+ in RG is $RG \cdot I(H)$.*

Proof.   Let T be a left transversal for H in G. Bearing in mind that

$$RG = \oplus_{t \in T} t(RH) \tag{2}$$

it follows that

$$RG \cdot I(H) = \oplus_{t \in T} t \cdot I(H) \tag{3}$$

By (2), a typical element x of RG can be written uniquely in the form $x = \sum_{t \in T} t x_t$ with $x_t \in RH$. Since $x_t H^+ = aug(x_t)H^+$, it follows that $xH^+ = 0$ if and only if $x_t \in I(H)$ for all $t \in T$. The required assertion now follows by virtue of (3). ∎

From now on, we write

$$\sigma_H : RG \to End_R((1_H)^G)$$

for the natural homomorphism of R-algebras given by

$$\sigma_H(x)(v) = xv \quad \text{for all} \quad x \in RG, v \in (1_H)^G \tag{4}$$

Given $x = \sum_{g \in G} x_g g \in RG$, $x_g \in R$, we also put

$$x^* = \sum_{g \in G} x_g g^{-1}$$

Then $x \mapsto x^*$ is obviously an antiautomorphism of RG. Finally, we put

$$N_{RG}(H^+) = \{x \in RG | H^+ x^* \in RG \cdot H^+\}$$

It is plain that $N_{RG}(H^+)$ is a subalgebra of RG such that if $H \lhd G$, then $N_{RG}(H^+) = RG$.

Consider the R-linear map

$$\omega_H : N_{RG}(H^+) \to End_{RG}((1_H)^G)$$

determined by

$$\omega_H(x)(gH^+) = gH^+x^* \quad (x \in N_{RG}(H^+))$$

It is obvious that

$$\omega_H(x)(v) = vx^* \quad \text{for all} \quad x \in N_{RG}(H^+), v \in (1_H)^G \qquad (5)$$

We now record some basic properties of the introduced map ω_H.

7.5. Proposition. *(Broúe and Robinson (1986)). The map*

$$\omega_H : N_{RG}(H^+) \to End_{RG}((1_H)^G)$$

satisfies the following properties:

(i) ω_H is a surjective homomorphism of R-algebras such that

$$Ker\omega_H = RG \cdot I(H)$$

(ii) $RN_G(H) \subseteq N_{RG}(H^+)$ and $\omega_H(g) = a_H(g^{-1})$ for all $g \in N_G(H)$.
(iii) $C_{RG}(H) \subseteq N_{RG}(H^+)$ and, for all $g \in G$, we have

$$\omega_H(Tr^H_{C_H(g)}(g)) = ((H \cap gHg^{-1}) : C_H(g))a_H(g^{-1}) \qquad (6)$$

In particular,

$$\omega_H(x) = Tr^G_H(\alpha_H(x^*)) \quad \text{for all} \quad x \in C_{RG}(H) \qquad (7)$$

(iv) $\omega_H(z) = \sigma_H(z^*)$ for all $z \in Z(RG)$, where σ_H is given by (4).
(v) If C is a conjugacy class of G, then

$$\omega_H((C^{-1})^+) = \sum_g |C \cap gH|a_H(g)$$

where g runs over a complete set of representatives of (H, H)-double cosets in G.

Proof. (i) If $x, y \in N_{RG}(H^+)$ and $g \in G$, we have

$$
\begin{aligned}
\omega_H(xy)(gH^+) &= gH^+(xy)^* = gH^+y^*x^* \\
&= \omega_H(x)\omega_H(y)(gH^+) \quad \text{(by (5))}
\end{aligned}
$$

which implies that ω_H is a homomorphism of R-algebras. Given $x \in RG$, we have $H^+x^* = (xH^+)^*$ and so $H^+x^* = 0$ if and only if $xH^+ = 0$. Thus, by Lemma 7.4, $Ker\omega_H = RG \cdot I(H)$.

Fix $g \in G$. By Theorem 2.7(ii), to prove surjectivity, it suffices to exhibit an element of $N_{RG}(H^+)$ whose image is $a_H(g)$. To this end, choose a left transversal $t_1, \ldots, t_n$ for $H \cap gHg^{-1}$ in H, and a right transversal $u_1, \ldots, u_n$ for $H \cap gHg^{-1}$ in H. Setting

$$X = \{t_i g^{-1} u_i | 1 \leq i \leq n\}$$

one readily verifies that

$$H^+ X^+ = X^+ H^+ = (Hg^{-1}H)^+$$

Thus $X^+ \in N_{RG}(H^+)$ and $\omega_H(X^+) = a_H(g)$, by virtue of the equality $a_H(g)(H^+) = (HgH)^+$.

(ii) If $g \in N_G(H)$, then $H^+g^* = H^+g^{-1} = g^{-1}H^+ \in RG \cdot H^+$ and therefore $RN_G(H) \subseteq N_{RG}(H^+)$. Again, suppose that $g \in N_G(H)$. Then $g^{-1} \in N_G(H)$, so $Hg^{-1}H = g^{-1}H$ and hence

$$\omega_H(g)(H^+) = H^+g^{-1} = g^{-1}H^+ = a_H(g^{-1})(H^+),$$

as asserted.

(iii) It is obvious that $C_{RG}(H) \subseteq N_{RG}(H^+)$. If $g \in G$, then

$$
\begin{aligned}
\omega_H(Tr^H_{C_H(g)}(g))H^+ &= H^+ Tr^H_{C_H(g)}(g^{-1}) \\
&= Tr^H_{C_H(g)}(H^+g^{-1}) \\
&= ((H \cap gHg^{-1}) : C_H(g))Tr^H_{H \cap gHg^{-1}}(H^+g^{-1}) \\
&= ((H \cap gHg^{-1}) : C_H(g))(Hg^{-1}H)^+,
\end{aligned}
$$

proving (6).

By Lemma 7.1(v), $a_H(g) = Tr^G_{H \cap gHg^{-1}}(\alpha_H(g))$ and so

$$
\begin{aligned}
\omega_H(Tr^H_{C_H(g^{-1})}(g^{-1})) &= ((H \cap gHg^{-1}) : C_H(g))Tr^G_{H \cap gHg^{-1}}(\alpha_H(g)) \\
&= Tr^G_{C_H(g)}(\alpha_H(g)) = Tr^G_H(Tr^H_{C_H(g)}(\alpha_H(g))) \\
&= Tr^G_H(\alpha_H(Tr^H_{C_H(g)}(g)),
\end{aligned}
$$

proving (7) for $x = Tr^H_{C_H(g^{-1})}(g^{-1})$. But the set of all such x forms an R-basis of $C_{RG}(H)$, hence (7) holds for all $x \in C_{RG}(H)$.

(iv) Given $z \in Z(RG)$ and $v \in (1_H)^G$, it follows from (5) that

$$\omega_H(z)v = vz^* = z^*v$$

and therefore $\omega_H(z) = \sigma_H(z^*)$.

(v) To calculate the coefficient of $a_H(g)$ in $\omega_H((C^{-1})^+)$, observe that

$$\begin{aligned}
\omega_H((C^{-1})^+)(H^+) &= H^+C^+ \\
&= \sum \lambda(C,g)(HgH)^+ \\
&= \sum \lambda(C,g)a_H(g)(H^+)
\end{aligned}$$

where $\lambda(C,g)$ is the coefficient of g in the element H^+C^+ (expressed as a linear combination of the natural basis elements of RG). Thus

$$\lambda(C,g) = |C \cap gH|$$

and the result follows. ■

7.6. Corollary. *(Broué and Robinson (1986)). Let R be an arbitrary commutative ring of prime characteristic p, let H be a p-subgroup of G and let C be an H-conjugacy class of G with $\omega_H(C^+) \neq 0$. Then for all $g \in C$*

$$C_H(g) = H \cap gHg^{-1}$$

and

$$\omega_H(C^+) = a_H(g^{-1})$$

Proof. Given $g \in C$, we have $C^+ = Tr^G_{C_H(g)}(g)$. Thus, by (6),

$$\omega_H(C^+) = ((H \cap gHg^{-1}) : C_H(g))a_H(g^{-1}) \neq 0$$

But H is a p-group and $char R = p$, hence $H \cap gHg^{-1} = C_H(g)$ and $\omega_H(C^+) = a_H(g^{-1})$, as asserted. ■

As a preliminary to the next result, we record the following observation.

7.7. Lemma.   *Let H be a subgroup of G, let V be an RH-module and let W be an RG-module.*

(i) $Hom_{RG}(V^G, W) \cong Hom_{RH}(V, W_H)$ as R-modules.

(ii) $Hom_{RG}((1_H)^G, W) \cong Inv(W_H)$ as R-modules.

where $Inv(W_H) = \{w \in W | hw = w$ for all $h \in H\}$.

(iii) If R is a field of prime characteristic p and H is a p-subgroup of G, then any simple RG-module is a homomorphic image of $(1_H)^G$.

Proof.   (i) Given $\lambda \in Hom_{RH}(V, W_H)$, we may define $\lambda^* \in Hom_{RG}(V$ by

$$\lambda^*(x \otimes v) = x(\lambda(v))   (v \in V, x \in RG)$$

The map $\lambda \mapsto \lambda^*$ is obviously R-linear. Furthermore, if $\lambda^* = 0$, then

$$\lambda^*(1 \otimes v) = \lambda(v) = 0   \text{ for all }   v \in V$$

and thus $\lambda = 0$. Finally, given $\psi \in Hom_{RG}(V^G, W)$, define $\varphi : V \to W$ by $\varphi(v) = \psi(1 \otimes v)$. Then $\varphi \in Hom_{RH}(V, W_H)$ and $\psi = \varphi^*$, as required.

(ii) By (i), it suffices to show that

$$Hom_{RH}(1_H, W_H) \cong Inv(W_H)$$

Given $f \in Hom_{RH}(1_H, W_H)$, we obviously have $f(1) \in Inv(W_H)$. Moreover, the map

$$\begin{cases} Hom_{RH}(1_H, W_H) & \to & Inv(W_H) \\ f & \mapsto & f(1) \end{cases}$$

is clearly R-linear and injective. Because, for any $w \in Inv(W_H)$, the map $f(r) = rw$ is an RH-homomorphism of 1_H into W_H, the desired assertion follows.

(iii) Let W be a simple RG-module. If V is a simple submodule of W_H, then $V \subseteq Inv(W_H)$, since R is a field of characteristic $p > 0$ and H is a p-group. Hence $Inv(W_H) \neq 0$ and, by (ii),

$$Hom_{RG}((1_H)^G, W) \neq 0$$

Since W is simple, this shows that W is a homomorphic image of $(1_H)^G$, as asserted. ∎

7.8. Proposition. *(Broué and Robinson (1986)). Let R be a field of prime characteristic p and let H be a p-subgroup of G.*

(i) $Ker\sigma_H \subseteq J(RG)$.

(ii) $Ker\omega_H \cap C_{RG}(H) \subseteq J(C_{RG}(H))$.

Proof. (i) If $x \in Ker\sigma_H$, then x annihilates $(1_H)^G$. Hence, by Lemma 7.7(iii), x annihilates every simple RG-module. Thus $x \in J(RG)$, as required.

(ii) It suffices to verify that $Ker\omega_H \cap C_{RG}(H)$ contains no nonzero idempotents. Our assumption on R and H guarantee that RH is local, hence any projective RH-module is free. Assume that e is a nonzero idempotent of $C_{RG}(H)$. Then $eRG \neq 0$ is a left projective (hence free) RH-module. Therefore H^+eRG is not zero : its dimension is the RH-rank of eRG. Thus

$$eH^+ = H^+e \neq 0$$

and so $e \notin Ker\omega_H$, as required. ∎

7.9. Corollary. *(Broué and Robinson (1986)). Let R be a field of characteristic $p > 0$ and let P be a normal p-subgroup of G. Then*

$$C_{RG}(P)/(Ker\omega_P \cap C_{RG}(P)) \cong R(C_G(P)/Z(P))$$

as R-algebras. In particular, $C_{RG}(P)$ is local if and only if $C_G(P)$ is a p-group.

Proof. Let C be a P-conjugacy class of G and let $g \in C$. Then, by Proposition 7.5(ii), (iii),

$$\omega_P(C^+) = \begin{cases} a_P(g^{-1}) & if \quad g \in C_G(P) \\ 0 & if \quad g \notin C_G(P) \end{cases}$$

Thus $\omega_P(C_{RG}(P)) = \omega_P(RC_G(P))$. Because the map $a_P(g^{-1}) \mapsto gZ(P)$, $g \in C_G(P)$ induces an R-algebra isomorphism

$$\omega_P(RC_G(P)) \to R(C_G(P)/Z(P)),$$

the first assertion is established. The second assertion follows from the first and Proposition 7.8(ii). ∎

Our next aim is to provide some general results linking the structure of $End_{RG}((1_H)^G)$ with control of fusion in finite groups.

Following Broué and Robinson (1986), we say that a subgroup H of G *controls* the G-fusion of its p-subgroups, p prime, if whenever P is a p-subgroup of H and g is an element of G such that

$$g^{-1}Pg \subseteq H,$$

then

$$g \in C_G(P)H$$

7.10. Theorem. *(Broué and Robinson (1986)). Let R be a field of characteristic $p > 0$ and let H be a subgroup of G. Then*

$$\omega_H(C_{RG}(H)) = End_{RG}((1_H)^G) \tag{8}$$

if and only if H controls the G-fusion of its p-subgroups.

Proof. By Proposition 7.5(iii) and Theorem 2.7, the equality (8) holds if and only if

(i) For any $g \in G$, there exists $g_1 \in HgH$ such that

$$p \nmid ((H \cap g_1 H g_1^{-1}) : C_H(g_1))$$

However, if $g_1 = h'gh$ for $h, h' \in H$, then

$$((H \cap g_1 H g_1^{-1}) : C_H(g_1)) = ((H \cap gHg^{-1}) : C_H(ghh'))$$

and so (i) is equivalent to :

(ii) For any $g \in G$, there exists $h \in H$ such that

$$p \nmid ((H \cap gHg^{-1}) : C_H(gh^{-1}))$$

We now show that (ii) is equivalent to :

(iii) For any $g \in G$ and any Sylow p-subgroup P of $H \cap gHg^{-1}$, there exists $h \in H$ such that $P \subseteq C_H(gh^{-1})$.

It is obvious that (iii) implies (ii). Conversely, suppose that (ii) holds. Then, given $g \in G$ and a Sylow p-subgroup P of $H \cap gHg^{-1}$, there exists $h \in H$ such that $C_H(gh^{-1})$ contains $h_1 P h_1^{-1}$ where $h_1 \in H \cap gHg^{-1}$. If $h_1 = gh_2g^{-1}$ with $h_2 \in H$, then

$$P \subseteq h_1^{-1} C_H(gh^{-1}) h_1 = C_H(gh_2^{-1}h^{-1}h_1)$$

which proves (iii). Thus we are left to show that H controls the G-fusion of its p-subgroups if and only if (iii) holds.

First suppose that (iii) holds. Let P be a p-subgroup of H and let $g^{-1}Pg \subseteq H$ for some $g \in G$. Then $P \subseteq H \cap gHg^{-1}$. Let Q denote a Sylow p-subgroup of $H \cap gHg^{-1}$ containing P. It follows from (iii) that

$$Q \subseteq C_H(gh^{-1}) \quad \text{for some} \quad h \in H$$

Setting $z = gh^{-1}$, we then have $z \in C_G(Q)$. Thus $z \in C_G(P)$ and $g = zh$, which shows that H controls the G-fusion of its p-subgroups.

Conversely, suppose that H controls the G-fusion of its p-subgroups. Let $g \in G$ and let P denote a Sylow p-subgroup of $H \cap gHg^{-1}$. Because $g^{-1}Pg \subseteq H$, we then have $g = zh$ for some $z \in C_G(P)$, $h \in H$. Since $z = gh^{-1}$, we infer that $P \subseteq C_H(gh^{-1})$, proving (iii) and hence the result. ∎

7.11. Corollary. *Let R be a field of characteristic $p > 0$ and let P be a p-subgroup of G which controls the G-fusion of its subgroups. Then $(1_P)^G$ is indecomposable if and only if $C_{RG}(P)$ is local.*

Proof. Apply Theorem 7.10 and Proposition 7.8(ii). ∎

8. An application : The Robinson's theorem

Let F be a field of characteristic $p > 0$ and let G be a finite group. If D is a defect group of a given block of FG and P any Sylow p-subgroup of G containing D, then a classical result of Green (1968, Theorem 3) asserts that there exists $x \in C_G(D)$ such that

$$D = P \cap xPx^{-1}$$

A remarkable improvement of the above was established by Robinson (1983, Corollary 1). Namely, Robinson demonstrated that there exists a p-regular element x of G with the following two properties:

(a) $D = P \cap xPx^{-1}$.

(b) D is a Sylow p-subgroup of $C_G(x)$.

Our aim here is to prove the above result as a consequence of some properties of G-algebras established in the previous section. In what follows, F is a field of characteristic $p > 0$, $Cl(G)$ the set of all conjugacy classes of G and, for any $C \in Cl(G)$, $\delta(C)$ is a defect group of C (with respect to p). We need a classical tool known as the Brauer correspondence. The following preliminary results will clear our path.

8.1. Lemma. *Let P be a p-subgroup of G and let $H = N_G(P)$.*
(i) If $C \in Cl(G)$ with $\delta(C) = P$, then $C \cap C_G(P)$ is a conjugacy class of H with defect group P.
(ii) If C is a conjugacy class of H with defect group P, then there exists $C \in Cl(G)$ such that $L = C \cap C_G(P)$ and $\delta(C) = P$.

Proof. (i) By definition, P is a Sylow p-subgroup of $C_G(g)$ for some $g \in C$. Hence $g \in C \cap C_G(P)$. If $g' \in C \cap C_G(P)$, then there is an $x \in G$ such that $xgx^{-1} = g'$. Since $g' \in C_G(P)$, P is a Sylow p-subgroup of $C_G(g') = xC_G(g)x^{-1}$. But xPx^{-1} is one also and so there is a $y \in C_G(g')$ with $yPy^{-1} = xPx^{-1}$; hence $y^{-1}x \in H$. Because y commutes with g', we have

$$g' = y^{-1}xgx^{-1}y$$

so that g' is conjugate to g in H. Finally, since $C_G(P) \lhd N_G(P)$ any element of H which is H-conjugate to g lies in $C \cap C_G(P)$.

(ii) By hypothesis, there is an element h in L such that P is a Sylow p-subgroup of $C_H(h)$. Hence $h \in C_G(P)$ and so $L \subseteq C \cap C_G(P)$, where C is the conjugacy class of G containing h. We now show that P is a defect group of C. It will then follow, by (i), that $L = C \cap C_G(P)$ and hence the proof will be complete.

Let Q be a Sylow p-subgroup of $C_G(h)$ with $P \subseteq Q$. Then Q is a defect group of C. If $Q \neq P$, chose a subgroup $P_1 \supseteq P$ of Q such that $(P_1 : P) = p$. Then $P \subseteq N_G(P) = H$ and $P_1 \subseteq C_H(h)$, contrary to the fact that P is a defect group of L. Thus $Q = P$ and the result is established. ∎

8.2. Lemma. *Let P be a p-subgroup of G, let $H = N_G(P)$ and let*

$$\pi : Z(FG) \to Z(FH)$$

be the Brauer homomorphism. Then π induces an injection of the block idempotents of FG having P as a defect group into nonzero idempotents of $Z(FH)$. *Moreover, for any block idempotent e of FG with defect group P, $Supp\,\pi(e)$ is a union of H-conjugacy classes having defect group P.*

Proof. Denote by e a block idempotent of FG with P as a defect group. By Proposition 2.12.19, there exist $C_1, \ldots, C_r$ in $Cl(G)$ which satisfy conditions (ii) and (iii) of Proposition 2.12.19 and such that

$$e = \lambda_1 C_1^+ + \cdots + \lambda_r C_r^+$$

for some nonzero λ_i in F. Hence, in the notation of Proposition 2.12.19, the conjugacy classes $C_1, \ldots, C_t$ have P as one of their defect groups, while defect groups of $C_{t+1}, \ldots, C_r$ have order less than $|P|$. Thus

$$C_k \cap C_G(P) = \emptyset \quad \text{for all} \quad k \in \{t+1, \ldots, r\}$$

and so $\pi(C_k^+) = 0$, $t+1 \leq k \leq r$, which proves that $\pi(e) = \lambda_1 \pi(C_1^+) + \cdots + \lambda_t \pi(C_t^+)$. Now

$$Supp\,\pi(C_i^+) = C_i \cap C_G(P) \qquad (1 \leq i \leq t)$$

and, by Lemma 8.1(i), the $C_i \cap C_G(P)$, $1 \leq i \leq t$, are distinct conjugacy classes of H with defect group P. Hence $\pi(e)$ satisfies the required property.

Assume that e' is another block idempotent of FG having P as a defect group, and let $\pi(e) = \pi(e')$. Then $Supp\,(e-e')$ is a union of conjugacy classes whose defect groups are conjugate to proper subgroups of P. Let γ be the irreducible representation of $Z(FG)$ associated with e. By Proposition 2.12.19 (iv), $\gamma(e - e') = 0$ and so $\gamma(e) = \gamma(e')$. But if e and e' were distinct, then $\gamma(e) = 1$ and $\gamma(e') = 0$. Hence $e = e'$ and the result follows. ∎

8.3. Lemma. *Let P be a normal p-subgroup of G. Then the kernel of the Brauer homomorphism*

$$\pi : Z(FG) \to Z(FG)$$

is a nilpotent ideal.

Proof. Let $C \in Cl(G)$ be such that $C \cap C_G(P) = \emptyset$. Owing to Theorem 2.11.9, it suffices to prove that C^+ is nilpotent. Because $FG \cdot I(P)$ is a nilpotent ideal of FG, the latter will follow provided we show that $C^+ \in FG \cdot I(P)$.

The group P acts by conjugation as a permutation group of the set C. Since $C \cap C_G(P) = \emptyset$, the size of each orbit under this action is divisible by p. But for all $x \in P$ and $g \in C$, we have

$$x^{-1}gx - g = x^{-1}g(x-1) + (x^{-1}-1)g \in FG \cdot I(P)$$

Thus $C^+ \in FG \cdot I(P)$, as required. ∎

In what follows, for any $C \in Cl(G)$, $I[C]$ denotes the ideal of $Z(FG)$ which is the F-linear span of all C_i^+, $C_i \in Cl(G)$, with $\delta(C_i) \subseteq_G \delta(C)$, (see Lemma 2.12.18).

8.4. Lemma. *Let P be a normal p-subgroup of G and let $C_1, \ldots, C_k$ (respectively $C_{k+1}, \ldots, C_t$) be all conjugacy classes of G with defect group P (respectively, all conjugacy classes of G whose defect groups are conjugate to proper subgroups of P). Denote by A the F-linear span of $C_1^+, \ldots, C_k^+$ and by B the F-linear span of $C_{k+1}^+, \ldots, C_t^+$. Then*

(i) A is a subalgebra of $Z(FG)$, B is a nilpotent ideal of $Z(FG)$ and for any $C \in \{C_1, \ldots, C_k\}$, $I[C] = A \oplus B$ (direct sum of F-spaces).

(ii) The defect groups of any block idempotent of FG contain P. If the order of P is p^d, then any block idempotent of FG of defect d lies in A.

Proof. (i) Fix $i, j \in \{1, \ldots, k\}$. By Lemma 2.12.18, we have

$$C_i^+ C_j^+ = \lambda_1 C_1^+ + \cdots + \lambda_k C_k^+ + \lambda_{k+1} C_{k+1}^+ + \cdots + \lambda_t C_t^+$$

where each λ_i is in F. Let

$$\pi : Z(FG) \to Z(FG)$$

be the Brauer homomorphism. Since $C_1, \ldots, C_k$ have P as their defect group, $C_i \subseteq C_G(P)$ for $i \in \{1, \ldots, k\}$ and so $\pi(C_i^+) = C_i^+$ for all $i \in \{1, \ldots, k\}$. On the other hand, because the defect groups of $C_{k+1}, \ldots, C_t$ are conjugate to proper subgroups of P, we have

$$C_{k+r} \cap C_G(P) = \emptyset \quad \text{for all} \quad r \in \{1, \ldots, t-k\}$$

Hence $\pi(C_{k+1}^+) = \cdots = \pi(C_t^+) = 0$ and so

$$C_i^+ C_j^+ = \pi(C_i^+)\pi(C_j^+) = \pi(C_i^+ C_j^+) = \sum_{r=1}^{k} \lambda_r C_r^+ \in A,$$

proving that A is a subalgebra.

By the definition of $I[C]$, $I[C] = A \oplus B$ as F-spaces. By Theorem 2.11.9, we also have $B = Ker\pi \cap I[C]$ and so B is an ideal. Finally, since by Lemma 8.3, $Ker\pi$ is nilpotent, so is B.

(ii) Let e be a block idempotent of FG with defect group D. By Proposition 2.12.19, $Supp\, e$ is a union of conjugacy classes whose defect groups are conjugate to subgroups of D. Let $C \in Cl(G)$ with $C \subseteq Supp\, e$. If $P \nsubseteq D$, then $C \cap C_G(P) = \emptyset$ and so, by Lemma 8.3, e is nilpotent, which is impossible. Consequently, $P \subseteq D$.

Finally, assume that e is of defect d. Then, by the above, P is a defect group of e. Hence, by (i) and Proposition 2.12.19, $e = a + b$ with $a \in A, b \in B$. But there is an integer n with $b^{p^n} = 0$. Thus

$$e = e^{p^n} = (a+b)^{p^n} = a^{p^n} + b^{p^n} = a^{p^n} \in A$$

and the result follows. ∎

We next record two useful facts pertaining to commutative F-algebras.

8.5. Lemma. *Let $\Gamma_i : A \to M_{n_i}(F)$, $i = 1, 2$, be two irreducible representations of a commutative F-algebra A. Then Γ_1 is equivalent to Γ_2 if and only if there exists an F-algebra isomorphism $f : \Gamma_1(A) \to \Gamma_2(A)$ such that $\Gamma_2 = f \circ \Gamma_1$.*

Proof. Assume that Γ_1 is equivalent to Γ_2. Then $n_1 = n_2$ and there is a non-singular $n_1 \times n_1$-matrix c such that $\Gamma_2(a) = c^{-1}\Gamma_1(a)c$ for all $a \in A$. Hence the map $f : \Gamma_1(A) \to \Gamma_2(A)$ defined by $f(x) = c^{-1}xc$ is an isomorphism of F-algebras such that $\Gamma_2 = f \circ \Gamma_1$.

To prove the converse, denote by e the block idempotent of A such that $\Gamma_1(e) = 1$. Then

$$\Gamma_2(e) = (f \circ \Gamma_1)(e) = f(1) = 1$$

and hence, by Lemma 2.12.6, Γ_1 is equivalent to Γ_2, as desired. ∎

8.6. Lemma. *Let $\rho : A \to B$ be a homomorphism of commutative F-algebras, let e be a block idempotent of A and let*

$$\rho(e) = u_1 + u_2 + \cdots + u_n$$

where the u_i are block idempotents of B. Denote by λ_i the irreducible representation of B associated with u_i and put $F_i = (\lambda_i\rho)(A)$, $1 \leq i \leq n$. Then, for all $i, j \in \{1, \ldots, n\}$, there exists an F-algebra isomorphism

$$\theta_{ij} : F_j \to F_i$$

such that

$$\lambda_i \circ \rho = \theta_{ij} \circ \lambda_j \circ \rho \quad (1 \leq i, j \leq n)$$

Proof. Fix $i, j \in \{1, \ldots, n\}$. Then, by Lemma 2.12.6,

$$\lambda_i\rho(e) = 1 = \lambda_j\rho(e) \tag{1}$$

Because $\lambda_i(B)$ is a finite field extension of F with $F_i \subseteq \lambda_i(B)$, it follows that F_i is a finite field extension of F. Thus, replacing F_i by its image in the regular representation of F_i, we may assume that $\lambda_i \circ \rho$ is an irreducible representation of A, $1 \leq i \leq n$. By (1) and Lemma 2.12.6, it follows that $\lambda_i \circ \rho$ and $\lambda_j \circ \rho$ are equivalent. The desired conclusion is therefore a consequence of Lemma 8.5. ∎

We are now ready to prove the following classical result.

8.7. Theorem. *(Brauer Correspondence). Let F be an arbitrary field of characteristic $p > 0$, let P be a p-subgroup of G and let $H = N_G(P)$. Then the Brauer homomorphism*

$$\pi : Z(FG) \to Z(FH)$$

determines a bijective correspondence between the block idempotents of
FG which have P as one of their defect groups and the block idempo-
tents of FH which have P as a (necessary unique) defect group.

Proof.   For the sake of clarity, we divide the proof into two steps.

Step 1.   Let e be a block idempotent of FG with defect group P.
Here we prove that $\pi(e)$ is a block idempotent of FH with defect group
P. Owing to Lemma 8.2, it suffices to show that $\pi(e)$ is a primitive
idempotent of $Z(FH)$. Assume by way of contradiction that

$$\pi(e) = e_1 + e_2 + \cdots + e_t \qquad (t > 1)$$

where the e_i are the orthogonal primitive idempotents of $Z(FH)$. Let
γ_i be the irreducible representation of $Z(FH)$ associated with e_i and
let F_i be the image of $Z(FG)$ under the homomorphism $\gamma_i \circ \pi, 1 \le i \le t$.
By Lemma 8.6, there exist F-isomorphisms $\theta_{ij} : F_j \to F_i$ such that

$$\theta_{ij} \circ \gamma_j \circ \pi = \gamma_i \circ \pi \qquad (1 \le i, j \le t) \qquad (2)$$

By Lemma 8.4(ii), the defect groups of each e_i contain P. Assume that
$k \in \{1, \ldots, t\}$ is such that the defect groups of e_k contain P properly.
Then, by Proposition 2.12.19, $\gamma_k(A) = 0$ where A is as in Lemma 8.4
with respect to FH. It follows that $(\gamma_k \circ \pi)(e) = 0$, which is impossible.
Hence all the e_i are in A.

Let $L_1, L_2, \ldots, L_r$ be all conjugacy classes of H with P as their
defect group. Then, by Lemma 8.1 (ii), there exist conjugacy classes
$C_1, \ldots, C_r$ of G with P as one of their defect groups and with $L_i = C_i \cap C_G(P), 1 \le i \le r$. Thus

$$e_i = \sum_{j=1}^{r} \lambda_{ij} \pi(C_j^+)$$

for some $\lambda_{ij} \in F$. Since $t > 1$, it follows from (2) that

$$\begin{aligned}
0 &= \gamma_2(e_1) = \gamma_2 \pi \Big(\sum_{j=1}^{r} \lambda_{1j} C_j^+ \Big) \\
&= \theta_{21} \gamma_1 \pi \Big(\sum_{j=1}^{r} \lambda_{ij} C_j \Big) = \theta_{21} \gamma_1(e_1) \\
&= 1,
\end{aligned}$$

a contradiction. Thus $\pi(e)$ is primitive.

Step 2. Let f be a block idempotent of FH with defect group P. By Lemma 8.2, it suffices to show that $\pi(e) = f$ for some block idempotent e of FG with defect group P.

By Lemma 8.4 (ii), f lies in A. Let γ be the irreducible representation of $Z(FH)$ associated with f. Since, by Lemma 8.1, every conjugacy class L of H with defect group P is of the form $C \cap C_G(P)$ for some $C \in Cl(G)$ with $\delta(C) = P$, A is contained in $\pi(Z(FG))$. Hence $\gamma\pi \neq 0$ and so $\gamma\pi$ is an irreducible representation of $Z(FG)$.

Let e be the block idempotent of FG associated with $\gamma\pi$ and write

$$e = \lambda_1 C_1^+ + \cdots + \lambda_t C_t^+ + \lambda_{t+1} C_{t+1}^+ + \cdots + \lambda_r C_r^+ \quad (\lambda_i \in F)$$

with the C_i as in Proposition 2.12.19. Then $(\gamma\pi)(C_{t+k}^+) = 0$ for all $k \in \{1,\ldots,r-t\}$, while $(\gamma\pi)(C_s^+) \neq 0$ for some $s \in \{1,\ldots,t\}$. Because $\pi(C_s^+) \neq 0$, we have $C_s \cap C_G(P) \neq \emptyset$. Thus the defect groups of C_s, and so of e, contain subgroups conjugate to P. But if the defect groups of e were to contain P properly, then by Proposition 2.12.19 and Lemma 8.3, $\gamma(A) = 0$, which contradicts $\gamma(f) = 1$. Thus the defect groups of e are conjugate to P. Therefore, by Step 1, $\pi(e)$ is primitive. But since $(\gamma\pi)(e) = \gamma(f)$, we have $\pi(e) = f$ as desired. ∎

We have now come to the demonstration for which this section has been developed.

8.8. Theorem. *(Robinson (1983)). Let F be an arbitrary field of characteristic $p > 0$, let D be a defect group of a block $B = B(e)$ of FG and let P be a Sylow p-subgroup of G containing D. Then there exists a p-regular element x of G with the following two properties:*
  (i) $D = P \cap xPx^{-1}$.
  (ii) D is a Sylow p-subgroup of $C_G(x)$.

Proof. For the sake of clarity, we divide the proof into two steps.

Step 1. Reduction to the case where D is normal in G. Assume that the result is true whenever $D \triangleleft G$. Let $H = N_G(D)$ and let S be a Sylow p-subgroup of H containing $P \cap H$. Then D is a defect group of some block of FH, by virtue of Theorem 8.7. Since $D \lhd H$, it follows from our assumption that $D = S \cap xSx^{-1}$ for some p-regular element x such that D is a Sylow p-subgroup of $C_H(x)$. Thus

$$D = (P \cap H) \cap x(P \cap H)x^{-1} = P \cap xPx^{-1} \cap H$$

If $P \cap xPx^{-1} \supsetneq D$, then $P \cap xPx^{-1} \cap H \supsetneq D$, a contradiction. Thus $P \cap xPx^{-1} = D$. Also if T is a Sylow p-subgroup of $C_G(x)$ with $T \supsetneq D$, then $T \cap H \supsetneq D$, whereas D is a Sylow p-subgroup of $C_H(x)$, a contradiction. Thus D is a Sylow p-subgroup of $C_G(x)$, as required.

Step 2. Here we complete the proof by treating the case where $D \lhd G$. Write $e = \sum_{i=1}^{n} \lambda_i C_i^{+}$ for some p-regular classes of G and some nonzero λ_i in F. Since D is a defect group of e and $D \lhd G$, it follows from Lemma 8.4 that each C_i has D as the defect group. Furthermore, by Proposition 7.8 (i), $\sigma_P(e) \neq 0$ and thus $\sigma_P(C^{+}) \neq 0$ for some $C \in \{C_1, \ldots, C_n\}$.

Since $\sigma_P(C^{+}) \neq 0$ and since $\sigma_P(C^{+}) = \omega_P((C^{-1})^{+})$ (by Proposition 7.5 (iv)), it follows from Corollary 7.6 that

$$C_P(x) = P \cap xPx^{-1} \qquad \text{for some} \quad x \in C^{-1}$$

Since D is the defect group of C^{-1}, we conclude that $D = P \cap xPx^{-1}$ and that D is a Sylow p-subgroup of $C_G(x)$. ∎

9.   The Brauer morphism

Throughout this section, p denotes a prime, G a finite group and R an arbitrary commutative ring.

Let A be a G-algebra over R. Given a subgroup K of G, we put

$$A(K) = A^K / (J(R)A^K + \sum_{H \subset K} A_H^K)$$

with the convention that for $K = 1, \sum_{H \subset K} A_H^K = 0$. The natural surjection

$$Br_K : A^K \to A(K)$$

is called the *Brauer morphism*. Note that if there is a proper subgroup H of K such that $(K : H)$ is a unit of R, then by Lemma 1.7, $A_H^K = A^K$

and so $A(K) = 0$. Therefore the notion of the Brauer morphism is useful only in the case $(K : H)$ is a nonunit of R, for any proper subgroup of H of K. For this reason, in the most important case where $R/J(R)$ is a field of characteristic p, K is usually assumed to be a p-subgroup of G, since otherwise $A(K) = 0$.

9.1. Lemma. *With the notation above, the following properties hold:*

(i) A^K is an $N_G(K)/K$-algebra and $J(R)A^K + \sum_{H \subset K} A_H^K$ is an $N_G(K)/K$-stable ideal of A^K.

(ii) $A(K)$ is an $N_G(K)/K$-algebra annihilated by $J(R)$.

(iii) The Brauer morphism Br_K is a homomorphism of $N_G(K)/K$-algebras.

Proof. (i) If $g \in N_G(K)$ and $a \in A^K$, then for any $x \in K$,

$$^x(^g a) = {}^{g(g^{-1}xg)}a = {}^g a$$

proving that A^K is an $N_G(K)$-algebra. Since K acts trivially on A^K, it follows that A^K is an $N_G(K)/K$-algebra in a natural way.

The second assertion is obvious for $K = 1$. Assume that H is a proper subgroup of K. Given $a \in A^H$ and $g \in N_G(K)$, it follows from Lemma 1.2 (iv) that

$$^g[Tr_H^K(a)] = Tr_{gHg^{-1}}^K(^g a) \in A_{gHg^{-1}}^K,$$

as required.

(ii) This is a direct consequence of (i).

(iii) Direct consequence of (i). ∎

9.2. Lemma. *Let A be a G-algebra over R and let K be a subgroup of G. Then*

(i) $Tr_K^{N_G(K)} \circ Br_K = Tr_1^{N_G(K)/K} \circ Br_K = Br_K \circ Tr_K^G$

(ii) $Br_K(A_K^G) = A(K)_K^{N_G(K)} = A(K)_1^{N_G(K)/K}$

Proof. (i) By Lemma 9.1 (ii), we may view $A(K)$ as an $N_G(K)$-algebra with the trivial action of K on $A(K)$. This obviously implies the first equality. To prove the second equality, it suffices to verify that

$$Tr_K^G(a) \equiv Tr_K^{N_G(K)}(a) \pmod{Ker(Br_K)}$$

for all $a \in A^K$. Let T be a full set of double coset representatives for $(N_G(K), K)$ in G containing 1. Then, by Lemma 1.3 (i),

$$Tr_K^G(a) = Tr_K^{N_G(K)}(a) + \sum_{t \in T - \{1\}} Tr_{tKt^{-1} \cap N_G(K)}^{N_G(K)}(^t a)$$

Now fix $t \in T - \{1\}$, put $L = tKt^{-1} \cap N_G(K)$ and denote by S a full set of double coset representatives for (K, L) in $N_G(K)$. Then each $sLs^{-1} \cap K$ is a proper subgroup of K and

$$Tr_L^{N_G(K)}(^t a) \in \sum_{s \in S} A_{sLs^{-1} \cap K}^K \subseteq Ker(Br_K^A),$$

as required.

(ii) This is a direct consequence of (i). ∎

9.3. Lemma. *Let R be a commutative local ring such that the field $F = R/J(R)$ is of prime characteristic p, and let P be a p-subgroup of G. Let A be a G-algebra over R such that A is a permutation P-algebra with an R-basis X permuted by P. If $x_1, x_2, \ldots, x_n$ are all distinct elements of X fixed by P, then $Br_P(x_1), \ldots, Br_P(x_n)$ is an F-basis for $A(P)$.*

Proof. By Lemma 9.1 (ii), $A(P)$ may be regarded as an F-algebra. Let $x_1, x_2, \ldots, x_{n+1}, \ldots, x_t$ be all representatives for the P-orbits of X, and let Q_i be the stabilizer of x_i in P, $1 \leq i \leq t$. By our choice of the x_i, we have $Q_1 = Q_2 = \cdots = Q_n = P$ and $Q_{n+1}, \ldots, Q_t$ are all proper subgroups of P. By Lemma 2.1 (i), $\{Tr_{Q_i}^P(x_i) | 1 \leq i \leq t\}$ is an R-basis for A^P. It therefore suffices to show that for any proper subgroup Q of P, $A_Q^P \subseteq M$ where M is the sum of $J(R)A^P$ and the R-linear span of $Tr_{Q_k}^P(x_k), n+1 \leq k \leq t$.

To this end, fix $x \in X$ and denote by H and L the stabilizers of x in Q and P, respectively. Since the elements $a \in A$ of the form $a = Tr_H^Q(x)$ form an R-basis for A^Q, it suffices to verify that for any such a, $Tr_Q^P(a) \in M$. Since

$$Tr_Q^P(Tr_H^Q(x)) = Tr_H^P(x) = Tr_L^P(Tr_H^L(x))$$

and $Tr_H^L(x) \in J(R)A^L$ whenever $L \neq H$, the required assertion is established. ∎

9.4. Lemma. *Let R be as in Lemma 9.3, let RX be the group algebra of a group X over R and let G act on X as a group of automorphisms of X.*

(i) RX is a permutation G-algebra with respect to the induced action of G on RX.

(ii) If P is a p-subgroup of G and H is the subgroup of all elements of X fixed by P, then the map

$$\begin{cases} RX(P) & \to & FH \\ \sum_{h \in H} \lambda_h Br_P(h) & \mapsto & \sum_{h \in H} \lambda_h h \qquad (\lambda_h \in F) \end{cases}$$

is an isomorphism of F-algebras.

Proof. (i) This is obvious and holds for an arbitrary commutative ring R.

(ii) By (i), RX is a permutation G-algebra and hence is a permutation P-algebra. Since the R-basis X of RX is permuted by P, it follows from Lemma 9.3 that $\{Br_P(h) | h \in H\}$ is an F-basis of $RX(P)$. Since Br_P preserves multiplication, the result follows. ∎

9.5. Corollary. *Let F be a field of characteristic $p > 0$, let P be a p-subgroup of G and let $H = C_G(P)$. Then, upon identification of $FG(P)$ with FH given by Lemma 9.4 (ii), the restriction of Br_P to $Z(FG)$ is the classical Brauer homomorphism (see Theorem 2.11.9).*

Proof. We apply Lemma 9.3 to the case where $R = F, X = G$ and G acts on itself by conjugation. Hence, by Lemma 9.3, $FG(P)$ and FH are identifiable in a natural way. Let $C_1, \ldots, C_n$ be all P-conjugacy classes of G. If $|C_i| > 1$, then $Br_P(C_i^+) = 0$ while $Br_P(h) = h$ for all $h \in H$. This obviously implies the required assertion. ∎

10.   Points and pointed groups

In what follows, R denotes a complete noetherian local ring and all R-algebras are assumed to be finitely generated as R-modules. Given an

R-algebra A, we write $U(A)$ for the unit group of A. Two elements x, y in A are said to be $U(A)$-*conjugate* (or simply *conjugate* if no confusion can arise) if $y = u^{-1}xu$ for some $u \in U(A)$. If $a \in A$, then the set

$$\{u^{-1}au \mid u \in U(A)\}$$

of all elements of A which are $U(A)$-conjugate to a is called a $U(A)$-*conjugacy class* of a. We begin by recording the following important result.

10.1. Theorem. *Let A be an R-algebra, let I be an ideal of A and, for each $x \in A$, let $\bar{x}$ be the image of x in $\bar{A} = A/I$.*

(i) Each idempotent $\varepsilon \in \bar{A}$ can be lifted to an idempotent $e \in A$, that is $\bar{e} = \varepsilon$. Moreover, for any idempotent e of A with $e \notin I$, $\bar{e}$ is primitive if and only if e is primitive.

(ii) If e_1, e_2 are primitive idempotents of A with $e_1, e_2 \notin I$, then $Ae_1 \cong Ae_2$ if and only if $\bar{A}\bar{e}_1 \cong \bar{A}\bar{e}_2$.

(iii) If e_1, e_2 are primitive idempotents of A, then $Ae_1 \cong Ae_2$ if and only if e_1 and e_2 are $U(A)$-conjugate.

(iv) If e_1, e_2 are primitive idempotents of A with $e_1, e_2 \notin I$, then e_1 and e_2 are $U(A)$-conjugate if and only if $\bar{e}_1$ and $\bar{e}_2$ are $U(\bar{A})$-conjugate.

Proof. We first note that (iv) follows from (i), (ii) and (iii). To prove (iii), assume that $e_2 = u^{-1}e_1 u$ for some $u \in U(A)$. Then, setting $e_{12} = u$ and $e_{21} = u^{-1}e_1$, we have $e_{12}e_{21} = e_1$ and $e_{21}e_{12} = e_2$. Hence, by Lemma 1.6.9, $Ae_1 \cong Ae_2$. The converse was established in the course of the proof of Proposition 1.5.9.

To prove (i) and (ii), we first treat the case where $I = J(A)$. By Lemma 3.4, A is complete in the $J(A)$-adic topology. Hence, if $I = J(A)$, then (i) and (ii) follow by repeating verbatim the arguments in the proof of Theorem 3.5.

Turning to the general case, let $\varepsilon \in \bar{A}$ be an idempotent and let $J \supseteq I$ be the ideal of A such that $J/I = J(A/I)$. Since, by Lemma 3.3(i), $A/J(A)$ is semiprimitive artinian, we may choose an ideal X of A with $X \supseteq J(A)$ such that

$$A/J(A) = J/J(A) \oplus X/J(A) \tag{1}$$

in which case

$$A/J \cong X/J(A) \quad \text{via} \quad x + J \to x + J(A) \quad (x \in X) \qquad (2)$$

By the case $I = J(A)$, ε can be lifted to an idempotent of A/J. Thus, by (2), ε can be lifted to an idempotent of $A/J(A)$ and hence to an idempotent e of A.

Assume that $e \notin I$. Since $J/I = J(A/I)$, we must have $e \notin J$. Hence, by (1) and Lemma 4.1, $e \in X$. By the case $I = J(A)$, e is primitive if and only if $e + J(A)$ is primitive. Hence, by (2), e is primitive if and only if $e + J$ is primitive. But, since $J/I = J(A/I)$, $e + J$ is primitive if and only if $e + I = \varepsilon$ is primitive, proving (i).

Assume that e_1, e_2 are primitive idempotents of A with $e_1, e_2 \notin I$. Then, by (i), $e_1, e_2 \in X$ and $\bar{e}_1, \bar{e}_2$ are primitive idempotents of A. By the case $I = J(A)$, we have

$$Ae_1 \cong Ae_2 \iff (A/J(A))(e_1 + J(A)) \cong (A/J(A))(e_2 + J(A))$$

Since $e_1, e_2 \in X$, it follows from (1) that

$$Ae_1 \cong Ae_2 \iff (X/J(A))(e_1 + J(A)) \cong (X/J(A))(e_2 + J(A))$$

and hence, by (2),

$$Ae_1 \cong Ae_2 \iff (A/J)(e_1 + J) \cong (A/J)(e_2 + J)$$

Therefore, by the case $I = J(A)$, $Ae_1 \cong Ae_2$ if and only if

$$(A/I)(e_1 + I) \cong (A/I)(e_2 + I)$$

Finally, since $(A/I)(e_i + I) \cong \bar{A}\bar{e}_i, i = 1, 2$, the result follows. ∎

10.2. Lemma. *Let A be an R-algebra. Then $J(A)$ is equal to the intersection of all maximal ideals of A.*

Proof. Let M be a maximal ideal of A. Then $J(A/M) = 0$ and so, by Corollary 1.4.6 (i), $J(A) \subseteq M$. Hence $J(A)$ is contained in the intersection of all maximal ideals of A.

Conversely, suppose that x belongs to all maximal ideals of A. Then its image $\bar{x}$ in $A/J(A)$ is contained in the intersection of all maximal ideals of $A/J(A)$. But, by Lemma 3.3 (i), $A/J(A)$ is semiprimitive artinian and hence $\bar{x} = 0$. Thus $x \in J(A)$, as required. ∎

10.3. Lemma. *Let M be a maximal ideal of an R-algebra A.*

(i) There exists a primitive idempotent e of A with $e \notin M$ and any two such primitive idempotents are $U(A)$-conjugate.

(ii) For any primitive idempotent e of A with $e \notin M$ and any ideal I of A,

$$e \notin I \quad \text{if and only if} \quad I \subseteq M$$

Proof. (i) Let ε be a primitive idempotent of the simple artinian ring A/M. By Theorem 10.1 (i), ε can be lifted to a primitive idempotent e of A, in which case $e \notin M$. Furthermore, if f is another primitive idempotent of A with $f \notin M$, then $\delta = f + M$ is a primitive idempotent of A/M, by Theorem 10.1 (i). Since A/M is simple artinian, ε and δ are conjugate. Hence, by Theorem 10.1 (iv), e and f are also conjugate.

(ii) If $I \subseteq M$, then obviously $e \notin I$. Conversely, assume that $e \notin I$. Then, by Lemma 4.1, $e \notin I + M$ and, by the maximality of M, we obtain $I \subseteq M$. ∎

Let A be an R-algebra. Following Puig (1981), by a *point* of A, we understand an $U(A)$-conjugacy class of primitive idempotents of A. In what follows, we write $pt(A)$ for the set of all points of A. If I is an ideal of A and $\alpha \in pt(A)$, then the subset $\alpha + I$ of A/I is defined by

$$\alpha + I = \{e + I | e \in \alpha\}$$

10.4. Lemma. *Let I be an ideal of an R-algebra A.*

(i) For any $\alpha \in pt(A)$, either $\alpha \subseteq I$ or $\alpha + I \in pt(A/I)$.

(ii) The map $\alpha \mapsto \alpha + I$ is a bijection from the set of $\alpha \in pt(A)$ with $\alpha \nsubseteq I$ onto $pt(A/I)$. In particular, $\alpha \mapsto \alpha + J(A)$ is a bijection of $pt(A)$ onto $pt(A/J(A))$.

Proof. (i) Assume that $\alpha \nsubseteq I$ and choose $e_\alpha \in \alpha$. Then $e_\alpha \notin I$ and so, by Theorem 10.1 (i), $e_\alpha + I \in pt(A/I)$. Since $A/J(A)$ is artinian, the natural map

$$U(A) \to U(A/I)$$

is surjective. Thus $\alpha + I \in pt(A/I)$.

(ii) This is a direct consequence of (i) and Theorem 10.1 (i), (iv). ∎

For each $\alpha \in pt(A)$, let $A(\alpha)$ denote a unique block of $A/J(A)$ containing $\alpha + J(A)$.

10.5. Corollary. *For each $\alpha \in pt(A)$, choose $e_\alpha \in \alpha$. Then*

(i) The map $\alpha \mapsto Ae_\alpha/J(A)e_\alpha$ induces a bijection from the set $pt(A)$ onto the set of isomorphism classes of simple A-modules.

(ii) The map $\alpha \mapsto A(\alpha)$ is a bijection from the set $pt(A)$ onto the set of all blocks of $A/J(A)$. In particular, $A/J(A) = \oplus_{\alpha \in pt(A)} A(\alpha)$ is a block decomposition of $A/J(A)$.

(iii) For any $\alpha \in pt(A)$, there exists a unique ideal M_α of A such that $\alpha \not\subseteq M_\alpha$. Furthermore, the map $\alpha \mapsto M_\alpha$ is a bijection from the set $pt(A)$ onto the set of maximal ideals of A.

Proof. (i) By Lemma 10.4 (ii), the map $\alpha \mapsto \alpha + J(A)$ is a bijection of $pt(A)$ onto $pt(A/J(A))$. Since

$$(A/J(A))(e_\alpha + J(A)) \cong Ae_\alpha/J(A)e_\alpha$$

it follows from Theorem 10.1 (iii) that the map

$$\alpha + J(A) \mapsto Ae_\alpha/J(A)e_\alpha$$

induces a bijection from $pt(A/J(A))$ onto the isomorphism classes of simple A-modules. This proves the required assertion.

(ii) This is a direct consequence of (i).

(iii) By Lemma 10.2, $e_\alpha \notin M_\alpha$ for some maximal ideal M_α of A, in which case $\alpha \not\subseteq M_\alpha$. If $e_\alpha \notin M$ for some other maximal ideal M of A, then by Lemma 10.3 (ii), $M \subseteq M_\alpha$ and so $M = M_\alpha$. The desired conclusion is now a consequence of Lemma 10.3 (i). ∎

10.6 Lemma. *Let A be an R-algebra, let $\alpha \in pt(A)$ and let $\pi_\alpha : A \to A(\alpha)$ be the canonical homomorphism.*

(i) For any ideal I of A,

$$\pi_\alpha(I) \neq 0 \quad \text{if and only if} \quad \alpha \subseteq I$$

(ii) For any primitive idempotent e of A,

$$\pi_\alpha(e) = 0 \quad \text{if} \quad e \notin \alpha$$

(iii) If an ideal J of A does not contain primitive idempotents of A, then $J \subseteq J(A)$.

Proof. (i) We first note that $\pi_\alpha(I) \neq 0$ if and only if

$$A(\alpha) \subseteq (I + J(A))/J(A)$$

which in turn is equivalent to

$$\alpha + J(A) \subseteq (I + J(A))/J(A)$$

Since the latter is equivalent to $\alpha \subseteq I + J(A)$ or, by Lemma 4.1, to $\alpha \subseteq I$, the required assertion follows.

(ii) If $\pi_\alpha(e) \neq 0$, then $(e + J(A))(f + J(A)) \neq 0$ for some primitive idempotent f of A such that $f + J(A) \in \alpha + J(A)$. Hence $e + J(A) \in \alpha + J(A)$ and so $e \in \alpha$.

(iii) The homomorphisms $\pi_\alpha, \alpha \in pt(A)$ induce an isomorphism

$$\begin{cases} A/J(A) & \xrightarrow{\pi} & \prod_{\alpha \in pt(A)} A(\alpha) \\ a + J(A) & \mapsto & (\pi_\alpha(a)) \end{cases}$$

Hence, by (i), if an ideal J of A does not contain primitive idempotents of A, then $J \subseteq J(A)$. ∎

10.7. Lemma. *Let I be an ideal of an R-algebra A, let Γ be the set of all points γ of A such that $\gamma \not\subseteq I$ and let $\bar{A} = A/I$.*

(i) $J(A) + I$ is the sum of all ideals J of A such that $\gamma \not\subseteq J$ for all $\gamma \in \Gamma$.

(ii) The homomorphisms $\pi_\gamma, \gamma \in \Gamma$, induce an R-algebra isomorphism

$$\bar{A}/J(\bar{A}) \to \prod_{\gamma \in \Gamma} A(\gamma)$$

Proof. (i) Let J be an ideal of A such that $\gamma \not\subseteq J$ for all $\gamma \in \Gamma$. Then, by Lemma 4.1, $\gamma \not\subseteq J(A) + I + J$ for all $\gamma \in \Gamma$. Assume that $e + I$

is a primitive idempotent of A/I such that $e + I \in (J(A) + I + J)/I$. Then, by Theorem 1.1 (i), there exists a primitive idempotent f of A such that $e + I = f + I$, in which case $f \notin I$ and $f \in J(A) + I + J$. But if $\gamma \in pt(A)$ is such that $f \in \gamma$, then $\gamma \not\subseteq I, \gamma \not\subseteq J$ and $\gamma \subseteq J(A) + I + J$, which contradicts Lemma 4.1. Thus $(J(A) + I + J)/I$ does not contain primitive idempotents of A/I. Hence, by Lemma 10.6 (iii), $(J(A)+I+J)/I \subseteq J(A/I) = (J(A)+I)/I$, where the last equality follows from Lemma 10.2. Thus $J \subseteq J(A) + I$, as required.

(ii) Since $J(A/I) = (J(A) + I)/I$ we have $\bar{A}/J(A) \cong A/(J(A) + I)$. The homomorphisms $\pi_\alpha, \alpha \in pt(A)$ induce an isomorphism

$$\left\{ \begin{array}{ccc} A/J(A) & \xrightarrow{\pi} & \prod_{\alpha \in pt(A)} A(\alpha) \\ a + J(A) & \mapsto & (\pi_\alpha(a)) \end{array} \right.$$

Hence, by (i) and Lemma 10.6 (i), π induces an isomorphism

$$\left\{ \begin{array}{ccc} A/(J(A) + I) & \to & \prod_{\gamma \in \Gamma} A(\gamma) \\ a + J(A) + I & \mapsto & (\pi_\gamma(a)) \end{array} \right.$$

thus completing the proof. ∎

From now on, we fix the following notation and assumptions:
G is a finite group.
K is a subgroup of G.
R is a complete noetherian local ring with $char\, R/J(R) = p > 0$.
A is a G-algebra over R.
$A(K) = A^K/(J(R)A^K + \sum_{H \subset K} A_H^K)$
with the convention that for $K = 1, \sum_{H \subset K} A_H^K = 0$.
$Br_K : A^K \to A(K)$ is the Brauer morphism.
$A(\alpha)$ is a unique block of $A/J(A)$ containing $\alpha + J(A)$, $\alpha \in pt(A)$.
$\pi_\alpha = \pi_\alpha(A) : A \to A(\alpha)$ is the canonical homomorphism of R-algebras.
$A \cdot \alpha \cdot A$ (or AeA if $e \in \alpha$) is the two-sided ideal of A generated by $\alpha \in pt(A)$.
Observe that, by Lemma 1.7,

$$A(K) = 0 \quad \text{unless } K \text{ is a } p\text{-subgroup of } G \qquad (3)$$

The following terminology is extracted from Puig (1981). A *pointed group* is a pair (H, α) consisting of a group H, which is a subgroup of

G, and a point α of A^H; this pair (H, α) will be denoted by H_α.

We say that a point α of A^H is *local* (or that the pointed group H_α is *local*), if $Br_H(\alpha) \neq 0$. Note that if α is a local point of A^H, then $A(H) \neq 0$ and hence, by (3), H is a p-subgroup of G. On the other hand, if $H = 1$, then $A(H) = A/J(R)A$ and hence any point of A^H is local.

10.8. Proposition. *Let P be a p-subgroup of G, let Γ be the set of all local points of A^P and, for each $\gamma \in \Gamma$, let $\pi_\gamma = \pi_\gamma(A^P)$.*

(i) A point γ of A^P is local if and only if $\pi_\gamma(A_Q^P) = 0$ for all proper subgroups Q of P.

(ii) The Brauer morphism $Br_P : A^P \to A(P)$ induces a bijection from Γ onto the set of all points of $A(P)$.

(iii) The homomorphisms $\pi_\gamma, \gamma \in \Gamma$, induce an R-algebra isomorphism

$$A(P)/J(A(P)) \to \prod_{\gamma \in \Gamma} A^P(\gamma)$$

Proof. (i) By Lemma 10.6 (i), $\pi_\gamma(A_Q^P) \neq 0$ implies $\gamma \subseteq A_Q^P$, in which case $Br_P(\gamma) = 0$. Conversely, if $Br_P(\gamma) = 0$, then by Lemma 4.1, there exists a proper subgroup Q of P such that $\gamma \subseteq A_Q^P$. Hence, by Lemma 10.6 (i), $\pi_\gamma(A_Q^P) \neq 0$.

(ii) By definition, a point γ of A^P is local if and only if $\gamma \subseteq Ker Br_P$. Now apply Lemma 10.4 (ii).

(iii) Apply Lemma 10.7 (ii) to the special case where A^P plays the role of A and $Ker Br_P$ the role of I. ∎

10.9. Corollary. *Let P be a p-subgroup of G and let R be a field of characteristic $p > 0$. Then there exists a canonical bijection between the local points of $C_{RG}(P)$ and the isomorphism classes of simple $RC_G(P)$-modules.*

Proof. Let A be the group algebra of G over R with the action of G on A given by conjugation. Then $A^P = C_{RG}(P)$ and, by Lemma 9.4 (ii), $A(P)$ is identifiable with $RC_G(P)$. Now apply Proposition 10.8 (ii) and Corollary 10.5 (i). ∎

Let H_α and S_β be two pointed groups. We write

$$S_\beta \subseteq H_\alpha$$

and say that S_β is *contained* in H_α if $S \subseteq H$ and, for any $e \in \alpha$, there exists $f \in \beta$ such that $ef = fe = f$. It is clear that the relation $\subseteq$ between pointed groups is transitive. Also, $H_\beta \subseteq H_\alpha$ if and only if $H_\beta = H_\alpha$.

10.10. Lemma. *Let H_α and S_β be two pointed groups. Then the following conditions are equivalent:*

(i) $S_\beta \subseteq H_\alpha$.

(ii) $S \subseteq H$ and $\pi_\beta(\alpha) \neq 0$.

(iii) $S \subseteq H$ and, for any $e \in \alpha$, there exists $f \in \beta$ such that $A^S f$ is a direct summand of $A^S e$.

Proof. We may obviously assume that $S \subseteq H$. If $S_\beta \subseteq H_\alpha$ and $e \in \alpha$, then there exists $f \in \beta$ such that $ef = fe = f$, in which case

$$A^S e = A^S f \oplus A^S(e - f)$$

Conversely, if $A^S e = A^S f \oplus M$ for some $f \in \beta$ and some submodule M of $A^S e$, then $A^S f = A^S f_1$ for some primitive idempotent f_1 of A^S such that $ef_1 = f_1 e = f_1$. Since $f_1 \in \beta$, this shows that (i) is equivalent to (iii).

To prove that (ii) is equivalent to (iii), assume that $e \in \alpha$ and let J be a set of pairwise orthogonal idempotents of A^S such that $e = \sum_{j \in J} j$. Then, by Lemma 10.6 (ii), $\pi_\beta(j) = 0$ for any $j \in J - \beta$. Hence $\pi_\beta(\alpha) \neq 0$ is equivalent to $J \cap \beta \neq \emptyset$, which is obviously equivalent to (iii). ∎

10.11. Lemma. *Let H be a subgroup of G, let V be a finitely generated RG-module and let $A = End_R(V)$. Then*

(i) The points of A^H are in bijective correspondence with the isomorphism classes of indecomposable RH-modules which are direct summands of V_H.

(ii) If S is a subgroup of H and $\alpha \in pt(A^H)$, $\beta \in pt(A^S)$, then $S_\beta \subseteq H_\alpha$ if and only if an indecomposable RS-module associated with β is a direct summand of W_S where W is an indecomposable RH-module associated with α.

Proof. (i) An idempotent e of $A^H = End_{RH}(V)$ is primitive if and only if the RH-module $e(V)$ is indecomposable. Furthermore, if e, f are primitive idempotents of A^H, then e and f are $U(A^H)$-conjugate if and only if the RH-modules $e(V)$ and $f(V)$ are isomorphic. Since every indecomposable direct summand of V_H is of the form $e(V)$ for some primitive idempotent e of A^H, the required assertion follows.

(ii) Apply (i) and Lemma 10.10. ∎

Let H be a subgroup of G and let α be a point of A^H. We define the *defect groups* of α as the subgroups P of H minimal subject to

$$\alpha \subseteq A_P^H$$

Let e be an arbitrary element of α. Then $\alpha \subseteq A_P^H$ if and only if $e \in A_P^H$. Considering A just as an H-algebra, it follows from Theorem 4.2(ii), that the defect groups of α are p-subgroups of H, which are unique up to H-conjugacy. For future use, we now record

10.12. Lemma. *Let P be a p-subgroup of G and let γ be a point of A^P. Then P is a defect group of γ if and only if γ is a local point of A^P.*

Proof. This is a direct consequence of Lemma 4.1 ∎

It is clear that the group G acts on the set of pointed groups via:

$$^g(H_\alpha) = (gHg^{-1})_\beta \quad \text{where} \quad \beta = \, ^g\alpha \quad (\alpha \in pt(A^H))$$

10.13. Lemma. *Let H_α and S_β be two pointed groups and let $g \in G$.*
(i) $S_\beta \subseteq H_\alpha$ if and only if $^g(S_\beta) \subseteq \, ^g(H_\alpha)$.
(ii) H_α is a local pointed group if and only if so is $^g(H_\alpha)$.

Proof. (i) Let $e \in \alpha$ and $f \in \beta$. Then

$$ef = fe = f \iff \, ^ge\,^gf = \, ^gf\,^ge = \, ^gf$$

and the required assertion follows.

(ii) We may assume that H is a p-subgroup of G. Hence, by Lemma

10.12, H_α is a local pointed group if and only if H is a defect group of α. Since H is a defect group of α if and only if gHg^{-1} is a defect group of $^g\alpha$, the result is established. ∎

We have now accumulated all the information necessary to prove the following important result.

10.14. Theorem. *(Puig (1981)). Let H_α be a pointed group and let P be a defect group of α.*

(i) There exists a local point γ of A^P such that

$$P_\gamma \subseteq H_\alpha \quad \text{and} \quad \alpha \subseteq Tr_P^H(A^P \cdot \gamma \cdot A^P)$$

(ii) If Q_δ is any local point group with $Q_\delta \subseteq H_\alpha$, then

$$Q_\delta \subseteq {}^h(P_\gamma) \quad \text{for some} \quad h \in H$$

(iii) The following conditions are equivalent:
(a) Q_δ is a maximal local pointed group with $Q_\delta \subseteq H_\alpha$.
(ii) Q_δ is a local pointed group with $Q_\delta \subseteq H_\alpha$ and $\alpha \subseteq A_Q^H$.
(c) $Q_\delta = {}^h(P_\gamma)$ for some $h \in H$.

(iv) The group H acts transitively on the set of maximal local pointed groups Q_δ with $Q_\delta \subseteq H_\alpha$. In particular, $N_H(P)$ acts transitively on the set of all local points δ of A^P such that $P_\delta \subseteq H_\alpha$.

Proof. (i) Let $e \in \alpha$ and let J be a set of pairwise primitive idempotents of A^P such that $e = \sum_{j \in J} j$. By definition, P is a minimal subgroup of H such that $e \in A_P^H$. Since $e = Tr_P^H(a)$ for some $a \in A^P$ and $e^2 = e$, we have

$$
\begin{aligned}
e &= Tr_P^H(a)e = Tr_P^H(ae) \in Tr_P^H(A^P e) \\
&\subseteq Tr_P^H(A^P e A^P) \\
&\subseteq \sum_{j \in J} Tr_P^H(A^P j A^P)
\end{aligned}
$$

But, for any $j \in J$, $Tr_P^H(A^P j A^P)$ is a two-sided ideal of the algebra A^H and hence, by Lemma 4.1,

$$e \in Tr_P^H(A^P f A^P) \quad \text{for some} \quad f \in J \tag{4}$$

Moreover, the minimal choice of P implies $f \notin A_Q^P$ for any proper subgroup Q of P. Hence, if γ is the point of A^P containing f, then P is a defect group of γ and, by Lemma 10.12, γ is a local point of A^P. Furthermore, $P_\gamma \subseteq H_\alpha$ (since $fe = ef = f$) and, by (4),

$$\alpha \subseteq Tr_P^H(A^P \cdot \gamma \cdot A^P),$$

as required.

(ii) Assume that Q_δ is a local pointed group contained in H_α. Since $Q_\delta \subseteq H_\alpha$, it follows from Lemma 10.10 that $\pi_\delta(e) \neq 0$. But $e \in Tr_P^H(A^P f A^P)$ and clearly

$$Tr_P^H(A^P f A^P) \subseteq \sum_{h \in H} Tr_{Q \cap hPh^{-1}}^P(A^{hPh^{-1}} \, {}^h f A^{hPh^{-1}})$$

Moreover, since δ is a local point of A^Q, we have

$$\pi_\delta(A_R^Q) = 0$$

for all proper subgroups R of Q (see Proposition 10.8 (i)). Thus there exists $h \in H$ such that $Q \subseteq hPh^{-1}$ and $\pi_\delta({}^h f) \neq 0$, which proves that $Q_\delta \subseteq {}^h(P_\gamma)$, by virtue of Lemma 10.10.

(iii) By (i) and Lemma 10.13, for each $h \in H$, ${}^h(P_\gamma)$ is a local pointed group contained in H_α. Hence (a) and (c) are equivalent, by virtue of (ii). Since, for any $h \in H$, $\alpha = {}^h\alpha \in {}^h(A_P^H) = A_{hPh^{-1}}^H$, (c) implies (b). Finally, assume that (b) holds. Since $\alpha \subseteq A_Q^H$, it follows from Theorem 4.2 (i) that $P \subseteq xQx^{-1}$ for some $x \in H$. But, by (ii), $Q_\delta \subseteq {}^h(P_\gamma)$ for some $h \in H$. Hence $Q_\delta = {}^h(P_\gamma)$, proving (c).

(iv) This is a direct consequence of (iii). ∎

If H is a subgroup of G and S_β is a pointed group, we denote by $N_H(S_\beta)$ the stabilizer of S_β in H, i.e.

$$N_H(S_\beta) = \{h \in H \mid {}^h(S_\beta) = S_\beta\}$$

It is clear that S is a normal subgroup of $N_H(S_\beta)$.

10.15. Lemma. *If Q_δ is a pointed group, then $A^Q(\delta)$ is endowed with the $(N_G(Q_\delta)/Q)$-algebra structure of A, i.e.*

$$\overline{{}^g Q} \pi_\delta(a) = \pi_\delta({}^g a) \qquad (g \in N_G(Q_\delta), a \in A^Q)$$

Proof. It clearly suffices to verify that $A^Q(\delta)$ is an $N_G(Q_\delta)$-algebra via

$$^g\pi_\delta(a) = \pi_\delta(^g a) \qquad (g \in N_G(Q_\delta), a \in A^Q)$$

or, equivalently, that $Ker\pi_\delta$ is an $N_G(Q_\delta)$-stable ideal of A^Q. Let $e + J(A^Q)$ be the identity element of $A^Q(\delta)$. Then $a \in Ker\pi_\delta$ if and only if $ae \in J(A^Q)$. But, for $g \in N_G(Q_\delta)$, we have $^g\delta = \delta$ and hence $^g e \equiv e\,(mod J(A^Q))$. Thus $ae \in J(A^Q)$ implies

$$0 \equiv {}^g(ae) \equiv {}^g ae\,(mod J(A^Q),$$

as required. ∎

10.16. Lemma. *Let Q_δ be a pointed group. Then, for any $g \in N_G(Q) - N_G(Q_\delta)$,*

$$\pi_\delta(^g\delta) = 0$$

Proof. By definition, $\pi_\delta(\delta) = \delta + J(A^Q)$ is a unique point of $A^Q(\delta)$. Assume that $\pi_\delta(^g\delta) \neq 0$. Then, by Lemma 10.4 (i), $\pi_\delta(^g\delta)$ is a point of $A^Q(\delta) = \pi_\delta(A^Q)$ and hence $\pi_\delta(\delta) = \pi_\delta(^g\delta)$. Therefore, by Lemma 10.4 (ii), $^g\delta = \delta$ and so $g \in N_G(Q_\delta)$, as required. ∎

10.17. Lemma. *Let $Q_\delta \subseteq H_\alpha$ be pointed groups. Then*

$$Ker\pi_\delta \cap A^H \subseteq Ker\pi_\alpha$$

Proof. Assume by way of contradiction that $Ker\pi_\delta \cap A^H \nsubseteq Ker\pi_\alpha$. Since $Ker\pi_\alpha$ is a maximal ideal of A^H, we then have

$$A^H = Ker\pi_\alpha + (Ker\pi_\delta \cap A^H)$$

Since $\alpha \subseteq A^H$ and $\alpha \nsubseteq Ker\pi_\alpha$, it follows from Lemma 4.1 that

$$\alpha \subseteq Ker\pi_\delta \cap A^H$$

which implies that $\pi_\delta(\alpha) = 0$, contrary to Lemma 10.10. ∎

10.18. Lemma. *Let B and S be artinian algebras over a commutative ring L, let S be simple and let I be an ideal of B. Then every* *surjective L-homomorphism* $f : I \to S$ *preserving multiplication can be extended to an L-algebra homomorphism* $B \to S$.

Proof. Put $J = I \cap J(B)$. If $f(J) \neq 0$, then $f(J) = S$ and so $f(J^n) = S$ for all $n \geq 1$. But J is a nilpotent ideal of I, so $S = 0$, a contradiction. Thus $J \subseteq Ker f$ and hence f induces a homomorphism $f^* : I/J \to S$ given by $f^*(x + J) = f(x)$ for each $x \in I$. Thus the map $(I + J(B))/J(B) \overset{g}{\to} S$ given by $g(x + J(B)) = f(x)$ for each $x \in I$ is a homomorphism. Since $(I + J(B))/J(B)$ is a direct summand of $B/J(B)$, there exists a homomorphism $g^* : B/J(B) \to S$ such that $g^*(x + J(B)) = f(x)$ for each $x \in I$. Thus the homomorphism $\pi : B \to S$ given by $\pi(b) = g^*(b + J(B))$ is such that for each $x \in I$, $\pi(x) = f(x)$. Finally, since π is surjective, it preserves identity elements. ∎

In what follows, $A^Q(\delta)$ is regarded as an $(N_H(Q_\delta)/Q)$-algebra in a manner described in Lemma 10.15. Also "Q_δ is maximal in H_α" means that Q_δ is a maximal local pointed group with $Q_\alpha \subseteq H_\alpha$.

10.19. Theorem. *(Puig (1981)). Let H_α be a pointed group, let Q_δ be a local pointed group contained in H_α and let $N = N_H(Q_\delta)/Q$. Then*

(i) $\pi_\delta(A_Q^H) = A^Q(\delta)_1^N$.

(ii) $\pi_\delta(Tr_Q^H(a)) = Tr_1^N(\pi_\delta(a))$ *for all* $a \in A^Q \cdot \delta \cdot A^Q$.

(iii) Q_δ *is maximal in* H_α *if and only if* $\pi_\delta(\alpha) \subseteq A^Q(\delta)_1^N$.
(iv) *If* Q_δ *is maximal in* H_α, *then there exists an algebra homomorphism*

$$\pi_\alpha^\delta : A^Q(\delta)^N \to A^H(\alpha)$$

such that

$$\pi_\alpha^\delta(\pi_\delta(a)) = \pi_\alpha(a) \quad \text{for all} \quad a \in A_Q^H$$

Proof. (i) and (ii). Given $a \in A^Q$, it follows from Lemma 1.3 (i) that

$$Tr_Q^H(a) = \sum_x Tr_{Q \cap xQx^{-1}}^Q(^x a) \tag{5}$$

where x runs over a complete set of representatives for the double cosets of H with respect to Q. Moreover, since δ is a local point of A^Q, we have

$$\pi_\delta(A_R^Q) = 0 \tag{6}$$

for all proper subgroups R of Q (see Proposition 10.8 (i)). Let T be a transversal for Q in $N_H(Q_\delta)$ and let S be a right transversal for $N_H(Q_\delta)$ in $N_H(Q)$. Then TS is a complete set of representatives of double cosets QxQ with $x \in N_H(Q)$. Thus, for all $a \in A^Q$,

$$
\begin{aligned}
\pi_\delta(Tr_Q^H(a)) &= \sum_{s \in S}(\sum_{t \in T} \pi_\delta(^{ts}a)) \quad \text{(by (5) and (6))} \\
&= \sum_{s \in S}(\sum_{t \in T} {}^{tQ}\pi_\delta(^s a)) \quad \text{(by Lemma 10.15)} \\
&= \sum_{s \in S} Tr_1^N(\pi_\delta(^s a)) \tag{7}
\end{aligned}
$$

and therefore

$$\pi_\delta(A_Q^H) \subseteq A^Q(\delta)_1^N \tag{8}$$

But, by Lemma 10.16, for any $s \in N_H(Q) - N_H(Q_\delta)$, we have $\pi_\delta(^s\delta) = 0$. Hence, for any $a \in A^Q \cdot \delta \cdot A^Q$,

$$\pi_\delta(^s a) \in \pi_\delta(A^Q \cdot {}^s\delta \cdot A^Q) = 0$$

and so, by (7),

$$\pi_\delta(Tr_Q^H(a)) = Tr_1^N(\pi_\delta(a)), \quad \text{for all} \quad a \in A^Q \cdot \delta \cdot A^Q \tag{9}$$

proving (ii). Moreover, since $\pi_\delta(A^Q \cdot \delta \cdot A^Q) = A^Q(\delta)$, it follows from (8) and (9) that

$$\pi_\delta(A_Q^H) = A^Q(\delta)_1^N,$$

proving (i).

(iii) Since $Q_\delta \subseteq H_\alpha$, it follows from Lemma 10.17 that

$$Ker\pi_\delta \cap A^H \subseteq Ker\pi_\alpha$$

Hence $\alpha \subseteq A_Q^H$ is equivalent to $\pi_\delta(\alpha) \subseteq \pi_\delta(A_Q^H)$, so by (i) and Theorem 10.14 (iii), Q_δ is maximal in H_α if and only if

$$\pi_\delta(\alpha) \subseteq A^Q(\delta)_1^N,$$

as required.

(iv) Assume that Q_δ is maximal in H_α. Then, by (iii), $\pi_\delta(\alpha) \subseteq A^Q(\delta)_1^N$ and so π_α and π_δ restricted to A_Q^H induce a surjective homomorphism

$$A^Q(\delta)_1^N \to A^H(\alpha)$$

Since $A^H(\alpha)$ is a simple algebra, it follows from Lemma 10.18 that this homomorphism can be extended to a homomorphism

$$A^Q(\delta)^N \to A^H(\alpha)$$

of R-algebras. This completes the proof of the theorem. ∎

10.20. Theorem. *(Puig (1981)). Let H_α, S_β and Q_δ be pointed groups such that $Q_\delta \subseteq H_\alpha \subseteq S_\beta$, Q_δ is local and $N_S(Q_\delta) \subseteq H$. Then Q_δ is maximal in H_α if and only if Q_δ is maximal in S_β.*

Proof. If Q_δ is maximal in S_β, then obviously Q_δ is maximal in H_α. Conversely, assume that Q_δ is maximal in H_α and set $N = N_S(Q_\delta)/Q$. Since $N_S(Q_\delta) \subseteq H$, we have $N = N_H(Q_\delta)/Q$. Also, by Theorem 10.19 (i) applied to $Q_\delta \subseteq S_\beta$, we have

$$\pi_\delta(A^S) \supseteq \pi_\delta(A_Q^S) = A^Q(\delta)_1^N$$

Since $Q_\delta \subseteq H_\alpha$, we also have $\pi_\delta(\alpha) \neq 0$ by Lemma 10.10. Thus, by Theorem 10.19 (iii),

$$0 \neq \pi_\delta(\alpha) \subseteq A^Q(\delta)_1^N \subseteq \pi_\delta(A^S) \tag{10}$$

Now, since $Q_\delta \subseteq S_\beta$, we have $\pi_\delta(\beta) \neq 0$ (Lemma 10.10). Hence, for any $e \in \beta$, $\pi_\delta(e)$ is a primitive idempotent of $\pi_\delta(A^S)$ (Theorem 10.1 (i)). Moreover, since $H_\alpha \subseteq S_\beta$, there exists $f \in \alpha$ such that $ef = fe = f$. Consequently, we have

$$\pi_\delta(e)\pi_\delta(f) = \pi_\delta(f)\pi_\delta(e) = \pi_\delta(f)$$

Since, by (10), $\pi_\delta(f)$ is an idempotent of $\pi_\delta(A^S)$, and $\pi_\delta(e)$ is a primitive idempotent of $\pi_\delta(A^S)$, we deduce that $\pi_\delta(e) = \pi_\delta(f)$. Hence, by (10),

$$\pi_\delta(\beta) \subseteq \pi_\delta(\alpha) \subseteq A^Q(\delta)_1^N$$

and so, by Theorem 10.19 (iii), Q_δ is maximal in S_β. ∎

10.21. Corollary. *(Puig (1981)). Let $Q_\delta \subset P_\alpha$ be local pointed groups. Then there exists a local pointed group S_β such that*

$$Q_\delta \subset S_\beta \subseteq P_\alpha \quad and \quad S \subseteq N_P(Q_\delta)$$

Proof. Put $K = N_P(Q_\delta)$. It is clear that there exists a point γ of A^K such that

$$Q_\delta \subseteq K_\gamma \subseteq P_\alpha$$

Since $Q_\delta \neq P_\alpha$, it follows from Theorem 10.20 that Q_δ is not maximal in K_γ. Choosing S_β to be maximal in K_γ, the result follows. ∎

As a preliminary to the next result, we record the following standard fact.

10.22. Proposition. *Let C be a finite-dimensional semisimple algebra over a field. If θ is an automorphism of C which is the identity mapping on the centre of C, then θ is an inner automorphism.*

Proof. See Bourbaki (1958). ∎

10.23. Lemma. *Assume that $F = R/J(R)$ is algebraically closed, let $Q_\delta \subseteq H_\alpha$ be pointed groups and put $S = A^Q(\delta), N = N_H(Q_\delta)/Q$. Then there exists a finite group $\hat{N}$ and a central p'-subgroup Z of $\hat{N}$ with $\hat{N}/Z \cong N$ such that the action of N on S can be lifted to a group homomorphism from $\hat{N}$ to $U(S)$.*

Proof. Since S is a simple algebra over an algebraically closed field F, there is a vector space V over F such that $S \cong End_F(V)$. Identifying S with $End_F(V)$, it follows from Proposition 10.22 that for any $x \in N$, there exists $u_x \in GL(V)$ such that

$$^x a = u_x a u_x^{-1} \quad \text{for all} \quad a \in End_F(V)$$

Then the map $N \to GL(V)$, $x \mapsto u_x$ is a projective representation of N. Hence, by the argument in the proof of Proposition 2.10.12 and Lemma 2.10.4 (iii), the required assertion follows. ∎

We are now ready to prove the final result of this section.

10.24. Theorem. *(Puig (1981)). Assume that $F = R/J(R)$ is algebraically closed, let H_α be a pointed group and let Q_δ be a maximal local pointed group contained in H_α. Assume further that*

$$A_Q^H = R \cdot 1_A + \sum_P A_P^H \tag{11}$$

where P runs over the set of proper subgroups of Q, and put $N = N_H(Q_\delta)/Q$. Then there exists a finite group $\hat{N}$, a central p'-subgroup Z with $\hat{N}/Z \cong N$ and a block B of $F\hat{N}$ with defect zero such that

$$B \cong A^Q(\delta) \qquad \text{as } N\text{-algebras}$$

Proof. Put $S = A^Q(\delta)$, let V be a vector space over F such that

$$S \cong End_F(V)$$

and let $\hat{N}$ be as in Lemma 10.23. Then, by Lemma 10.23, the action of N on S can be lifted to a group homomorphism from $\hat{N}$ to $U(S)$. Thus V becomes an $F\hat{N}$-module in a natural way.

For any proper subgroup P of Q, we have

$$A_P^H \subseteq \sum_{h \in H} A_{Q \cap hPh^{-1}}^Q \tag{12}$$

by virtue of Corollary 1.5 (i). Since Q_δ is local, it follows from (12) and Proposition 10.8 (i) that $\pi_\delta(A_P^H) = 0$. Hence, by (11) and Theorem 10.19 (i),

$$\pi_\delta(A_Q^H) = S_1^N = F \cdot 1_S$$

Consequently, $1_V = Tr_1^{\hat{N}}(f)$ for some $f \in End_F(V)$ and $End_{F\hat{N}}(V) = F \cdot 1_V$. The latter implies that V is simple, while the former, by Theorem 5.3, implies that V is a projective $F\hat{N}$-module (hence V comes from a projective $R\hat{N}$-module U such that $End_{R\hat{N}}(U) = R \cdot 1_U$). Thus V belongs to a block B of $F\hat{N}$ of defect zero and we have $B \cong End_F(V)$ as N-algebras. ∎

11. Interior G-algebras

Let G be a finite group and let R be a commutative ring. In most G-algebras A, which occur in the applications, the action of G on A is induced by a group homomorphism from G to $U(A)$ or, equivalently, by the restriction to G of an R-algebra homomorphism from RG to A. This special case deserves, therefore, a separate treatment and, following Puig (1981) and Ikeda (1986, 1987), we introduce the following definitions.

An *interior G-algebra* is a pair (A, ρ) consisting of an R-algebra A and an R-algebra homomorphism.

$$\rho : RG \to A$$

An interior G-algebra (A, ρ) has an evident G-algebra structure, namely, the one defined by

$$^{g}a = \rho(g)a\rho(g)^{-1} \quad \text{for all} \quad a \in A, g \in G$$

Note also that A is an $R(G \times G)$-module via

$$(g_1, g_2)a = \rho(g_1)a\rho(g_2)^{-1} \quad \text{for all} \quad g_1, g_2 \in G, a \in A$$

If ρ is an epimorphism, then we refer to the interior G-algebra (A, ρ) as an *epimorphic interior G-algebra*. The interior G-algebra (A, ρ) is called a *local interior G-algebra* if A^G is a local ring.

Note that if (A, ρ) is an interior G-algebra, then for any subgroup H of G, we have

$$A^H = C_A(\rho(RH)) = C_A(\rho(H)) \tag{1}$$

In particular, if (A, ρ) is an epimorphic interior G-algebra, then

$$A^G = Z(A) \tag{2}$$

Let (A, ρ) and (A', ρ') be interior G-algebras. By a *homomorphism* from (A, ρ) to (A', ρ'), we understand an R-algebra homomorphism

$$f : A \to A'$$

such that
$$\rho' = f \circ \rho$$

If, in addition, f is an isomorphism, then we say that f is an *isomorphism* of (A, ρ) onto (A', ρ'). Two interior G-algebras (A, ρ) and (A', ρ') are said to be *isomorphic* if there exists an isomorphism of (A, ρ) onto (A', ρ').

We warn the reader that if (A, ρ) and (A', ρ') are interior G-algebras, then any homomorphism f from (A, ρ) to (A', ρ') is still a homomorphism for the associated G-algebra structures, but the converse need not be true.

Let (A, ρ) be an interior G-algebra. If ρ is not relevant to the discussion, we shall refer to A itself as an interior G-algebra and put

$$x \cdot a \cdot y = \rho(x)a\rho(y) \qquad \text{for all} \quad x, y \in RG, a \in A$$

11.1. Lemma.   *Let A and A' be interior G-algebras and let $f : A \to A'$ be an R-algebra homomorphism. Then the following conditions are equivalent:*

(i) f is a homomorphism of interior G-algebras.
(ii) $f(x \cdot a \cdot y) = x \cdot f(a) \cdot y$ for all $x, y \in RG, a \in A$.
(iii) f is a homomorphism of $R(G \times G)$-modules.
(iv) $f(g \cdot 1_A) = g \cdot 1_{A'}$ for all $g \in G$.

Proof.   (i) $\Rightarrow$ (ii): Let ρ and ρ' be the homomorphisms attached to A and A', respectively. By hypothesis, $\rho' = f \circ \rho$. Hence, given $x, y \in RG$ and $a \in A$, we have

$$f(x \cdot a \cdot y) = f(\rho(x)a\rho(y)) = \rho'(x)f(a)\rho'(y) = x \cdot f(a) \cdot y,$$

as required.

(ii) $\Rightarrow$ (iii): Given $(g_1, g_2) \in G \times G$ and $a \in A$, we have

$$f((g_1, g_2)a) = f(g_1 \cdot a \cdot g_2^{-1}) = g_1 \cdot f(a) \cdot g_2^{-1} = (g_1, g_2)f(a),$$

as asserted.

(iii) $\Rightarrow$ (iv): Since $(g, 1)1_A = g \cdot 1_A$, we have

$$f(g \cdot 1_A) = (g, 1)f(1_A) = g \cdot 1_{A'},$$

as required.

(iv) $\Rightarrow$ (i): It suffices to show that for any $g \in G, \rho'(g) = f(\rho(g))$. Since $\rho(g) = g \cdot 1_A$, we have $f(\rho(g)) = f(g \cdot 1_A) = g \cdot 1_{A'} = \rho'(g)$. $\blacksquare$

From now on, we assume that R is a *complete noetherian local ring with residue class field $R/J(R)$ of characteristic $p > 0$ and all R-algebras are finitely generated as R-modules.* Owing to Theorem 4.2, these assumptions ensure that defect groups of any primitive idempotent of A^G are conjugate p-subgroups of G.

For a local interior G-algebra A, a *defect group* of A is defined to be a defect group of the primitive idempotent 1 of A^G. Thus, by definition, a defect group of a local interior G-algebra A is a minimal element in the set of all subgroups H of G such that

$$A_H^G = A^G \tag{3}$$

Owing to (1), condition (3) is equivalent to the requirement that the map

$$Tr_H^G : C_A(\rho(H)) \to C_A(\rho(G))$$

is surjective.

11.2. Lemma. *Let (A, ρ) be an epimorphic interior G-algebra. Then the following conditions are equivalent:*
(i) (A, ρ) is a local interior G-algebra.
(ii) The R-algebra A is indecomposable.
(iii) A is an indecomposable $R(G \times G)$-module.

Proof. By (2), (i) is equivalent to the requirement that $Z(A)$ is a local ring. By Corollary 3.6, the latter is equivalent to the condition that 1 is the only nonzero idempotent of $Z(A)$, i.e. that the R-algebra A is indecomposable. Finally, (iii) is equivalent to (ii), since the $R(G \times G)$-submodules of A are precisely the two-sided ideals of A, by virtue of the assumption that (A, ρ) is epimorphic. $\blacksquare$

Let e be a block idempotent of RG and let $B = RGe$ be the corresponding block of RG. The block B is an R-algebra with the identity element e and B is an epimorphic local interior G-algebra by the R-algebra homomorphism

$$\begin{cases} RG & \to & B \\ x & \mapsto & xe \end{cases}$$

Then the defect groups of B in the sense of interior G-algebras are the defect groups of B in the classical sense defined in Sec. 4.

We next provide some other important examples of interior G-algebras. In what follows, all RG-modules are assumed to be finitely generated.

11.3. Example. *Let V be an RG-module and let*

$$\rho_V : RG \to End_R(V)$$

be the corresponding homomorphism of R-algebras given by $\rho_V(a)(x) = ax$, for all $a \in RG, x \in V$. Then

(i) $(End_R(V), \rho_V)$ is an interior G-algebra such that the corresponding action of G on $End_R(V)$ coincides with the classical action defined in Sec. 1.

(ii) The interior G-algebra $(End_R(V), \rho_V)$ is local if and only if V is indecomposable.

(iii) If V is indecomposable, then the defect groups of the local interior G-algebra $(End_R(V), \rho_V)$ conicide with the vertices of V.

Proof. (i) This is a direct consequence of the definitions.

(ii) By (i), $End_R(V)^G = End_{RG}(V)$. Now apply Corollary 3.8.

(iii) Apply (i) and Proposition 6.1 (i). ■

11.4. Example. *Assume that R is a splitting field for RG with $char R = p > 0$. If V is a simple RG-module and $\rho_V : RG \to End_R(V)$ is the corresponding homomorphism of R-algebras, then $(End_R(V), \rho_V)$ is an epimorphic local interior G-algebra. Furthermore, the defect groups of $(End_R(V), \rho_V)$ coincide with the vertices of the simple RG-module V.*

Proof. Since V is simple and R is a splitting field for RG, the homomorphism ρ_V is surjective. Now apply Example 11.3. ■

Let (A, ρ) be an epimorphic local interior G-algebra. Then, by Lemma 11.2, A is an indecomposable $R(G \times G)$-module and it is natural to investigate the vertex of A. To do this, we first establish some preliminary results.

Let (A, ρ) be an interior G-algebra and let H be a subgroup of G. Since A is an $R(H \times G)$-module, the R-linear map

$$\rho_H : R(H \times G) \to End_R(A)$$

determined by

$$\rho_H(h, g)(a) = \rho(h)a\rho(g)^{-1} \qquad (h \in H, g \in G, a \in A)$$

is a homomorphism of R-algebras. This shows that $(End_R(A), \rho_H)$ is an interior $H \times G$-algebra. It follows from the definition of the action of $H \times G$ on $End_R(A)$ that $End_R(A)^{H \times G}$ consists precisely of all elements f in $End_R(A)$ for which

$$f(\rho(h)a\rho(g)^{-1}) = \rho(h)f(a)\rho(g)^{-1} \qquad (4)$$

for all $h \in H, g \in G$.

11.5. Lemma. *Let (A, ρ) be an interior G-algebra, let H be a subgroup of G and let $E = End_R(A)$.*

(i) The map $\psi_H : A^H \to E^{H \times G}$, $b \mapsto l_b$, where $l_b(a) = ba$ for all $a \in A$, is an injective homomorphism of R-algebras. Furthermore, if (A, ρ) is epimorphic, then ψ_H is an isomorphism of R-algebras.

(ii) If $Tr_H^G : A^H \to A^G$ and $Tr_{H \times G}^{G \times G} : E^{H \times G} \to E^{G \times G}$ are trace maps, then

$$\psi_G \circ Tr_H^G = Tr_{H \times G}^{G \times G} \circ \psi_H$$

Proof. (i) Since $b \in A^H$, we have $\rho(h)b = b\rho(h)$ for all $h \in H$. Hence, for all $h \in H, g \in G$,

$$l_b(\rho(h)a\rho(g)^{-1}) = b\rho(h)a\rho(g)^{-1} = \rho(h)l_b(a)\rho(g)^{-1},$$

proving, by (4), that $l_b \in E^{H \times G}$. It is now clear that ψ_H is an injective homomorphism of R-algebras.

Assume, further, that (A, ρ) is epimorhic. If $f \in E^{H \times G}$ then, by (4), with $h = g, a = 1_A$, we have

$$f(1_A) = \rho(h)f(1_A)\rho(h)^{-1} = {}^h f(1_A) \quad \text{for all} \quad h \in H$$

and so $f(1_A) \in A^H$. By (4) (with $a = 1_A, h = 1$), we have

$$f(\rho(g)) = f(1_A)\rho(g) \quad \text{for all} \quad g \in G$$

and hence $f(\rho(x)) = f(1_A)\rho(x)$ for all $x \in RG$. But ρ is surjective, so $f(a) = f(1_A)a$ for all $a \in RG$ and therefore $f = \psi(f(1_A))$, as required.

(ii) Let T be a left transversal for H in G and let $a \in A^H$. Then, for all $x \in A$,

$$
\begin{aligned}
\left[\psi_G(Tr_H^G(a))\right](x) &= \left[\psi_G(\sum_{t \in T} \rho(t)a\rho(t)^{-1})\right](x) \\
&= \sum_{t \in T} \rho(t)a\rho(t)^{-1}x = (\sum_{t \in T} {}^{(t,1)}\psi_H(a))(x) \\
&= Tr_{H \times G}^{G \times G}(\psi_H(a))(x),
\end{aligned}
$$

as required. ∎

Given an RG-module V, we write $Inv(V)$ for the R-module of G-invariant elements of V, i.e.

$$Inv(V) = \{v \in V | gv = v \quad \text{for all} \quad g \in G\}$$

Hence, if H is a subgroup of G, then $Inv(V_H)$ denotes the R-module of H-invariant elements of V. If T is a left transversal for H in G, then the map

$$Tr_H^G : Inv(V_H) \to Inv(V)$$

given by

$$Tr_H^G(v) = \sum_{t \in T} tv$$

is an R-homomorphism which is independent of the choice of T. This is of course a generalization of the trace map defined previously for G-algebras. The reader may easily verify that many properties of the trace map proved for G-algebras hold in a more general context of RG-modules. Given subgroups $H \subseteq K$ of G, we put

$$V_H^K = Tr_H^K(Inv(V_H))$$

and

$$V(K) = Inv(V_K)/(\sum_{H \subset K} V_H^K + J(R)Inv(V_K))$$

11.6. Lemma.   *Let H be a subgroup of G and let V be an H-projective RG-module. Then, for any subgroup K of G, $V(K) = 0$ unless K is G-conjugate to a subgroup of H.*

Proof.   Since V is H-projective, it follows from Theorem 5.3 that

$$Tr_H^G(End_{RH}(V)) = End_{RG}(V)$$

Hence, by Corollary 1.5 (i),

$$1_V \in Tr_H^G(End_{RH}(V)) \subseteq \sum_{g \in G} Tr_{K \cap gHg^{-1}}^K(End_{R(K \cap gHg^{-1})}(V))$$

For each $g \in G$, we may therefore choose $\psi_g \in End_{R(K \cap gHg^{-1})}(V)$ such that $1_V = \sum_{g \in G} Tr_{K \cap gHg^{-1}}^K(\psi_g)$ and thus

$$Inv(V_K) \subseteq \sum_{g \in G} Tr_{K \cap gHg^{-1}}^K(\psi_g)(Inv(V_K))$$

But, for any $v \in Inv(V_K)$, we have $\psi_g(v) \in Inv(V_{K \cap gHg^{-1}})$ and

$$Tr_{K \cap gHg^{-1}}^K(\psi_g)(v) = Tr_{K \cap gHg^{-1}}^K(\psi_g(v))$$

which implies that

$$Inv(V_K) \subseteq \sum_{g \in G} V_{K \cap gHg^{-1}}^K,$$

thus completing the proof. ∎

Let R be a commutative ring. Then R is called a *discrete valuation ring* if it is a principal ideal domain that has a unique prime ideal P.

Such P is necessarily a maximal ideal and the field R/P is called the *residue class field* of R. We say that a discrete ring is *complete* if it is complete as a noetherian local ring.

From now on, we assume that R is a complete discrete valuation ring such that the residue class field F of R is of prime characteristic p and such that F is a splitting field for FG (we will also allow the case where $R = F$). Any R-algebra below is R-free of finite rank and any RG-module is also R-free of finite rank.

Let B be a block of the group algebra RG and let (A, ρ) be a local interior G-algebra. Following Ikeda (1987), we say that (A, ρ) *belongs to B*, if $\rho(B) \neq 0$.

11.7.   Lemma.   *Let (A, ρ) be an epimorphic local interior G-algebra belonging to a block B of RG. Assume that A is a projective RG-module, where the RG-module structure on A is defined by*

$$xa = \rho(x)a \qquad \text{for all} \quad x \in RG, a \in A$$

Then ρ restricts to an isomorphism of B onto A.

Proof.   We first claim that there is no loss of generality in assuming that $R = F$. Indeed, suppose that the result holds for $R = F$, put $\bar{A} = A/J(R)A$ and let $\bar{\rho} : FG \to A$ be the homomorphism of F-algebras induced by ρ. Then $(\bar{A}, \bar{\rho})$ is an epimorphic local interior G-algebra such that $\bar{A}$ is a projective FG-module where the FG-module structure on $\bar{A}$ is defined via $\bar{\rho}$. Furthermore, $\bar{A}$ belongs to the block $\bar{B} = B/J(R)B$ of FG. Hence, by our assumption, the map $\bar{\rho} : \bar{B} \to \bar{A}$ is an isomorphism of F-algebras. In particular, $dim_F \bar{B} = dim_F \bar{A}$ and so $rank_R(B) = rank_R(A)$. Furthermore $\rho(B) = A$ since $\bar{\rho}(\bar{B}) = \bar{A}$. Thus ρ restricts to an isomorphism of B onto A, as claimed.

By the foregoing, we may assume that $R = F$. Note that, since (A, ρ) is an epimorphic local interior G-algebra, we have $\rho(B) = A$. Thus we are left to verify that $dim_F A = dim_F B$. Let e be a primitive idempotent of FG such that $\rho(e) \neq 0$. Then, by Theorem 2.1 (i), $\rho(e)$ is a primitive idempotent of A. Thus $A\rho(e)$ is an indecomposable A-module and hence is an indecomposable FG-module. We now claim that

$$A\rho(e) \cong FGe \qquad \text{as } FG\text{-modules} \tag{5}$$

Since A is a projective FG-module, the module $A_{G\times 1}$ is projective and hence the regular module ${}_A A$ is a projective FG-module. Since the A-module $A\rho(e)$ is an indecomposable direct summand of the regular module ${}_A A$, the FG-module $A\rho(e)$ is projective. On the other hand, the restriction

$$\rho : FGe \to A\rho(e)$$

is a surjective homomorphism of FG-modules. Thus the FG-module $A\rho(e)$ is isomorphic to a direct summand of the indecomposable FG-module FGe, proving (5).

Let $V_1, \ldots, V_r$ be all nonisomorphic simple A-modules and let $W_1, \ldots, W$ be the corresponding simple FG-modules defined via ρ. Let $W_1, \ldots, W_r$, $W_{r+1}, \ldots, W_t$ be all nonisomorphic simple FG-modules belonging to B, choose primitive idempotents $e_1, \ldots, e_t$ of FG such that

$$FGe_i/J(FG)e_i \cong W_i \qquad (1 \leq i \leq t)$$

and put $f_i = \rho(e_i), 1 \leq i \leq t$. Then $f_i = 0$ for $i \geq r+1$ and, by (5),

$$Af_i \cong FGe_i \qquad (1 \leq i \leq r) \tag{6}$$

as FG-modules. Under the above arrangement of the indices i, let

$$C_A = (c'_{ij}) \quad \text{and} \quad C_B = (c_{ij})$$

be the Cartan matrices of the F-algebras A and B, respectively. Then C_A is an $r \times r$ matrix, while C_B is a $t \times t$ matrix. The A-modules Af_i and $\oplus_{i=1}^r c'_{ij} V_j$ have the same composition factor (counting multiplicity) and the same is true for the FG-modules FGe_i and $\oplus_{j=1}^t c_{ij} W_j$. By (6), we have $c'_{ij} = c_{ij}$ for $i \in \{1, 2, \ldots, r\}$. Since $f_i = 0$ for $i \geq r+1$, we also have $c_{ij} = 0$ for $j \geq r+1$. Since, by Theorem 2.6.1 (v), the matrix C_B is symmetric, it follows that

$$C_B = \begin{bmatrix} C_A & 0 \\ 0 & * \end{bmatrix}$$

Since B is a block of FG and F is a splitting field for FG, the reader may easily verify that the above implies $r = t$ and $C_A = C_B$. Since F is a splitting field for FG, we have

$$A \cong \oplus_{i=1}^r (dim_F V_i) Af_i$$

and

$$B \cong \oplus_{i=1}^{r} (dim_F W_i) FGe_i$$

But $dim_F V_i = dim_F W_i$, $1 \leq i \leq r$, hence by (6), $A \cong B$ as FG-modules. It follows that $dim_F A = dim_F V$, as required. ∎

We have now come to the demonstration for which this section has been developed. In what follows, we put $\Delta(D) = \{(g,g)|g \in D\}$ and, for any subgroups X, Y of $G \times G$, $X \subseteq_{G \times G} Y$ (respectively, $X =_{G \times G} Y$) means that X is $G \times G$-conjugate to a subgroup of Y (respectively, X and Y are $G \times G$-conjugate).

11.8. Theorem. *(Ikeda (1987)). Let (A, ρ) be an epimorphic local interior G-algebra belonging to a block B of RG, let D be a defect group of A and let Q be a vertex of the indecomposable $R(G \times G)$-module A (see Lemma 11.2). Then*

(i) $\Delta(D) \subseteq_{G \times G} Q \subseteq_{G \times G} D \times D$.

(ii) $\Delta(D) =_{G \times G} Q$ *if and only if the restriction $\rho : B \to A$ is an isomorphism of R-algebras.*

(iii) If $\Delta(D) =_{G \times G} Q$, then D is a defect group of the block B.

Proof. (i) By definition, D is a minimal element in the set of all subgroups H of G such that $A_H^G = A^G$. Let V be the restriction of the $R(G \times G)$-module A to $\Delta(G)$. Then $V(D) = V(\Delta(D)) \neq 0$. Hence, by Lemma 11.6,

$$\Delta(D) \subseteq_{G \times G} Q$$

To prove the second containment, it suffices to show that the $R(G \times G)$-module A is $D \times D$-projective. To this end, we first note that by Lemma 11.5 (ii),

$$\psi_G(A_D^G) = Tr_{D \times G}^{G \times G}(\psi_D(A^D)) \subseteq E_{D \times G}^{G \times G}$$

Since D is a defect group of A, the identity element 1_A is contained in A_D^G and so the identity map id_A is contained in $E_{D \times G}^{G \times G}$. Hence, by Theorem 5.3, the $R(G \times G)$-module A is $D \times G$-projective. A similar argument shows that the $R(G \times G)$-module A is $G \times D$-projective. Hence, by Mackey Decomposition Theorem (see Curtis and Reiner (1961)), it follows that A is $gDg^{-1} \times D$-projective for some $g \in G$, and hence A is $D \times D$-projective, proving (i).

(ii) and (iii) Assume that the restriction $\rho : B \to A$ is an isomorphism of R-algebras. Then the $R(G \times G)$-module A is isomorphic to the $R(G \times G)$-module B. Hence D is a defect group of B and so, by Corollary 6.5, $\Delta(D)$ is a vertex of A. Thus Q and $\Delta(D)$ are $G \times G$-conjugate.

Conversely, assume that Q and $\Delta(D)$ are $G \times G$-conjugate. Then $\Delta(D)$ is a vertex of A and so the $R(G \times G)$-module A is $\Delta(G)$-projective. Therefore, by Mackey Decomposition Theorem, A is a projective RG-module, where the RG-module structure on A is defined via ρ. Invoking Lemma 11.7, we conclude that ρ restricts to an isomorphism of B onto A, as desired. ∎

As a preliminary to the next result, let us recall the following piece of information. Let V be an RG-module and let $V^* = Hom_R(V, R)$. Then V^* is an RG-module via

$$(gf)(v) = f(g^{-1}v) \qquad (v \in V, g \in G, f \in V^*)$$

We refer to V^* as the *contragredient* of V. In case $V \cong V^*$, we say that V is *self-contragredient*. A routine verification shows that if H is a subgroup of G and V is an RH-module, then $(V^*)^G \cong (V^G)^*$. Accordingly, if $V = 1_H$ is the trivial RH-module, then obviously $V \cong V^*$ and hence $(1_H)^G$ is self-contragredient. In particular, since

$$RG \cong (1_{\Delta(G)})^{G \times G}$$

we deduce that the $R(G \times G)$-module RG is *self-contragredient*.

Let I be an indecomposable two-sided ideal of RG and let

$$RG = B_1 \oplus \cdots \oplus B_n$$

be the block decomposition of RG. Since

$$I = (I \cap B_1) \oplus \cdots \oplus (I \cap B_n),$$

it follows that $I \subseteq B_k$ for exactly one $k \in \{1, \ldots, n\}$. It is therefore natural to enquire about the necessary and sufficient conditions under which I is a block of RG. Since I is an indecomposable submodule of the $R(G \times G)$-module RG, it would be desirable to provide such a criterion in terms of the vertex of the $R(G \times G)$-module I. This is achieved by the following result.

11.9. Theorem. *(Ikeda (1987)). Let I be an indecomposable two-sided ideal of RG. Then I is a block of RG if and only if a vertex of the $R(G \times G)$-module I is contained in $\Delta(G)$.*

Proof. If I is a block of RG with defect group D, then by Corollary 6.5, $\Delta(D)$ is a vertex of I. Conversely, assume that a vertex Q of I is contained in $\Delta(G)$. Since the $R(G \times G)$-module RG is self-contragredient, there exists a two-sided ideal I_0 of RG such that

$$(RG/I_0)^* \cong I \quad \text{as } R(G \times G)\text{-modules} \tag{7}$$

Put $A = RG/I_0$ and let $\rho : RG \to A$ be the natural homomorphism. Since A is an indecomposable $R(G \times G)$-module, it follows from Lemma 11.2 that A is an epimorphic local interior G-algebra. Now $Q \subseteq \Delta(G)$ and, by (7), Q is a vertex of A. Hence, by Theorem 11.8, there exists a block B of RG such that the restriction $\rho : B \to A$ is an isomorphism of R-algebras. But then ρ is obviously an isomorphism of $R(G \times G)$-modules. Hence, since $I \cong A^* \cong B^*$, it follows that I is a direct summand of A. Thus I is a block of A and the result follows. ∎

12. Bilinear forms on G-algebras

In this section, we develop some general properties of bilinear forms on G-algebras. An application to block theory of group algebras is also provided. The material presented is based on an important work of Broué and Robinson (1986) which contains a number of further applications and demonstrates how the techniques developed unify some diverse results on height 0 characters, multiplicities of Scott modules, etc.

Let V be a vector over a field F and let

$$f : V \times V \to F$$

be a symmetric bilinear form. Then the *rank* of f, written *rank* f, is defined by

$$rank\, f = dim_F(V/V^\perp)$$

where

$$V^\perp = \{x \in V | f(x, V) = 0\}$$

It is clear that the rank of f is equal to the rank of the matrix (a_{ij}), $1 \leq i, j \leq n$, where $a_{ij} = f(v_i, v_j)$ and $v_1, \ldots, v_n$ is a basis of V.

12.1. Lemma.   *Let $f : V \times V \rightarrow F$ and $f_i : V_i \times V_i \rightarrow F$ be symmetric bilinear forms, $1 \leq i \leq n$. Assume that for each $i \in \{1, \ldots, n\}$, there exists an F-linear map $\pi_i : V \rightarrow V_i$ and a nonzero $\mu_i \in F$ such that*

(i) The induced map $\prod_{i=1}^n \pi_i : V \rightarrow \prod_{i=1}^n V_i, v \mapsto (\pi_i(v))$ is surjective.

(ii) $f(x, y) = \sum_{i=1}^n \mu_i f_i(\pi_i(x), \pi_i(y))$ for all $x, y \in V$.
Then

$$rank\, f = \sum_{i=1}^n rank\, f_i$$

Proof.   We may clearly assume that each $\mu_i = 1$. Define the symmetric bilinear form $g : \prod_{i=1}^n V_i \times \prod_{i=1}^n V_i \rightarrow F$ by

$$g((v_i), (v_i')) = \sum_{i=1}^n f_i(v_i, v_i') \quad (v_i, v_i' \in V_i)$$

and put $\pi = \prod_{i=1}^n \pi_i$. Then, by (ii),

$$f(x, y) = g(\pi(x), \pi(y)) \qquad \text{for all} \quad x, y \in V$$

and therefore, by (i)

$$f(x, V) = 0 \iff g(\pi(x), \pi(V)) = 0 \iff g(\pi(x), \prod_{i=1}^n V_i) = 0 \quad (x \in V)$$

Thus π induces an F-isomorphism $V/V^\perp \rightarrow \prod_{i=1}^n V_i/(\prod_{i=1}^n V_i)^\perp$. Hence

$$rank\, f = dim_F(V/V^\perp) = \sum_{i=1}^n rank\, (f_i),$$

as required. ∎

From now on, G denotes a finite group and F an algebraically closed field of characteristic $p > 0$. We also fix a finite-dimensional G-algebra A over F and assume further that A is endowed with an F-linear map

$$\lambda : A \to F$$

which is *symmetric* in the sense that $\lambda(ab) = \lambda(ba)$ for all $a, b \in A$ and also *G-stable* in the sense that

$$\lambda({}^g a) = \lambda(a) \qquad \text{for all} \quad a \in A, g \in G$$

In what follows, P denotes a p-subgroup of G and

$$Br_P : A^P \to A(P)$$

is the corresponding Brauer morphism. Recall, from Lemma 9.1 (ii), that $A(P)$ is an $N_G(P)/P$-algebra.

12.2. Lemma. *We have $Ker\,Br_P \subseteq Ker\,\lambda$ and thus*
(i) λ induces an F-linear map $\lambda_P : A(P) \to F$ given by

$$\lambda_P(Br_P(a)) = \lambda(a) \qquad \text{for all} \quad a \in A^P \tag{1}$$

(ii) λ_P is symmetric and $N_G(P)/P$-stable.

Proof. Let Q be a proper subgroup of P and let $a = Tr_Q^P(x)$ for some $x \in A^Q$. It suffices to show that $\lambda(a) = 0$. Let T be a left transversal for Q in P. Then, since λ is G-stable,

$$\lambda(a) = \lambda(\sum_{t \in T} {}^t x) = |T|\lambda(x) = 0,$$

as required. ∎

Following Broué and Robinson (1986), we define the bilinear form

$$f_{P,G}^{A,\lambda} : A_P^G \times A_P^G \to F$$

as follows: if $a, b \in A_P^G$ with $b = Tr_P^G(b'), b' \in A^P$, then

$$f_{P,G}^{A,\lambda}(a, b) = \lambda(ab')$$

12.3. Lemma.   *The form $f_{P,G}^{A,\lambda}$ is well defined, symmetric, and associative.*

Proof.   Let $a = Tr_P^G(a')$ and $b = Tr_P^G(b')$ for some $a', b' \in A^P$. Then, by (1), we have

$$\lambda(ab') = \lambda_P(Br_P(ab')) = \lambda_P(Br_P(a)Br_P(b')) \qquad (2)$$

Note also that, by Lemma 9.2 (i),

$$Br_P \circ Tr_P^G = Tr_1^{N_G(P)/P} \circ Br_P$$

Hence we must also have

$$Br_P(a) = Tr_1^{N_G(P)/P}(Br_P(a'))$$

and

$$Br_P(b) = Tr_1^{N_G(P)/P}(Br_P(b')) \qquad (3)$$

Replacing G, P, A by $N_G(P)/P, 1, A(P)$, we may therefore assume that $P = 1$. Since λ is G-stable, we have $\lambda(^g(a')b) = \lambda(a'\,^{g^{-1}}b)$ for all $g \in G$ and hence

$$\lambda(Tr_1^G(a')b') = \lambda(a'Tr_1^G(b'))$$

Thus if $Tr_1^G(b') = Tr_1^G(b'')$, then $\lambda(ab') = \lambda(ab'')$ and so the form is well defined.

Since $\lambda(ab') = \lambda(b'a)$, the form is symmetric. Furthermore, since $cb = Tr_1^G(cb')$ for all $c \in A^G$, we have

$$f_{1,G}^{A,\lambda}(ac, b) = \lambda(acb') = f_{1,G}^{A,\lambda}(a, cb),$$

proving that the form is also associative. ∎

By Lemmas 12.2 and 12.3, we may now define the bilinear form

$$f_{1,N_G(P)/P}^{A(P),\lambda_P} : A(P)_1^{N_G(P)/P} \times A(P)_1^{N_G(P)/P} \to F$$

with the role of P, G, A and λ played by 1, $N_G(P)/P$, $A(P)$ and λ_P, respectively.

12.4. Corollary. *For all $a, b \in A_P^G$, we have*

$$f_{P,G}^{A,\lambda}(a, b) = f_{1,N_G(P)/P}^{A(P),\lambda_P}(Br_P(a), Br_P(b))$$

Proof. This is a direct consequence of (2) and (3). ∎

12.5. Lemma. *For any $n \geq 1$, the ordinary trace map $M_n(F) \to F$ is the only (up to scalar multiples) symmetric F-linear map.*

Proof. Let V be the F-space of all symmetric F-linear maps $M_n(F) \to F$. We must show that $dim_F V = 1$. But for any $\varphi \in V$,

$$[M_n(F), M_n(F)] \subseteq Ker\varphi$$

and

$$dim_F M_n(F)/[M_n(F), M_n(F)] = 1,$$

hence the result. ∎

For any point α of A, let χ_α denote the character of A afforded by the corresponding simple A-module $Ae/J(A)e$ where $e \in \alpha$. We let $\lambda(\alpha)$ denote $\lambda(e)$ (which depends only on α, not on the particular e chosen).

12.6. Lemma. *Assume that $J(A) \subseteq Ker\lambda$. Then*

$$\lambda = \sum_{\alpha \in pt(A)} \lambda(\alpha)\chi_\alpha$$

Proof. Because $J(A) \subseteq Ker\lambda$ and, by Lemma 12.4, there is (up to scalar multiples) only one symmetric F-linear map on a matrix algebra, we may write

$$\lambda = \sum_{\alpha \in pt(A)} \mu_\alpha \chi_\alpha \qquad (\mu_\alpha \in F)$$

It is easy to see that $\chi_\alpha(\beta) = \delta_{\alpha\beta}$ for all $\alpha, \beta \in pt(A)$, so $\mu_\alpha = \lambda(\alpha)$ for each $\alpha \in pt(A)$, as required. ∎

Given a point α of A^P, we denote by V_α the associated simple A^P-module and by

$$\sigma_\alpha : A^P \to End_F(V_\alpha)$$

the associated homomorphism. We also let

$$tr_\alpha : End_F(V_\alpha) \to F$$

be the ordinary trace map and denote by $N(P, \alpha)$ the stabilizer of α in $N_G(P)/P$.

12.7. Lemma.   *Assume that $J(A) \subseteq Ker\lambda$. Then*
(i) $End_F(V_\alpha)$ is an $N(P, \alpha)$-algebra via

$$^x(\sigma_\alpha(a)) = \sigma_\alpha(^xa) \quad \text{for all} \quad a \in A^P, x \in N(P, \alpha) \tag{4}$$

(ii) The F-linear map tr_α is both symmetric and $N(P, \alpha)$-stable.

Proof.   (i) Since $Ker\sigma_\alpha$ is a maximal ideal of A^P with $\alpha \not\subseteq Ker\sigma_\alpha$ and since each element of $N(P, \alpha)$ stabilizes α, it follows from Corollary 10.5 (iii) that $Ker\sigma_\alpha$ is an $N(P, \alpha)$-stable ideal of A^P. Hence $End_F(V_\alpha)$ becomes an $N(P, \alpha)$-algebra in a desired manner.

(ii) It is clear that tr_α is symmetric. By Proposition 10.22, the group $N(P, \alpha)$ acts on $End_F(V_\alpha)$ via inner automorphisms. Hence tr_α is $N(P, \alpha)$-stable. ∎

Owing to Lemmas 12.7 and 12.3, if $J(A) \subseteq Ker\lambda$, then we may define the bilinear form

$$f_{1,N(P,\alpha)}^{End_F(V_\alpha),tr_\alpha} : End_F(V_\alpha)_1^{N(P,\alpha)} \times End_F(V_\alpha)_1^{N(P,\alpha)} \to F$$

with the role of P, G, A and λ played by $1, N(P, \alpha), End_F(V_\alpha)$ and tr_α.

12.8. Proposition.   *(Broué and Robinson (1986)). Assume that $J(A) \subseteq Ker\lambda$.*
(i) For all $a, b \in A_P^G$, we have

$$f_{P,G}^{A,\lambda}(a, b) = \sum_{\alpha \in S} \lambda(\alpha) f_{1,N(P,\alpha)}^{End_F(V_\alpha),tr_\alpha}(\sigma_\alpha(a), \sigma_\alpha(b))$$

where S denotes a set of all representatives for the $N_G(P)/P$-orbits of local points of A^P such that $\lambda(\alpha) \neq 0$.

(ii) We have

$$rank\ f_{P,G}^{A,\lambda} = \sum_{\alpha \in S} dim_F[((End_F(V_\alpha))_1^{N(P,\alpha)}]$$

Proof. (i) By Proposition 10.8 (ii), the Brauer morphism $Br_P : A^P \to A(P)$ induces a bijection from the set of all local points of A^P onto the set of all points of $A(P)$. Hence, by Corollary 12.4, we are reduced to the case where $P = 1$ (replacing A, G, P, λ by $A(P)$, $N_G(P)/P$, 1, λ_P, respectively). In that case $G(\alpha) = N(1, \alpha)$ is the stabilizer of α in G.

Let $b = Tr_1^G(b')$ be an element of A_1^G and for each $\alpha \in pt(A_1^G)$, let $T(\alpha)$ be a right transversal for $G(\alpha)$ in G. Then, for any given $\alpha \in pt(A_1^G)$, it follows from Lemma 1.3 (i) (with $K = G$, $D = G(\alpha)$ and $H = 1$) that

$$b = \sum_{t \in T(\alpha)} Tr_1^{G(\alpha)}(^t(b')) = Tr_1^{G(\alpha)}(\sum_{t \in T(\alpha)} {}^t(b'))$$

Hence, by (4), we have

$$\sigma_\alpha(b) = Tr_1^{G(\alpha)}(\sum_{t \in T(\alpha)} \sigma_\alpha(^t(b')))$$

By the definition of the form $f_{1,G}^{A,\lambda}$, we then have to prove that for all $a \in A_1^G$,

$$\lambda(b'a) = \sum_{\alpha \in S} \lambda(\alpha) tr_\alpha(\sum_{t \in T(\alpha)} \sigma_\alpha(^t(b')) \sigma_\alpha(a)),$$

that is

$$\lambda(b'a) = \sum_{\alpha \in S} \lambda(\alpha) \chi_\alpha(\sum_{t \in T(\alpha)} {}^t(b')a) \tag{5}$$

By Lemma 12.6, we know that

$$\lambda = \sum_{\alpha \in pt(A)} \lambda(\alpha) \chi_\alpha = \sum_{\alpha \in S} \lambda(\alpha) \sum_{t \in T(\alpha)} \chi_{(t_\alpha^{-1})} \tag{6}$$

On the other hand, for any given $\alpha \in pt(A)$, we have

$$
\begin{aligned}
\sum_{t \in T(\alpha)} \chi_{(t_\alpha^{-1})}(b'a) &= \sum_{t \in T(\alpha)} \chi_\alpha({}^t(b'a)) \\
&= \sum_{t \in T(\alpha)} \chi_\alpha({}^t(b')a) \\
&= \chi_\alpha\left(\sum_{t \in T(\alpha)} {}^t(b')a \right) \qquad (7)
\end{aligned}
$$

This proves (5), by applying (6) and (7).

(ii) Again we may reduce to the case $P = 1$, and we do so. If $\varphi \in End_F(V_\alpha)_1^{G(\alpha)}$ is such that $tr_\alpha(\varphi\psi) = 0$ for all $\psi \in End_F(V_\alpha)$, then $\varphi = 0$. Hence

$$
rank\ f_{1,G(\alpha)}^{End_F(V_\alpha),tr_\alpha} = dim_F(End_F(V_\alpha))_1^{G(\alpha)} \qquad (8)
$$

Moreover, the map

$$
\prod_{\alpha \in S} \sigma_\alpha : A_1^G \rightarrow \prod_{\alpha \in S} [End_F(V_\alpha)]_1^{G(\alpha)}
$$

is obviously surjective. The desired conclusion is therefore a consequence of (i), (8) and Lemma 12.1. ∎

Assume that A^G is local. Then a *defect group* of A is defined to be a defect group of the primitive idempotent 1 of A^G.

12.9. Lemma. *Suppose that A^G is local and let P be a defect group of A. Then*
(i) There exists a local point γ of A^P such that $1 \in Tr_P^G(A^P \cdot \gamma \cdot A^P)$.
(ii) Any other local point of A^P is in the $N_G(P)/P$-orbit of γ.

Proof. This is a special case of Theorem 10.14 (i), (iv) in which $H = G$ and $\alpha = \{1\}$. ∎

Assume that A^G is local and that P is a defect group of A. Then any local point γ of A^P such that

$$
1 \in Tr_P^G(A^P \cdot \gamma \cdot A^P)
$$

is called a *source* of the G-algebra A. The existence of a source is guaranteed by Lemma 12.9 (i). In what follows, P denotes a p-subgroup of G.

12.10. Theorem. *(Broué and Robinson (1986)). Assume that $J(A) \subseteq \operatorname{Ker} \lambda$ and that A^G is local.*

(i) If $f_{P,G}^{A,\lambda} \neq 0$, then P is contained in a defect group of A.

(ii) If P is a defect group of A and γ is a source of A, then $f_{P,G}^{A,\lambda} \neq 0$ if and only if $\lambda(\gamma) \neq 0$.

(iii) If P is a defect group of A and γ is a source of A with $\lambda(\gamma) \neq 0$, then

$$\operatorname{rank} f_{P,G}^{A,\lambda} = \dim_F[(\operatorname{End}_F(V_\gamma))_1^{N(P,\gamma)}]$$

where $N(P,\gamma)$ is the stabilizer of γ in $N_G(P)/P$.

(iv) If $A^G \subseteq Z(A)$, then $f_{P,G}^{A,\lambda} \neq 0$ if and only if P is a defect group of A and $\lambda(\gamma) \neq 0$ for γ a source of A.

(v) If $A^G \subseteq Z(A)$, P is a defect group of A and $\lambda(\gamma) \neq 0$ for γ a source of A, then

$$\operatorname{rank} f_{P,G}^{A,\lambda} = 1$$

Proof. (i) Let D be a defect group of A. By Corollary 1.5 (i), we have

$$1 \in A_D^G \subseteq \sum_{g \in G} A_{gDg^{-1} \cap P}^P$$

Hence if P is not G-conjugate to the subgroup of D, i.e. if all $gDg^{-1} \cap P$ are proper subgroups of P, then by Lemma 4.1, $1 \in A_H^P$ for some proper subgroup H of P. But then $A(P) = 0$ and so, by Corollary 12.4, $f_{P,G}^{A,\lambda} = 0$.

(ii) and (iii) Suppose that P is a defect group of A and let γ be a source of A. By Lemma 12.9 (ii), there exists exactly one $N_G(P)/P$-orbit of local points of A^P, namely the orbit of γ. Hence, by Proposition 12.8 (ii),

$$\operatorname{rank} f_{P,G}^{A,\lambda} = \begin{cases} \dim_F[(\operatorname{End}_F(V_\gamma))_1^{N(P,\gamma)}] & \text{if} \quad \lambda(\gamma) \neq 0 \\ 0 & \text{if} \quad \lambda(\gamma) = 0 \end{cases}$$

and, in particular, if $f_{P,G}^{A,\lambda} \neq 0$ then $\lambda(\gamma) \neq 0$.

Conversely, assume that $\lambda(\gamma) \neq 0$. By Theorem 10.19 (ii),

$$\sigma_\gamma(Tr_P^G(x)) = Tr_1^{N(P,\gamma)}(\sigma_\gamma(x)) \quad \text{for all} \quad x \in A^P \cdot \gamma \cdot A^P$$

But $1 \in A_P^G$, so we may take x such that $Tr_P^G(x) = 1$ and therefore

$$1 = Tr_1^{N(P,\gamma)}(\sigma_\gamma(x)) \in (End_F(V_\gamma))_1^{N(P,\gamma)},$$

proving that $f_{P,G}^{A,\lambda} \neq 0$.

(iv) and (v) Assume that $A^G \subseteq Z(A)$. By (i) and (ii), to prove (iv), it suffices to show that if P is strictly contained in a defect group of A, then $f_{P,G}^{A,\lambda} = 0$. But if P is strictly contained in a defect group of A, then $1 \notin A_P^G$ and so $A_P^G \subseteq J(A^G)$, since A^G is local (see Lemma 10.6 (iii)). It follows that each element of A_P^G is central and nilpotent. Using the fact that λ vanishes on each nilpotent element (Lemma 12.6) and the definition of $f_{P,G}^{A,\lambda}$, we therefore deduce that $f_{P,G}^{A,\lambda} = 0$.

Since A^G is central in A, it is mapped into the center of $End_F(V_\gamma)$ by the homomorphism σ_γ. Thus $\sigma_\gamma(A_P^G) = 0$ or F and, on the other hand, $\sigma_\gamma(A_P^G) \neq 0$ by virtue of (ii) and Proposition 12.8 (i). Hence $\sigma_\gamma(A_P^G) = F$ and so, by Proposition 12.8 (i),

$$rank \, f_{\gamma,G}^{A,\lambda} = 1$$

proving (v). ■

For the rest of this section, G denotes a finite group and R a complete discrete valuation ring of characteristic 0 such that the residue class field $F = R/J(R)$ of R is algebraically closed of prime characteristic p.

For each $x = \sum x_g g \in RG$, $x_g \in R, g \in G$, put

$$\bar{x} = \sum \bar{x}_g g$$

where $\bar{x}_g = x_g + J(R)$. Then the map

$$\begin{cases} RG & \to & FG \\ x & \mapsto & \bar{x} \end{cases}$$

is a surjective homomorphism with kernel $J(R)G$. For any R-linear map $\psi : RG \to R$, we denote by $\bar{\psi}$ the reduction of ψ modulo $J(R)$, i.e. $\bar{\psi}$ is the F-linear map $FG \to F$ given by

$$\bar{\psi}(\bar{x}) = \overline{\psi(x)} \quad \text{for all} \quad x \in RG$$

From now on, we fix an F-linear map

$$\lambda : FG \to F$$

which is a class function on G. Then, regarding FG as a G-algebra in a natural way, λ is symmetric and G-stable. In what follows, we also assume that

$$J(FG) \subseteq Ker\,\lambda$$

12.11. Lemma. *There exists an R-linear combination ψ of characters of R-free RG-modules such that $\lambda = \bar{\psi}$, where $\bar{\psi}$ is the reduction of ψ modulo $J(R)$.*

Proof. By Lemma 12.6, λ is an F-linear combination of the characters of simple FG-modules. Since $FG \cong RG/J(R)G$, the required assertion follows. ∎

Let e be a block idempotent of RG with defect group P and let χ be an R-linear combination of characters RG-modules. By Proposition 4.4, $\bar{e}$ is a block idempotent of FG. Since the restriction of $\bar{\chi}$ to $FG\bar{e}$ (which we also denote by $\bar{\chi}$) is symmetric and G-stable, we may define the bilinear form

$$f_{P,G}^{FG\bar{e},\bar{\chi}} : (FG\bar{e})_P^G \times (FG\bar{e})_P^G \to F$$

Furthermore, since $(FG\bar{e})^G = Z(FG\bar{e})$ is local, we may define a source γ of $FG\bar{e}$ (with respect to P).

12.12. Lemma. *With the assumptions and notation above, the following conditions are equivalent:*
(i) $\chi(e)/(G : P)$ is invertible in R.
(ii) $\bar{\chi}(\gamma) \neq 0$.
(iii) $f_{P,G}^{FG\bar{e},\bar{\chi}} \neq 0$.

Proof. We obviously have $J(FG) \subseteq Ker\bar{\chi}$. Since, by Proposition 4.4 (iv), P is a defect group of $\bar{e}$, it follows from Theorem 12.10 (ii) that (ii) is equivalent to (iii).

We now prove that (i) is equivalent to (iii). To this end, we first write $e = Tr_P^G(x)$ for some $x \in (RGe)^P = (RG)^P e$. Then

$$\chi(e) = (G:P)\chi(x) \tag{9}$$

and

$$f_{P,G}^{FG\bar{e},\bar{\chi}}(\bar{e},\bar{e}) = \bar{\chi}(\bar{x}) = \overline{\chi(x)} \tag{10}$$

By (9), condition (i) is equivalent to $\overline{\chi(x)} \neq 0$. Hence, by (10), it suffices to check that $f_{P,G}^{FG\bar{e},\bar{\chi}}$ is nonzero if and only if it is nonzero on $(\bar{e},\bar{e})$. Since

$$Z(FG\bar{e}) = (FG\bar{e})_P^G = F \cdot \bar{e} \oplus J(Z(FG\bar{e}))$$

and

$$f_{P,G}^{FG\bar{e},\bar{\chi}}(J(Z(FG)\bar{e}),x) = 0$$

for all $x \in Z(FG\bar{e})$, the result follows. ∎

12.13. Corollary. *(Broué and Robinson (1986)). Let χ be an R-linear combination of characters of RG-modules, let P be a p-subgroup of G and let $e_1,\ldots,e_n$ be all block idempotents e of RG with defect group P and such that $\chi(e)/(G:P)$ is invertible in R. Then*

$$rank\, f_{P,G}^{FG,\bar{\chi}} = n$$

Proof. Let $e_1, e_2, \ldots, e_n, e_{n+1}, \ldots e_m$ be all block idempotents of RG. Then

$$
\begin{aligned}
rank\, f_{P,G}^{FG,\bar{\chi}} &= \sum_{i=1}^{m} rank\, f_{P,G}^{FG\bar{e_i},\bar{\chi}} \quad \text{(by Lemma 12.1)} \\
&= \sum_{i=1}^{n} rank\, f_{P,G}^{FG\bar{e_i},\bar{\chi}} \quad \text{(by Lemma 12.12)} \\
&= n \quad \text{(by Lemma 12.12 and Theorem 12.10 (v))}
\end{aligned}
$$

as required. ∎

12.14. Lemma. *Let Q be a Sylow p-subgroup of G and let χ_p be the character of the RG-module $(1_Q)^G$. If e is a block idempotent of RG with defect group P, then $\chi_P(e)/(G:P)$ is invertible in R.*

Proof. By Lemma 12.12, it suffices to show that $f_{P,G}^{FG\bar{e},\bar{\chi}_p}$ is nonzero. Write $e = \sum e_g g$ with $e_g \in R$, $g \in G$ so that $\bar{e} = \sum \bar{e}_g g$. Since $Br_P(\bar{e}) \neq 0$, there exists a p-regular element t of $C_G(P)$ such that $\bar{e}_{t^{-1}} \neq 0$. Taking into account that

$$f_{P,G}^{FG\bar{e},\bar{\chi}_p}(\bar{e}, Tr_P^G(t)) = \bar{\chi}_p(\bar{e}t)$$

it therefore suffices to show that

$$\bar{\chi}_p(\bar{e}t) = (G:Q)\bar{e}_{t^{-1}} \tag{11}$$

Let $RG_{p'}$ be the R-linear span of the set $G_{p'}$ of all p-regular elements of G. Then the multiplication by e preserves $(RG_{p'})^G$ (Broué (1978, p.91)), and hence

$$
\begin{aligned}
\chi_p(et) &= \frac{1}{(G:C_G(t))}\chi_p(eTr_{C_G(t)}^G(t)) \\
&= (G:Q)e_{t^{-1}},
\end{aligned}
$$

which clearly implies (ii). ∎

12.15. Corollary. *(Broué and Robinson (1986)). Let P be a p-subgroup of G, let χ_p be the character of the RG-module $(1_Q)^G$ where Q is a Sylow p-subgroup of G and let P be a p-subgroup of G. Then the rank of the form $f_{P,G}^{FG,\bar{\chi}_p}$ is equal to the number of blocks of FG with defect group P.*

Proof. This is a direct consequence of Corollary 12.13 and Lemma 12.14. ∎

For more explicit information about the rank of $f_{P,G}^{FG,\bar{\chi}_p}$, we refer to the original paper of Broué and Robinson (1986). An alternative approach can be found in papers of Robinson (1983), Broué (1984) and Külshammer (1984b). A detailed account of Külshammer's approach together with a number of important applications of Robinson's theorem (Robinson (1983)) is contained in Karpilovsky (1987).

Bibliography

Alperin, J.L. and Broué, M.
(1979)   Local methods in block theory, Ann. of Math. 110, 143-157.

Asano, S.
(1961a)   On the radical of quasi-Frobenius algebras, Kodai Math. Sem. Rep. 13, 131-151.

(1961b)   Remarks concerning two quasi-Frobenius rings with isomorphic radicals, Kodai Math. Sem.Rep. 13, 224-226.

(1961c)   Note on some generalizations of quasi-Frobenius rings, Kodai Math. Sem. Rep. 13, 227-234.

Azumaya, G.
(1948)   On almost symmetric algebras, Jap. J. Math. 19, 329-343.

(1959)   A duality theory for injective modules (Theory of quasi-Frobenius modules), Amer. J. Math. 81, 249-278.

(1981)   Fixed subalgebra of a commutative Frobenius algebra, Proc. Amer. Math. Soc. 81, No. 2, 213-216.

Bongale, P. R.
(1967)   Filtered Frobenius algebras, Math. Z. 97, 320-325.

(1968a)   Filtered Frobenius algebras, J. Algebra 9, 79-93.

(1968b)   Filtered quasi-Frobenius rings, Math. Z. 106, 191-196.

Bourbaki, N.
(1958)   Algébre, Ch. 8: Modules et anneaux semi-simples, Hermann, Paris.

(1959)   Algébre, Ch. 5: Hermann, Paris.

(1961)   Algébre Commutative, Ch. 2: Hermann, Paris.

(1964)   Algébre Commutative, Ch. 7: Hermann, Paris.

(1972) Commutative algebra (English translation), Addison-Wesley, Reading, Massachusetts.

(1974) Algebra I, Chapter 1-3 (English translation), Addison-Wesley, Reading, Massachusetts.

Brandt, J.
(1982) A lower bound for the number of irreducible characters in a block, J. Algebra 74, 509-514.

Brauer, R.
(1935) Über die Darstellungen von Gruppen in Galoischen Feldern, Act. Sci. Ind. 195, Paris.

(1968) On blocks and sections in finite groups, II, Amer. J. Math. 90, 895-925.

(1969) Defect groups in the theory of representations of finite groups, Illin. J. Math. 13, 53-73.

Brauer, R. and Feit, W.
(1959) On the number of irreducible characters of finite groups in a given block, Proc. Nat. Acad. Sci. USA, 45, 361-365.

Brauer, R. and Nesbitt, C. J.
(1941) On a problem of E. Artin, Ann. of Math. (2) 68, 713-720.

Broué, M.
(1978) Radical, hauteurs, p-sections et blocs, Ann. of Math. 107, 89-107.

(1984) On a theorem of G. Robinson, J. London Math. Soc. (2) 29, 425-434.

(1985) On Scott modules and p-permutation modules: An approach through the Brauer morphism, Proc. Amer. Math. Soc. 93, 401-408.

Broué, M. and Puig, L.
(1980) Characters and local structure in G-algebras, J. Algebra 63, 306-317.

(1981) A Frobenius theorem for blocks, Invent. Math. 56, 117-128.

Broué, M. and Robinson, G. R.
(1986) Bilinear forms on G-algebras, J. Algebra 104, 377-396.

Burnside, W.
(1911)   Theory of groups of finite order, Cambridge Univ. Press, Cambridge.

Byrd, K. A.
(1970)   Some characterizations of uniserial rings, Math. Ann. 186, 163-170.

Chlebowitz, M. and Külshammer, B.
(1989)   Symmetric local algebras with 5-dimensional center, preprint.

Clifford, A. H.
(1937)   Representations induced in an invariant subgroup, Ann. of Math. 38, 533-550.

Curtis, C. W.
(1959)   Quasi-Frobenius rings and Galois theory, Illin. J. Math. 3, 134-144.

Curtis, C. W. and Reiner, I.
(1962)   Representation theory of finite groups and associative algebras, Interscience, New York.

(1981)   Methods of representation theory with applications to finite groups and orders, Vol. 1, Interscience, New York.

Dade, E. C.
(1966)   Blocks with cyclic defect groups, Ann. of Math. 84, 20-48.

(1968)   Degrees of modular irreducible representations of p-solvable groups, Math. Z. 104, 141-143.

(1980)   Group-graded rings and modules, Math. Z. 174, 241-262.

Dieudonné, J.
(1958)   Remarks on quasi-Frobenius rings. Illin. J. Math. 12, 346-354.

Eilenberg, S. and Nakayama, T.
(1955)   On the dimension of modules and algebras II, Frobenius algebras and quasi-Frobenius rings, Nagoya Math. J., 1-16.

Endo, S.
(1967)   Complete faithful modules and quasi-Frobenius algebras, J.
           Math. Soc. Japan 19, 437-456.

Enochs, E.
(1968)   A note on quasi-Frobenius rings, Pacif. J. Math. 24, 69-70.

Faith, C.
(1966)   Rings with ascending condition on annihilators, Nagoya Math.
           J. 27, 179-191.

Feller, E. H.
(1967)   A type of quasi-Frobenius ring, Canad. Math. Bull. 10, 19-27.

Fossum, T. V.
(1970)   Characters and centers of symmetric algebras, J. Algebra 16,
           4-13.
(1971)   Characters and orthogonality in Frobenius algebras, Pacif. J.
           Math. 36, No. 1, 123-131.

Fried, E.
(1964)   Beiträge zur Theorie der Frobenius-Algebren, Math. Ann. 155,
           265-269.

Garotta, O.
(1988)   Suites presque scindées de G-algébres intérieures et G-algébres
           intérieures de suites presque scindées, thése, Ecole Normale
           Sup., Paris.

Gow, R.
(1978)   A note on p-blocks of a finite group, J. London Math. Soc. (2)
           18, 61-64.

Green, J. A.
(1962)   Blocks of modular representations, Math. Z. 79, 100-115.
(1968)   Some remarks on defect groups, Math. Z. 107, 133-150.
(1978)   On the Brauer homomorphism, J. London Math. Soc. 17, 58-66.

(1983)   Multiplicities, Scott modules and lower defect groups, J. London
                     Math. Soc. 28, 282-292.

(1985)   Functors on categories of finite group representations, J. Pure
                     Appl. Algebra 37, 265-298.

Hall, M. Jr.
(1959)   The theory of groups, Macmillan, New York.

Hannula, A. T.
(1972)   The Morita context and the construction of QF rings, Lecture
                     Notes in Mathematics, Vol. 353, 113-130, Springer, Berlin.

(1973)   On the construction of quasi-Frobenius rings, J. Algebra 25,
                     403-414.

Harada, M.
(1983)   A characterization of QF-algebras, Osaka J. Math. 20, No. 1,
                     1-4.

Harris, M. E.
(1988)   On classical Clifford theory, Trans. Amer. Math. Soc. 309, No.
                     2, 831-842.

Hauger, G. and Zimmermann, W.
(1973)   Quasi-Frobenius-Moduln, Arch. Math. (Basel) 24, 379-386.

Higman, D. G.
(1954)   Indecomposable representations at characteristic p, Duke Math.
                     J. 21, 377-381.

(1955)   Induced and produced modules, Canad. J. Math. 7, 490-508.

Hirata, K.
(1959)   On relative homological algebra of Frobenius extensions, Nagoya
                     Math. J. 15, 17-28.

Horn, P. J.
(1974)   A note on weakly symmetric rings, Canad. Math. Bull. 17, No.
                     4, 531-533.

Iizuka, K.
(1961)   On Brauer's theorem on sections in the theory of blocks of group
                     algebras, Math. Z. 75, 299-304.

Iizuka, K. and Ito, Y.
(1972)   A note on blocks of defect groups of a finite group, Kumamoto J. Sci. (Math) 9, 25-32.

Iizuka, K. and Watanabe, A.
(1973)   On the number of blocks of irreducible characters of a finite group with a given defect group, Kumamoto J. Sci. (Math), 55-61.

Ikeda, M.
(1951)   Some generalizations of quasi-Frobenius rings, Osaka Math. J. 3, 227-239.
(1952)   A characterization of quasi-Frobenius rings, Osaka Math. J. 4, 203-210.
(1953)   On a theorem of Gaschütz, Osaka Math. J. 5, 53-58.

Ikeda, M. and Nakayama, T.
(1954)   On some characteristic properties of quasi-Frobenius and regular rings, Proc. Amer. Math. Soc. 5, 15-25.

Ikeda, T.
(1984)   Corestriction and p-subgroups, Hokkaido Math. J. 13, 285-298.
(1986)   Some properties of interior G-algebras, Hokkaido Math. J., 15, No. 3, 453-467.
(1987)   A characterization of blocks with vertices, J. ALgebra 105, No. 2, 344-350.

Itoh, S.
(1983)   On weak normality and symmetric algebras, J. Algebra 85, 40-50.

Jacobson, N.
(1956)   Structure of rings, Amer. Math. Soc. Colloq. Publ., Vol. 37.

Jans, J. P.
(1957)   Indecomposable representations of algebras, Ann. of Math. 66, 418-429.
(1959)   On Frobenius algebras, Ann. of Math. 69, No. 2, 392-407.
(1960)   Some remarks on symmetric and Frobenius algebras, Nagoya Math. J. 16, 65-71.

(1967)   On orders in quasi-Frobenius rings, J. Algebra 7, 35-43.

Johns, B.
(1977)   Annihilator condition in Noetherian rings, J. Algebra 49, No. 1, 222-224.

Karpilovsky, G.
(1985)   Dimensions of irreducible modules over twisted group algebras, Arch. Math. (Basel), 45, 431-441.

(1987)   Structure of blocks of group algebras, Pitman Monographs and Surveys in Pure and Applied Mathematics, Vol. 33, Longman Scientific and Technical.

(1989)   A survey on the Jacobson radical of group algebras, Niew Arch. Voor Wiskunde, No. 1-2, 15-38.

Kasch, F.
(1954)   Grundlagen einer Theorie der Frobeniuserweiterungen, Math. Ann. 127, 453-474.

(1961a)   Ein Satz über Frobenius-Erweiterungen, Arch. Math. 12, 102-104.

(1961b)   Dualitätseigenschaften von Frobenius-Erweiterungen, Math. Z. 77, 219-227.

Kato, T.
(1968)   Some generalizations of QF-rings, Proc. Japan Acad. 44, 114-119.

Kawada, Y.
(1966)   On blocks of group algebras of finite groups, Sci. Rep. Tokyo Kyoiku Daigaku, A. 9, 87-110.

Kirichenko, V. V.
(1978)   Quasi-Frobenius rings and Gorenstein orders, Trudy Mat. Inst. Steklov 148, 168-174.

Kitamura, Y.
(1971/1972)   On quasi-Frobenius extensions, Math. J. Okayama Univ. 15, 41-48.

Külshammer, B.

(1981)   Bemerkungen über die Gruppenalgebra als symmetrische Algebra, J. Algebra 72 (1), 1-7.

(1982)   Bemerkungen über die Gruppenalgebra als symmetrische Algebra, II, J. Algebra 75(1), 59-69.

(1984a)   Symmetric local algebras and small blocks of finite groups, J. Algebra 88(1), 190-195.

(1984b)   Bemerkungen über die Gruppenalgebra als symmetrische Algebra, III, J. Algebra 88(1), 279-291.

(1985)   Bemerkungen über die Gruppenalgebra als symmetrische Algebra, IV, J. Algebra 93(2), 310-323.

Kupisch, H.

(1959)   Beiträge zur Theorie nichtalbeinfacher Ringe mit Minimal bedingung, J. Reine Angew. Math. 201, 100-112.

(1964)   Zu einem Satz von Nakayama und Ikeda, Arch. Math. 15, 144-145.

(1965)   Symmetrische Algebren mit endlich vielen unzerlegbaren Darstellungen I, J. Reine Angew. Math. 219, 1-25.

(1968)   Projective Moduln endlicher Gruppen mit zyklischer p-Sylow Gruppe, J. Algebra 10, 1-7.

(1969)   Unzerlegbare Moduln endlicher Gruppen mit zyklischer p-Sylow Gruppe, Math. Z. 108, 77-104.

(1970)   Symmetrische Algebren, II, J. Reine Angew. Math. 245, 1-14.

(1975)   Quasi-Frobenius algebras of finite representation type, Lecture Notes in Mathematics 488, 184-200, Springer-Verlag, Berlin.

(1978)   Basisalgebren symmetrischer Algebren und eine Vermutung von Gabriel, J. Algebra 55, 58-73.

Kupisch, H and Scherzler, E.

(1980)   Symmetric algebras of finite representation type, Lecture Notes in Mathematics 832, 328-368, Springer Verlag, Berlin-Heidelberg-New York.

Kupisch, H and Waschbüsch, J
(1984)   On multiplicative bases in quasi-Frobenius algebras, Math. Z.,
                 186, 401-405.

Kurata, V.
(1958)   Some remarks on quasi-Frobenius modules, Osaka Math. J. 10,
                 213-220.

Landrock, P.
(1981)   On the number of irreducible characters in a 2-block, J. Algebra
                 68, 426-442.

Lang, S.
(1984)   Algebra, Addison-Wesley, Reading, Massachusetts, Second Edi-
                 tion.

Lawrence, J.
(1977)   A countable self-injective ring is quasi-Frobenius, Proc. Amer.
                 Math. Soc. 65, No. 2, 217-220.

Mackey, G. W.
(1958)   Unitary representations of group extensions: I, Acta Math. 99,
                 265-311.

Mano, T.
(1982)   On modules over a serial ring whose endomorphism rings are
                 quasi-Frobenius, Tokyo Math. J. 5, No. 2, 441-456.

Michler, G. O.
(1972)   Blocks and centers of group algebras, Lecture Notes in Mathe-
                 matics, Vol 246, 429-563, Springer, Berlin.

Ming, R. Y. C.
(1976)   On annihilators and quasi-Frobenius rings, Bull. Soc. Math.
                 Belg. 28, No. 2, 115-120.

Morita, K.
(1951)   On group rings over a modular field which possesses radicals
                 expressible as principal ideals, Sci. Rep. Tokyo Bunrika Daikagu
                 (A)4, 177-194.

(1969a)   Duality in QF-3 rings, Math. Z. 108, 237-252.

(1969b)   A theorem on Frobenius extensions, Sci. Rep. Tokyo Kyoiku Daigaku, Sect. A, 10, 79-87.

Morita, K. and Tachikawa, H.
(1956)   Character modules, submodules of a free module and quasi-Frobenius rings, Math. Z. 65, 414-428.

Müller, B. J.
(1974)   The structure of quasi-Frobenius rings, Can. J. Math. 26, No. 5, 1141-1151.

Müller, W.
(1974)   Symmetrische Algebren mit injektivem zentrum, Manuscripta Math. 11, 283-289.

Murase, I.
(1969a)   On Cartan invariants of quasi-Frobenius rings, Sci. Papers College Gen. Ed. Univ. Tokyo 19, 1-11.

(1969b)   A remark on quasi-Frobenius rings and Frobenius algebras, Sci. Papers College Ged. Ed. Univ. Tokyo 19, 109-120.

(1970)   Quasi-matrix algebras and symmetric algebras, Sci. Papers College Gen. Ed. Univ. Tokyo 20, 13-20.

Nakayama, T.
(1939)   On Frobenius algebras I, Ann. of Math. 40, 611-633.

(1940)   Algebras with anti-isomorphic left and right ideal lattices, Proc. Imp. Acad. Tokyo 17, 53-56.

(1941)   On Frobenius algebras II, Ann. of Math. 42, 1-21.

(1942)   On Frobenius algebras III, Jap. J. Math. 18, 49-65.

(1949)   Supplementary remarks on Frobenius algebras I, Proc. Japan. Acad. 25, No. 7, 45-50.

(1952)   Orthogonality relations for Frobenius and quasi-Frobenius algebras, Proc. Amer. Math. Soc. 3, 183-195.

(1957)   On the complete cohomology theory of Frobenius algebras, Osaka Math. J. 9, 165-187.

(1958)   On algebras with complete homology, Abh. Math. Sem. Univ. Hamburg 22, 300-307.

Nakayama, T. and Ikeda, M.
(1950)   Supplementary remarks on Frobenius algebras II, Osaka Math.
                   J. 2, 7-12.

Nakayama, T. and Nesbitt, C.
(1938)   Note on symmetric algebras, Ann. of Math., 659-668.

Nakayama, T. and Tsuzuku, T.
(1959)   A remark on Frobenius extensions and endomorphism rings,
                   Nagoya Math. J. 15, 9-16.
(1960a)   On Frobenius extensions I, Nagoya Math. J. 17, 89-110.
(1960b)   On Frobenius extensions II, Nagoya Math. J. 19, 127-148.

Nesbitt, C. J. and Thrall, R. M.
(1946)   Some ring theorem with applications to modular representa-
                   tions, Ann. of Math. 47, 551-567.

Okuyama, T.
(1980)   Some studies of group algebras, Hokkaido Math. J. 9, 217-221.

Osima, M.
(1951)   Some studies on Frobenius algebras, Jap. J. Math. 21, 170-190.
(1952)   On the Schur relations for the representations of a Frobenius
                   algebra, J. Math. Soc. Japan 4, 1-13.
(1953)   Supplementary remarks on the Schur relations for a Frobenius
                   algebra, J. Math. Soc. Japan 5, 24-28.
(1955)   Note on blocks of group characters, Math. J. Okayama Univ. 4,
                   175-188.

Osofsky, L.
(1966)   A generalization of quasi-Frobenius rings, J. Algebra 4, 373-387.

Pareigis, B.
(1964)   Einige Bemerkungen über Frobenius-Erweiterungen, Math.
                   Ann. 153, 1-13.

Passman, D. S.
(1969)   Central idempotents in group rings, Proc. Amer. Math. Soc.
                   22, 555-556.

Pollingher, A and Zaks, A.
(1968)   Some remarks on quasi-Frobenius rings, J. Algebra 10, 231-239.

Puig, L.
(1979a)   Sur un théorém de Green, Math. Z, 166, 117-129.
(1979b)   Structure locale et caractéres, J. Algebra 56, 24-42.
(1981)   Pointed groups and construction of characters, Math. Z. 176, 265-292.
(1986)   Local fusions in block source algebras, J. Algebra 104, 358-369.
(1988)   Pointed groups and construction of modules, J. Algebra 116, 7-129.

Reynolds, W. F.
(1972)   Sections and ideals of centers of group algebras, J. Algebra 20, 176-181.

Robinson, G. R.
(1983)   The number of blocks with a given defect group, J. Algebra 84, 493-502.

Rutter, E. A. Jr.
(1969)   Two characterizations of quasi-Frobenius rings, Pacif. J. Math. 30, 777-784.

Skornjakov, L. A.
(1973)   More on quasi-Frobenius rings, Mat. Sb. (N.S.) 92(134), 518-529.

Storrer, H. H.
(1969)   A note on quasi-Frobenius rings and ring epimorphisms, Canad. Math. Bull. 12, 287-292.

Swan, R. G.
(1963)   The Grothendieck group of a finite group, Topology 2, 85-110.

Tachikawa, H.
(1962)   A characterization of QF-3 algebras, Proc. Amer. Math. Soc. 13, 701-703.

Thévenaz, J.

(1988a)   Some remarks on G-functors and the Brauer morphism, J. Reine Angew. Math. 384, 24-56.

(1988b)   Duality in G-algebras, Math. Z. 200, 47-85.

(1988c) G-algebras, Jacobson radical and almost split sequences, Invent. Math. 93, 131-159.

(1989)   Defect theory for maximal ideals and simple factors, preprint.

Thrall, R. M.

(1948)   Some generalizations of quasi-Frobenius algebras, Trans. Amer. Math. Soc. 64, 173-183.

Tol'skaya, T. S.

(1965)   Injectivity and freeness, Sibirsk. Matem. Zh. 6, 1202-1207.

Tsushima, Y.

(1978)   On the p'-section sum in a finite group ring, Math. J. Okayama Univ. 20, 83-86.

Utumi, Y.

(1960)   A remark on quasi-Frobenius rings, Proc. Japan Acad. 36, 15-17.

(1967)   Self-injective rings, J. Algebra 6, 56-64.

Wada, T.

(1977)   On the existence of p-blocks with given defect groups, Hokkaido Math. J. 6, 243-248.

Waschbüsch, J.

(1981)   Symmetrische Algebren von endlichen Modultyp, J. Reine Angew. Math. 321, 78-98.

Zaks, A.

(1970)   Quasi-Frobenius X-rings, Canad. Math. Bull. 13, 23-30.

(1973)   Isobaric QF-rings, Portugal Math. 32, 39-47.

Index